ELEMENTARY ALGEBRA FOR COLLEGE STUDENTS

FOURTH EDITION

IRVING DROOYAN
LOS ANGELES PIERCE COLLEGE
WILLIAM WOOTON

JOHN WILEY & SONS, INC.
New York • London • Sydney • Toronto

Library of Congress Cataloging in Publication Data:

Drooyan, Irving.
Elementary algebra for college students.
Includes index.
1. Algebra. I. Wooton, William, joint author.
II. Title.
QA152.D72 1976 512.9'042 75-6554
ISBN 0-471-22253-4

Printed in the United States of America

10 9 8 7 6 5 4

PREFACE

This edition of *Elementary Algebra for College Students* retains the basic point of view of the previous three editions with regard to subject matter and pedagogy. Portions of the text have been rewritten to enhance clarity and exercise sets have been reviewed with an eye to greater effectiveness. We have also made organizational changes. Chapters 1 and 2 now give more attention to the fundamental operations. We have introduced metric units of measurement, and have altered Section 5.13 on ratio and proportion to include work on the use of proportions in making English-metric and metric-English measurement conversions. Numerical evaluation of geometric formulas, formerly in Chapter 1, has been transferred to Chapter 3 and now appears in conjunction with applications. We have deleted the material on the solution of quadratic equations by graphing and the original Chapter 10 on number systems. Increased emphasis is given to alerting students to common errors.

The following excerpts from the prefaces of previous editions remain applicable to this volume.

This textbook has been written for students who are beginning their study of algebra at the college level and who are scheduled to complete two semesters of high-school work in one semester. The general organization of the material is traditional. Algebra is developed as a generalized arithmetic, and the assumptions underlying the operations of both arithmetic and algebra are stressed. No notion of a mathematical proof has been introduced in this book, since experience dictates that students at this level profit more from an intuitive approach.

The textual material is brief. A large number of sample problems are included, however, and are placed directly before exercises of a similar type. Word problems are introduced carefully, with emphasis on methods of solution. The line graph is used frequently in the early chapters as a basis for the concept of order and as a valuable background for the treatment of graphing equations in two variables in Chapter 6. Graphing an equation is accomplished directly from ordered pairs that satisfy the equation. We have found that our students use this procedure quite successfully and, furthermore, that this approach provides a stronger background for future work in mathematics.

Subject matter is continually reviewed through the use of both chapter and cumulative reviews at the end of each chapter. At the end of the book there are ten cumulative reviews.

We have retained the functional use of a second color to stress certain routine procedures that are used to simplify fractions, evaluate expressions, and transform equations.

A glossary, a list of definitions and processes, a table of metric conversions, and a table of squares, square roots, and prime factors are placed at the end of the book for the convenience of the student.

Answers are provided for the odd-numbered exercises in Chapter 1 to 9 and for all the exercises in the chapter reviews and in the cumulative reviews.

In addition to the answers to problems at the end of the book, a separate solutions booklet is available which contains completely worked-out solutions to the even-numbered problems whose answers are not included here. This booklet contains 1595 sample problems in addition to those provided in the text. Hence, when it is used in conjunction with the textbook, the material is almost self-teaching.

We wish to thank Harold S. Engelsohn of Kingsborough Community College, Mary McCammon of Pennsylvania State University, Larry Hayman of Mt. San Antonio College, Ralph C. Williams of Pasadena City College, and Richard Darmody of Ferris State College for their many good suggestions for improving this edition; and to Nancy Halloran for her assistance in preparing the manuscript.

Irving Drooyan
William Wooton

CONTENTS

1. Whole Numbers and Their Representation

1.1	Numbers and Their Graphical Representation	1
1.2	Fundamental Operations	5
1.3	Prime Factors; Exponential Notation	9
1.4	Order of Operations	12
1.5	Numerical Evaluation	13
1.6	Algebraic Expressions	16
1.7	Sums Involving Variables	18
1.8	Differences Involving Variables	21
1.9	Products Involving Variables	22
1.10	Quotients Involving Variables	24
	Chapter Review	28

2. The Integers—Signed Numbers

2.1	Integers and Their Graphical Representation	30
2.2	Sums of Integers	33
2.3	Sums Involving Variables	36
2.4	Differences of Integers	37
2.5	Differences Involving Variables	40
2.6	Products of Integers	43
2.7	Quotients of Integers	47
2.8	Numerical Evaluation	49
	Chapter Review	51
	Cumulative Review	52

3. First-Degree Equations

3.1 Equations as Symbolic Sentences 54
3.2 Solutions of Equations 56
3.3 Solutions of Equations Using Addition and Subtraction Properties 57
3.4 Solution of Equations Using the Division Property 60
3.5 Solution of Equations Using the Multiplication Property 63
3.6 Further Solution of Equations 65
3.7 Literal Equations—Formulas 67
3.8 Applications 69
3.9 Applications from Geometry 72
Chapter Review 77
Cumulative Review 79

4. Products and Factors

4.1 The Distributive Law 80
4.2 Factoring Monomials from Polynomials 82
4.3 Binomial Products I 84
4.4 Factoring Trinomials I 86
4.5 Binomial Products II 90
4.6 Factoring Trinomials II 91
4.7 Factoring the Difference of Two Squares 94
4.8 Equations Involving Parentheses 95
4.9 Applications 97
Chapter Review 103
Cumulative Review 104

5. Fractions

5.1 Fractions and Their Graphical Representation 106
5.2 Reducing Fractions to Lower Terms 109
5.3 Quotients of Polynomials 112
5.4 Lowest Common Denominator 117
5.5 Building Fractions 119
5.6 Sums of Fractions with Like Denominators 125
5.7 Sums of Fractions with Unlike Denominators 129

5.8 Products of Fractions 134
5.9 Quotients of Fractions 138
5.10 Complex Fractions 141
5.11 Fractional Equations 144
5.12 Applications 148
5.13 Ratio and Proportion 150
Chapter Review 155
Cumulative Review 156

6. First-Degree Equations in Two Variables

6.1 Solutions of Equations in Two Variables 158
6.2 Graphs of Ordered Pairs 162
6.3 Graphing First-Degree Equations 164
6.4 Intercept Method of Graphing 168
6.5 Direct Variation 170
6.6 Graphical Solution of Systems of Linear Equations 174
6.7 Algebraic Solution of Systems I 176
6.8 Algebraic Solution of Systems II 179
6.9 Solving Word Problems Using Two Variables 182
Chapter Review 185
Cumulative Review 186
Review of Factoring 186

7. Quadratic Equations

7.1 Solution of Equations in Factored Form 188
7.2 Solution of Incomplete Quadratic Equations by Factoring 191
7.3 Solution of Complete Quadratic Equations by Factoring 194
7.4 Applications 199
Chapter Review 203
Cumulative Review 203

8. Radical Expressions

8.1 Radicals 205
8.2 Irrational Numbers 207

8.3 Simplification of Radical Expressions—Monomials 210
8.4 Simplification of Radical Expressions—Polynomials 213
8.5 Products of Radical Expressions 217
8.6 Quotients of Radical Expressions 219
Chapter Review 223
Cumulative Review 224

9. Solution of Quadratic Equations by Other Methods

9.1 Extraction of Roots 226
9.2 Completing the Square 231
9.3 Quadratic Formula 234
9.4 Graphing Quadratic Equations in Two Variables 238
9.5 The Pythagorean Theorem 241
Chapter Review 246
Cumulative Review 246

Final Cumulative Reviews 248
Definitions and Processes 259
Glossary 264
Odd-Numbered Answers 271
Table of Metric Conversions 320
Table of Squares, Square Roots and Prime Factors 321
Index 323

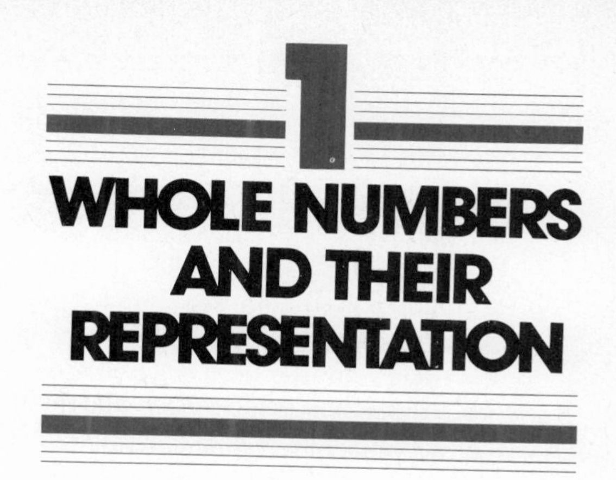

1 WHOLE NUMBERS AND THEIR REPRESENTATION

In this book, the things upon which we intend to focus our attention are numbers. We shall use the same procedures and symbols that we used in arithmetic, together with certain new symbols. The vocabulary used in arithmetic will apply in algebra, but we shall also need a number of new words. In short, we shall be studying arithmetic, but from a different and more general point of view.

1.1 NUMBERS AND THEIR GRAPHICAL REPRESENTATION

The numbers that we use to count things are called **natural numbers.** 1, 2, 3, 4, 5, . . . (and so forth), are natural numbers, whereas $2/3$, 3.141, $\sqrt{2}$, etc., are not.

A **prime number** is a natural number that is exactly divisible by itself and 1 only, that is, a multiple of no natural number other than itself and 1. For example, 2, 3, 5, 7, 11, and 13 are prime numbers. We exclude 1 from the set of prime numbers for reasons that will be noted on page 10.

When the number 0 is included with the natural numbers, the numbers in the enlarged collection 0, 1, 2, 3, . . . are sometimes called **whole numbers.**

Statements about numbers such as $4 = 2 \times 2$, $7 - 2 = 5$, and $6 + 5 = 11$ are called **equality statements,** and are interpreted to mean that the symbols on the left-hand side of the **equals** symbol, =, name the same number as the symbols on the right-hand side. Thus, "4" and "2×2" name the same natural number, as do "$7 - 2$" and "5."

The whole numbers are ordered; that is, it is always possible to say that one whole number is *greater than,* equal to, or *less than* another. Because of this property, we can use a **line graph** or **number line** to represent the relative order of a set of whole numbers. To do this we proceed as follows:

1. Draw a straight line.
2. Decide on a convenient unit of scale and mark off units of this length on the line, beginning on the left.
3. Label, on the bottom side of the line, enough of these units to establish the scale, usually two or three points.
4. Label, on the top side of the line, those points which represent the numbers to be graphed, and represent these points by heavy dots.

As an example, the graph of the prime numbers less than 8 appears in Figure 1.1.

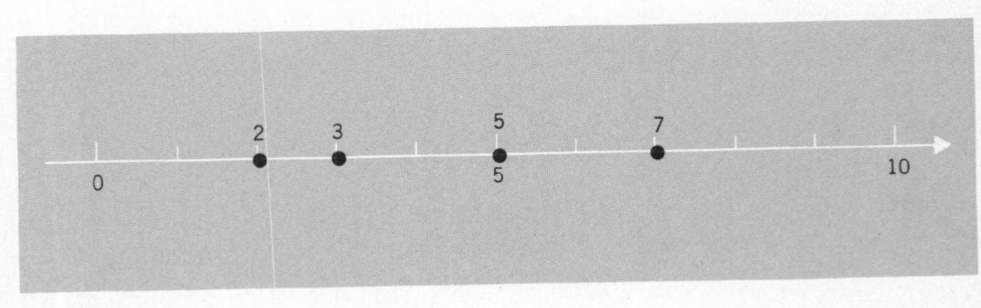

Figure 1.1

When representing the whole numbers on a line graph, we place a small arrow on the right to indicate that the numbers continue indefinitely to the right but no further to the left. The point representing 0 is called the **origin.**

The line graph tells us immediately whether one number is *less than* or *greater than* another. Of any two numbers, the number whose graph is to the left is less than the number whose graph is to the right.

The numbers 2, 4, 6, . . . are called **even numbers** and the numbers 1, 3, 5, . . . are called **odd numbers.** For example, the even numbers less than 9 are

2, 4, 6, and 8,

and the odd numbers greater than 9 are

11, 13, 15, ...

EXERCISES 1.1

■ *Which of the following are prime numbers?*

Sample Problems

a. 5 b. 15 c. 29

Ans.

a. 5 is prime (because it is exactly divisible only by 1 and itself).

b. 15 is not prime (because it is divisible by 3 and 5).

c. 29 is prime (because it is exactly divisible only by 1 and itself).

1. a. 7 b. 12 c. 21
2. a. 14 b. 19 c. 27
3. a. 24 b. 32 c. 37
4. a. 17 b. 43 c. 49

■ *List all prime numbers between the given numbers.*

5. 1 and 10 **6.** 11 and 20 **7.** 21 and 30
8. 31 and 40 **9.** 40 and 60 **10.** 60 and 100

■ *Graph the following numbers on a line graph (use a separate line graph for each problem).*

Sample Problem

The first four natural numbers divisible by 2.

Ans.

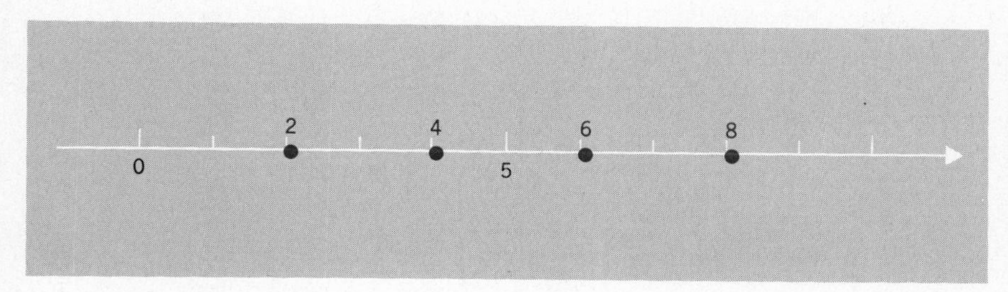

11. The natural numbers greater than 5 and less than 15.

12. The odd natural numbers greater than 7 and less than 12.

13. The natural numbers exactly divisible by 3 and less than 19.

14. The natural numbers exactly divisible by 4 and less than 19.

15. All prime numbers less than 10.

16. All prime numbers between 10 and 20.

17. The first four odd natural numbers.

18. The first four even natural numbers.

19. The first four natural numbers exactly divisible by 3.

20. The first six natural numbers not exactly divisible by 3.

Sample Problems

a. The prime numbers between 50 and 60.

b. The odd natural numbers.

Ans.

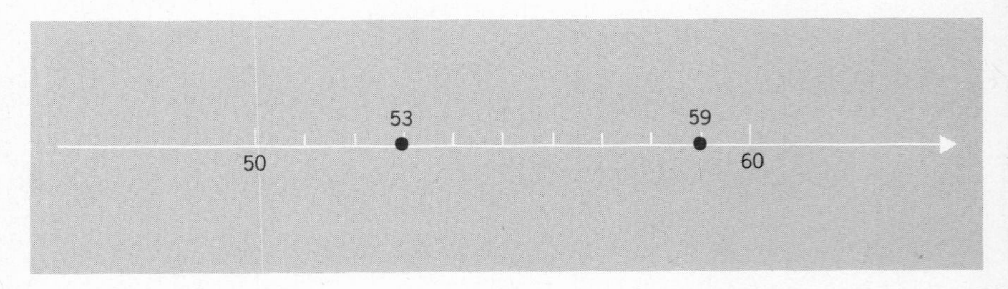

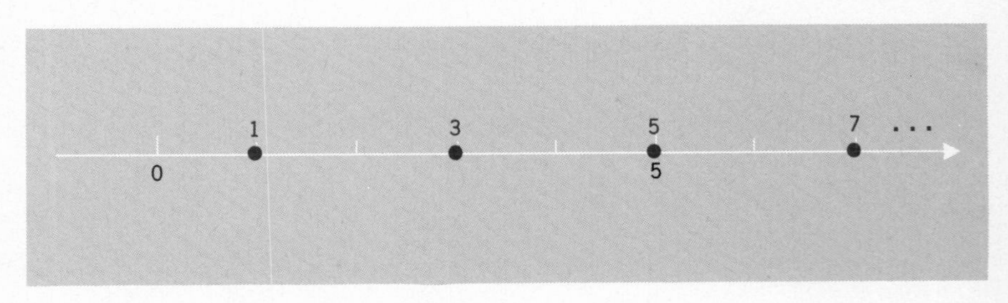

21. The prime numbers between 60 and 80.

22. The prime numbers between 80 and 100.

23. All natural numbers exactly divisible by 3.

24. All natural numbers exactly divisible by 4.

25. All natural numbers.

26. All whole numbers.

1.2 FUNDAMENTAL OPERATIONS

Mathematics can be described as a language. As such, it shares a number of characteristics with any other language. For instance, in mathematics, we find such things as verbs, nouns, pronouns, phrases, sentences, and many other concepts that are normally associated with a language. They have different names in mathematics, but the ideas are very comparable.

In a language, we use pronouns such as *he, she,* or *it* to stand in the place of nouns. In mathematics, we use symbols such as x, y, z, a, b, c, and the like, to stand in the place of numbers, where the letters used in this way are called **variables.** These symbols are sometimes also called *literal numbers* or *unknowns*. In this chapter they will always represent whole numbers.

In a language, the verbs are action words, expressing what happens to nouns. In mathematics, operations such as addition, multiplication, subtraction, or division express an action involving numbers. The symbols we use for these operations, and the properties these operations have, are the same in algebra as in arithmetic.

The addition of two numbers a and b is expressed by $a + b$, called the **sum** of a and b. The numbers a and b are called the **addends** of the sum. Since we use the symbol x so frequently in algebra as a variable, multiplication is indicated by using either a dot between the numbers or by enclosing one or both of the numbers in parentheses. For example,

$$2 \cdot 3, \qquad 2(3), \qquad \text{and} \qquad (2)(3)$$

represent the **product** of 2 and 3. The numbers 2 and 3 are called **factors** of the product. Multiplication of variables may be indicated in the same way or may be indicated by writing the symbols side by side. For example,

ab means the number a times the number b;

$3x$ means the number 3 times the number x;

abc means the number a times the number b times the number c.

A basic property of sums and products, called the **commutative law,** asserts the following:

The order in which the addends of a sum (or factors of a product) are considered does not change the sum (or product).

Thus it is always true that

$$a + b = b + a$$

and

$$a \cdot b = b \cdot a.$$

For example,

$$5 + 3 = 3 + 5,$$

and

$$5 \cdot 3 = 3 \cdot 5.$$

In subtracting one number from another, say in finding the difference $5 - 3$, we are seeking a number 2, which when added to 3 equals 5. More generally,

$a - b$ is the number which, when added to b, equals a.

For example,

$$9 - 5 = 4 \quad \text{because} \quad 5 + 4 = 9,$$

and

$$20 - 13 = 7 \quad \text{because} \quad 13 + 7 = 20.$$

In dividing a number (the dividend) by another (the divisor), we obtain a third number, the **quotient.** The quotient is defined as the number that, when multiplied by the divisor, gives the dividend. In general:

$a \div b$ or $\frac{a}{b}$ is the number q, such that $(b)(q) = a$.

For example,

$$12 \div 4 = 3 \quad \text{or} \quad \frac{12}{4} = 3 \quad \text{because} \quad 4 \cdot 3 = 12,$$

and

$$18 \div 2 = 9 \quad \text{or} \quad \frac{18}{2} = 9 \quad \text{because} \quad 2 \cdot 9 = 18.$$

Note the case where the divisor is 0. A symbol such as $\frac{5}{0}$ is meaningless since there is no number which, when multiplied by 0, gives 5. Neither do we use the symbol $\frac{0}{0}$ to represent a number because the product of 0 and any number is 0.

Division by 0 is undefined.

In a language, we use punctuation marks such as commas, periods, and semicolons to group words. In mathematics, we use symbols such as parentheses $()$, or brackets $[]$, or braces $\{\}$ to group numbers. Thus we can express the product of a number 3 and the sum $x + 5$ by

$$3(x + 5),$$

which shows that 3 multiplies the sum of x and 5. In this case it is helpful to view $x + 5$ as a single number.

Another useful property of sums and products, called the **associative law,** asserts the following:

The way in which three addends in a sum (or three factors in a product) are grouped for addition (or multiplication) does not change the sum (or product).

Thus it is always true that

$$(a + b) + c = a + (b + c)$$

and

$$a \cdot (b \cdot c) = (a \cdot b) \cdot c.$$

For example,

$$(2 + 3) + 4 = 2 + (3 + 4)$$

and

$$2 \cdot (3 \cdot 4) = (2 \cdot 3) \cdot 4.$$

If we are asked to represent symbolically the phrase "divide the sum of x and y by 7," we may write

$$(x + y) \div 7 \qquad \text{or} \qquad 7\overline{)x + y},$$

which shows that 7 is to be divided into the sum of x and y and not merely into one or the other. In algebra, however, we prefer to represent division by the use of the fraction, and in this case we have

$$\frac{x + y}{7},$$

where the line beneath the $x + y$ indicates grouping in the same sense as parentheses.

In a language, we can express ideas by using phrases; in mathematics, we can express ideas by using *algebraic expressions*. Thus, in the example just cited, $\frac{x+y}{7}$ is an algebraic expression.

EXERCISES 1.2

1. Each of the following phrases indicates the performance of a mathematical operation. Separate the phrases into four groups so that those in one group indicate addition, those in another subtraction, in another multiplication, and in another division.

Take away, increased by, less than, divide, times, add, difference, decreased by, multiply, more than, product, diminished by, exceeded by, sum, quotient, subtract.

2. Each of the following expressions indicates the performance of a mathematical operation. Separate them into four groups as in Exercise 1.

$(3)(4)$, $6 \div 3$, $4 - 3$, $3\overline{)4}$, $3 \cdot 4$, $4 + 3$, $\frac{6}{3}$,

$a \cdot b$, $a \div b$, $a - b$, $\frac{a}{b}$, ab, $b\overline{)a}$, $a + b$

■ *Express the following word phrases by using algebraic expressions.*

Sample Problems

a. Sum of three and four.

b. Six divided by two.

c. Product of 2 and x.

Ans.

a. $3 + 4$ or $4 + 3$

b. $\frac{6}{2}$ or $6 \div 2$

c. $2 \cdot x$, $2x$, $2(x)$, or $(2)(x)$

3. Add three and five.

4. Sum of six and four.

5. Product of six and nine.

6. Four times eight.

7. Nine diminished by five.

8. Eight divided by four.

9. x multiplied by y.

10. Product of x and y.

11. Subtract b from a.
12. a divided by b.
13. Add five to s.
14. t less r.
15. s diminished by t.
16. s increased by six.
17. Five less y.
18. Product of four and x.
19. From h subtract five.
20. Twice t.
21. Principal (P) times rate (r).
22. Distance (d) divided by time (t).
23. Sum of length (l) and width (w).
24. Cost (c) less five.
25. Five more than the cost (c).
26. Five times the cost (c).

Sample Problems

a. The product of x and the sum of y and three.
b. The sum of x and 3 less the product of 4 and y.

Ans.

a. $x(y + 3)$ or $(y + 3)x$

b. $(x + 3) - 4y$

27. Two times the sum of five and y.
28. Ten divided by the sum of x and four.
29. Five plus the product of four and x.
30. Nine less the quotient of b divided by four.
31. y less the sum of four and x.
32. y diminished by the product of five and t.
33. Product of b and the sum of c and d.
34. Divide the sum of x and y by five.
35. Sum of p and q less the product of four and y.
36. Product of five and x added to the sum of three and z.
37. Sum of r and s divided by the product of six and z.
38. Subtract the sum of x and y from the quotient of y divided by four.

1.3 PRIME FACTORS; EXPONENTIAL NOTATION

If we multiply 3 by 4, we obtain 12. We might also obtain 12 by multiplying the natural numbers 2 and 6, or 12 and 1, or 2, 2, and 3. In this book we are going to be interested primarily in the **prime factors** of a number, that is, factors that are themselves prime numbers. If we now ask for the prime factors of 12, we are restricted to the single set 2, 2, and 3. This is the reason we do not include 1 in the set of prime numbers. If 1 were included, another set of prime factors of 12 would be 1, 2, 2, and 3.

Many times, as in the factors of 12, the same factor occurs more than once in a product. We have a shorthand way of writing such products by using **exponents.** An exponent is a number written to the right and a little above a factor to indicate the number of times this factor occurs in a product. The product is then referred to as a **power** of the factor. Thus,

5^2 means (5) (5); read "five squared" or "5 to the second power."

x^3 means xxx; read "x cubed" or "x to the third power."

2^4 means (2)(2)(2)(2); read "two to the fourth power."

In general

$$a^n = aaa \cdot \cdots \cdot a \ (n \text{ factors}).$$

The number to which an exponent is attached is called the **base.** In the foregoing examples, 5^2, x^3, and 2^4, the bases are 5, x, and 2, respectively. It is to be understood that the exponent is attached only to the base and not to other factors in the product. Thus,

$3x^2$ means $3xx$; read "three x squared."

$5x^2y^3$ means $5xxyyy$; read "five x squared y cubed."

$2x^3$ means $2xxx$; read "two x cubed."

$(2x)^3$ means $(2x)(2x)(2x)$; read "the quantity $2x$ cubed."*

In the event that we write a variable such as x with no exponent indicated, it is to be understood that the exponent 1 is intended; that is,

$$x = x^1$$

EXERCISES 1.3

■ *Write in exponential form.*

Sample Problems

a. $3 \cdot 3 \cdot 3 \cdot xxyyy$

Ans. $3^3x^2y^3$

b. $(x-2)(x-2)(x-2)$

Ans. $(x-2)^3$

*The last two examples show that, in general $2x^3$ does not equal $(2x)^3$. In the expression $2x^3$, the exponent "3" applies only to the factor x.

1. $4 \cdot 4$ **2.** $3 \cdot 3 \cdot 3 \cdot 3$ **3.** xxx **4.** $y \cdot y \cdot y \cdot y$
5. $abbccc$ **6.** $aabbbc$ **7.** $3 \cdot 4 \cdot 4xyy$ **8.** $2yyzzz$
9. $(3x)(3x)$ **10.** $4(2y)(2y)$ **11.** $(x+3)(x+3)$ **12.** $y(y-1)(y-1)$

Sample Problems

a. $xx + xyyy$ b. $3xyy - 2 \cdot 2xxxy$

Ans. $x^2 + xy^3$ *Ans.* $3xy^2 - 2^2x^3y$

13. $xxy + xyy$ **14.** $xyyy - xxy$ **15.** $2abbb - bb$
16. $3aab + 4aaabb$ **17.** $(2x)(2x) - xx$ **18.** $y(3y)(3y) - (2y)(2y)$
19. $xx + yy - zzz$ **20.** $xyy - xxy - yyyy$
21. $2 \cdot 2 \cdot 2xx + 3 \cdot 3y + 4yyy$ **22.** $(x-y)(x-y) + (-y)(-y)$
23. $(2x)(2x)(2x) - (x-2)(x-2)$ **24.** $(3y)(3y) + xxx(x-y)(x-y)$

■ *Write in completely factored form without exponents.**

Sample Problems

a. $6x^3y^2$ b. $12(3x)^2$

Ans. $2 \cdot 3 \cdot xxxyy$ *Ans.* $2 \cdot 2 \cdot 3(3x)(3x)$

25. 8 **26.** 24 **27.** x^4 **28.** y^3
29. x^2y^4 **30.** x^3y **31.** $15ab^2c$ **32.** $4a^2b^2c^2$
33. $(2x)^3(2y)^3$ **34.** $(2x^3)(2y^3)$
35. $5(3a)^2(4b)^2$ **36.** $5(3a^2)(4b^2)$

Sample Problems

a. $4(2x+1)^2$ b. $y^2(y-2)^3$

Ans. $2 \cdot 2(2x+1)(2x+1)$ *Ans.* $yy(y-2)(y-2)(y-2)$

37. $(a-4)^3$ **38.** $(2a+1)^2$ **39.** $y^3(3y+4)^2$
40. $x^2(2x-1)^3$ **41.** $x^2y^3(x-y)^2$ **42.** $x^3y(y+2x)^3$

*A table of prime factors of the natural numbers from 2 to 100 appears on page 321.

1.4 ORDER OF OPERATIONS

The expression $4 + 6 \cdot 2$ could be interpreted in more than one way. We might look at it as meaning either

$$(4 + 6) \cdot 2 \qquad \text{or} \qquad 4 + (6 \cdot 2)$$

in which case the result is either

$$10 \cdot 2 = 20 \qquad \text{or} \qquad 4 + 12 = 16.$$

This is clearly an undesirable situation. To avoid such ambiguities, we shall make some agreements relative to the use of parentheses and fraction bars and with respect to the order of performing mathematical operations. Let us agree to accomplish any sequence of mathematical operations in the following order:

1. Perform any operations inside parentheses, or above or below a fraction bar.
2. Compute all indicated powers.
3. Perform all other multiplication operations and any division operations in the order in which they occur from left to right.
4. Perform additions and subtractions in order from left to right.

Thus, in the example above,

$$\begin{aligned} 4 + 6 \cdot 2 &= 4 + (6 \cdot 2) \\ &= 4 + 12 \\ &= 16 \end{aligned}$$

EXERCISES 1.4

■ *Simplify.*

Sample Problem

$2^3 + 3(4 + 1)$

Simplify quantity in the parentheses.

$2^3 + 3(5)$

Compute power.

$8 + 3(5)$

Multiply.

$8 + 15$

Add.

Ans. 23

1. $(3)(4) - 5$ **2.** $(6)(2) - 7$ **3.** $3 + (3)(5)$

4. $9 - (4)(2)$ **5.** $6(0) + 2$ **6.** $4 - 5(0)$

7. $(5)(1) + 3^2$ **8.** $(2)(7) + 2^3$ **9.** $2^3(2^2) - 8$

10. $(4)(2^2) - 2(2^3)$ **11.** $5(6 + 3)$ **12.** $2(7 - 3)$

13. $(5 - 3)(4)$ **14.** $(8 + 6)(2)$ **15.** $(6 - 1)(5 + 3)$

16. $(4 - 2)(7 + 1)$ **17.** $2(5 - 3)^3$ **18.** $4(3 + 2)^2$

Sample Problem

$$\frac{2+7}{5-2} - \frac{2^2+2^2}{4}$$

Simplify the numerators and denominators.

$$\frac{9}{3} - \frac{8}{4}$$

Divide as indicated.

$$3 - 2$$

Subtract

Ans. 1

19. $\frac{4(3)}{2} - 3$ **20.** $8 - \frac{12(3)}{9}$ **21.** $\frac{5+7}{4} - 2$

22. $\frac{16-4}{6} + 3$ **23.** $\frac{3(8)}{12} - \frac{6+4}{5}$ **24.** $\frac{5+9}{7} + \frac{2(8)}{4}$

25. $\frac{3^2+5}{2} + \frac{5^2-4}{3}$ **26.** $\frac{2^3}{4} + \frac{5+3^3}{8}$ **27.** $\frac{5^2-1}{6} + \frac{2(3^2)}{6}$

28. $\frac{2^3-1}{7} + \frac{3(2^3)}{4}$ **29.** $\frac{(2^2)(3^2)}{5+1} - \frac{7^2-6^2}{5+8}$ **30.** $\frac{4^2-2^2}{(4-2)^2} + \frac{3^3-2^4}{4+7}$

31. $\frac{2(3)+4}{6-1} - \frac{8(3)}{3(4)}$ **32.** $\frac{3^3+3}{5(2)} + \frac{2+2^3}{5(2)} - \frac{8^2}{16}$

33. $\frac{4^2-3^2}{7} + \frac{2(3^2)+2}{2(5)} - \frac{26}{3(5)-2}$ **34.** $\frac{26-2(3)^2}{4^2-3(4)} + \frac{4+6^2}{3(5)-7} - \frac{5+5^2}{6+3^2}$

35. $3\left(\frac{5^3-100}{3+2}\right)\left(\frac{2^5+4}{15-3^2}\right)$ **36.** $4\left(\frac{8^2-2(3)^2}{5^2-2}\right)\left(\frac{6^3-4(5^2)}{5^2+4}\right)$

1.5 NUMERICAL EVALUATION

Any meaningful collection of numbers, variables, and signs of operation such as $2xy + y$ is called an **algebraic expression** or, simply, an **expression.**

In the algebraic expression $x + 3$, the letter x represents some number. Suppose now, that we are supplied with the information that x represents the number 7. We may remove the letter and substitute in its place 7, which gives us $7 + 3$ or 10. We have made a **numerical evaluation** of $x + 3$ when x is 7. This process is particularly useful in evaluating **formulas,** which express relationships between physical quantities. For example, at an average rate r, in time t an automobile will travel a distance d given by

$$d = rt.$$

If $r = 50$ miles per hour and $t = 3$ hours, then, in miles,

$$d = (50)(3) = 150.$$

EXERCISES 1.5

■ *If $x = 3$, find the value of each expression.*

Sample Problems

a. $4 + 2x^2$ b. $\frac{4x + 6}{2}$

Substitute 3 for x.

$4 + 2 \cdot 3^2$ $\frac{4(3) + 6}{2}$

Simplify.

$4 + 2 \cdot 9$ $\frac{12 + 6}{2}$

$4 + 18$ $\frac{18}{2}$

Ans. 22 *Ans.* 9

1. $5x^2$ **2.** $8x^2$ **3.** $2x + 6$ **4.** $2x - 1$

5. $2(x + 6)$ **6.** $2(x - 1)$ **7.** $2 + 4x$ **8.** $5 + 3x$

9. $2x^2$ **10.** $(2x)^2$ **11.** $3x^2$ **12.** $(3x)^2$

13. $4x^2 - 3$ **14.** $(4x)^2 - 3$ **15.** $(4x)^2 - 3^2$ **16.** $(4x - 3)^2$

17. $\frac{3 + 6x}{3}$ **18.** $\frac{3(1 + 2x)}{3}$ **19.** $\frac{(2x - 2)^2}{2}$ **20.** $\frac{4(x - 1)^2}{2}$

■ *If $x = 2$ and $y = 3$, find the value of each expression.*

Sample Problems

a. $x(y+1)+3$	b. x^3y+xy^2	
$2(3+1)+3$	$(2)^3(3)+(2)(3)^2$	Substitute values.
$2(4)+3$	$8(3)+2(9)$	Simplify.
$8+3$	$24+18$	
Ans. 11	*Ans.* 42	

21. xy^2 **22.** $(xy)^2$ **23.** x^2y **24.** x^2y^2

25. $2(y-x)$ **26.** $2y-x$ **27.** $3(y+x)$ **28.** $3y+x$

29. $2y-2x$ **30.** $3y+3x$ **31.** $\dfrac{3x+y}{y}$ **32.** $\dfrac{x+2y}{x}$

■ *If $a=1$, $b=3$, $c=2$, find the value of each of the following.*

33. abc **34.** abc^2 **35.** ab^2c **36.** a^2bc

37. a^2b^2c **38.** $a(bc)^2$ **39.** $(abc)^2$ **40.** $ab+c$

41. $a(b+c)$ **42.** $a+(b-c)$ **43.** $(a+b)^2+c$ **44.** $a+(b-c)^2$

45. $(a+b+c)^2$ **46.** $a(b-c)^2$ **47.** $a^2(b+c)$ **48.** $\dfrac{a+b}{c}$

49. $\dfrac{(b-a)^2}{c}+\dfrac{(b+a)^2}{c}$ **50.** $\left(\dfrac{b^2-c^3}{a}\right)+\left(\dfrac{b^2+c^3}{a}\right)$

51. $\left(\dfrac{bc+b^2}{5a}\right)^2+\left(\dfrac{c^3+4b}{c^2}\right)^2$ **52.** $\left(\dfrac{a^2+b^2+c^2}{b^2-2a}\right)^3+\left(\dfrac{b^2-c^2-a^2}{c^2-2a}\right)^3$

■ *Evaluate each formula.*

Sample Problem

$P=\dfrac{k}{V}$; for P, where $k=88$ and $V=11$.

$P=\dfrac{88}{11}$

Ans. $P=8$

53. $W=F\cdot d$; for W, where $F=2000$ and $d=6$.

54. $P=2l+2w$; for P, where $l=64$ and $w=41$.

55. $V = a + 16t$; for V, where $a = 12$ and $t = 4$.

56. $A = 12(b + c)$; for A, where $b = 16$ and $c = 4$.

57. $A = \frac{bh}{2}$; for A, where $b = 500$ and $h = 200$.

58. $V = lwh$; for V, where $l = 14$, $w = 6$ and $h = 2$.

59. $A = \frac{h(b + c)}{2}$; for A, where $h = 6$, $b = 4$ and $c = 6$.

60. $l = a + (n - 1)d$; for l, where $a = 3$, $n = 7$ and $d = 2$.

1.6 ALGEBRAIC EXPRESSIONS

Any single collection of factors such as xyz, 2, or $2x^2y$ is called a **term.** An expression containing one term, such as

$$x, \quad y^2z, \quad 3x^2y^3, \quad \text{or} \quad x^2y$$

is called a **monomial.** An expression containing two terms, such as

$$2x + y, \quad 2a - 3b, \quad \text{or} \quad x^2 + y^2$$

is called a **binomial.** An expression containing three terms, such as

$$x + y + z, \quad 2a + 3b + 5c, \quad \text{or} \quad x^2 - 3x + 4$$

is called a **trinomial.** Any term or sum of terms is called a **polynomial.** Thus, all of the above monomials, binomials, and trinomials are also polynomials.

Any collection of factors in a term is called the **coefficient** of the remaining factors in the term. Thus, in the term $3xy$, 3 is the coefficient of xy, x is the coefficient of $3y$, y is the coefficient of $3x$, and $3x$ is the coefficient of y. In the event that we wish to refer to the numerical part of the term only, we speak of the **numerical coefficient.** For example, in $3xy$, 3 is the numerical coefficient. In a term such as xy, the numerical coefficient is understood to be 1.

The exponent on the variable in a term containing only one variable is called the **degree** of the term. For example,

$3x^2$ is of second degree,

$2y^3$ is of third degree, and

$4z$ is of first degree,

where the exponent on z is understood to be 1.

The degree of a polynomial is the degree of the term of highest degree.

For example,

$2x + 1$ is of first degree,

$3y^2 - 2y + 4$ is of second degree, and

$y^5 - 3y^2 + y$ is of fifth degree.

EXERCISES 1.6

■ *a. Identify each expression as monomial, binomial, or trinomial.*
b. Write each term separately and state its numerical coefficient.

Sample Problem

$4y^3 + 2y^2 + y$

Ans.

a. trinomial	b. $4y^3$ coefficient: 4	$2y^2$ coefficient: 2	y coefficient: 1

1. $3y^3 + 2y$ **2.** $4x^2 + 3x$ **3.** $2x^3 + x$

4. $x + x^2$ **5.** $4y^3$ **6.** $5x$

7. $7x^2$ **8.** y^4 **9.** $y^5 + 2x^3$

10. $x^3 + 3y$ **11.** $6x^3 + 2y^2$ **12.** $5y^4 + x$

13. $2x^4 + 2$ **14.** $4y^3 + 4$ **15.** $3x^2 + 3y + 4z$

16. $3y^3 + 2y + x$ **17.** $x^3 + 4x$ **18.** $x^3 + 4x^2 + 1$

19. $3xy^2 + y$ **20.** $x^2y + yz^2$ **21.** 3

22. 7 **23.** $7xyz + 3x$ **24.** $7z + 4xy^2z$

25. $3x^2y + x^2y^2$ **26.** $3xy + 2xz + yz$ **27.** $4x^2yz^3$

28. $x^5y^3z^2$

■ *Write each term separately and state the degree of the term.*

Sample Problem

$4x^2 + 2x^3 + x$

Ans. $4x^2$ degree: 2	$2x^3$ degree: 3	x degree: 1

29. $4y^2$ **30.** $5x^3$ **31.** $2x^4 + x$

32. $5y^3 + y^2$ **33.** $x^2 + 3x$ **34.** $y^5 + 4y$

35. $2x^2 + x + 4x$ **36.** $4y^2 + 2y + y$ **37.** $y^4 + 2y^3 + y$

38. $4x^3 + 2x^2 + 2x$ **39.** $z^5 + 2z^2 + z$ **40.** $z^6 + 3z^3 + z$

■ *Give the degree of the polynomial in each of the following.*

Sample Problem

$$3y^4 + 2y^2 + 1$$

Ans. Fourth degree (the term of highest degree, $3y^4$, is of fourth degree).

41. Exercise 35 above
42. Exercise 36
43. Exercise 37
44. Exercise 38
45. Exercise 39
46. Exercise 40

1.7 SUMS INVOLVING VARIABLES

In adding natural numbers, a counting procedure can be used to arrive at a sum. If we wish to add 3 to 5, we can first count out five units, then starting with the next unit, count out three more, yielding the number 8 as the sum. Suppose now, we wish to add three 2's to five 2's, that is, $5(2) + 3(2)$. These like quantities can be added by counting five 2's, arriving at the number 10, and then counting three more 2's, to make a total of eight 2's or 16. This addition is demonstrated on a line graph in Figure 1.2.

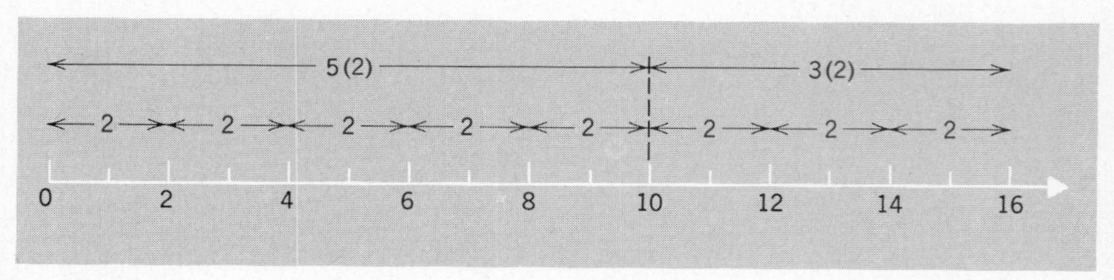

Figure 1.2

In algebra, where terms are usually made up of both numerals and variables, we have to decide what constitutes like quantities in order that we may apply the idea of addition just developed. We could add 2's as we did because they represented a common unit in each number. For variables, we are sure that $x = x$ or that $ab = ab$ regardless of the numbers that these letters represent. It is also apparent that, in general, $x \neq x^2$ (the symbol $\neq$ is read "is not equal to"), $a^3 \neq a^2$, $x \neq xy$, *etc.* We therefore define **like terms** to be any terms which are exactly alike in their variable factors. Like terms may differ only in numerical coefficients. Thus $2x$ and $3x$ are like terms; $2x$ and $3x^2$ are unlike terms.

With this definition for like terms and in view of the discussion above, we state the rule:

To add like terms, add their numerical coefficients.

For example,

$$5(2) + 3(2) = 8(2), \qquad 2x + 3x = 5x,$$

$$3ay^2 + 2ay^2 + 5ay^2 = 10ay^2, \qquad \text{and} \qquad xy + xy = 2xy.$$

We can illustrate the addition of like terms on a line graph by considering the unit of distance to be equal to the variable part of each term. Thus, $2x + 3x = 5x$ can be represented as in Figure 1.3.

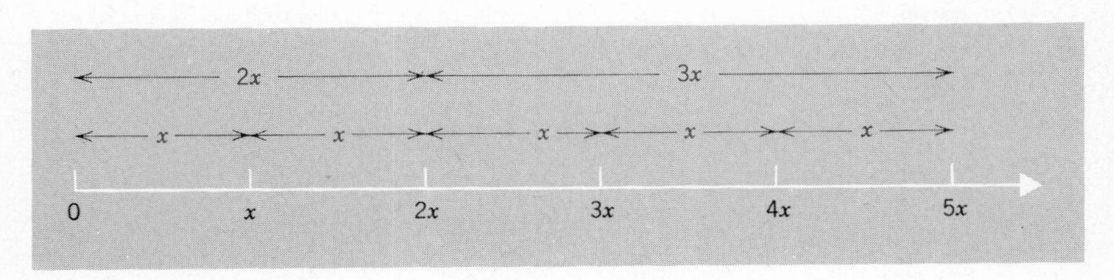

Figure 1.3

The examples above illustrate a basic property of numbers called the **distributive law:**

$$b \cdot a + c \cdot a = (b + c) \cdot a.$$

Using this property we can write

$$\begin{aligned} 2x + 3x &= (2 + 3)x \\ &= 5x. \end{aligned}$$

We shall have numerous opportunities to use different forms of the distributive law throughout this book.

Algebraic expressions having identical values for all substitutions for any variable or variables they contain are called **equivalent expressions.** Thus,

$$2x + 3x \qquad \text{and} \qquad 5x$$

are equivalent expressions since their values are the same for every number substituted for x.

If we have an algebraic expression containing unlike terms, we cannot combine the numerical coefficients of the terms. In the event that we have both like and unlike terms in the same expression, we can add the like terms and leave the unlike terms in the same form in which they originally appear. Thus, $2x + 3y + 4x$ can be simplified to $6x + 3y$ or $3y + 6x$, but the x and y terms cannot be combined. The expressions $6x + 3y$ and $3y + 6x$ are **equivalent,** that is, *they equal the same number for all replacements of the variables.* The fact that

$$6x + 3y = 3y + 6x$$

reflects the commutative property for sums mentioned in Section 1.2.

EXERCISE 1.7

■ *Simplify.*

Sample Problem

$3x^2 + 2x + 5x^2 + x$

Add the coefficients of like terms.

Ans. $8x^2 + 3x$

ASSIGN
ODD 1-39

1. $4x + 3x$ **2.** $2y + 6y$ **3.** $5a + a$
4. $7b + b$ **5.** $3x^2 + 5x^2$ **6.** $2y^2 + 4y^2$
7. $2x^2 + 3x^2 + 5x$ **8.** $4y + 5y^2 + y^2$ **9.** $5a^2 + a + 3a^2$
10. $a^2 + 7a^2 + a$ **11.** $2b^3 + b^3 + 2b^2 + b$ **12.** $5b^3 + b^2 + 2b^2 + 3b$

Sample Problem

$3x^2y + 2xy^2 + x^2y^2 + 4x^2y^2 + 5xy^2$

$3x^2y + 2xy^2 + x^2y^2 + 4x^2y^2 + 5xy^2$

Add the coefficients of like terms.

Ans. $3x^2y + 7xy^2 + 5x^2y^2$

13. $2xy + 3xy + xy$ **14.** $4xy + xy + 7xy$

15. $xy^3 + 3xy^3 + 5xy^3$

16. $2x^2y + 3x^2y + 8x^2y$

17. $x^2y + 5x^2y + 2xy^2$

18. $3xy^2 + 4x^2y + x^2y$

19. $2x + 2xy + 5xy + 4y$

20. $5x^2y + 3xy + 2x^2y + xy^2$

21. $x + 3y + z + 3x + y$

22. $2x + y + 4z + 2x + 5z$

23. $x^2y + xy + xy^2 + xy^2 + xy$

24. $3x^2y + 4x + 2z + 7x^2y + 5x$

25. $2x + 3x + 4x^2 + x^3 + 2x^3$

26. $x + 2x^2 + x^3 + 2x + x^2$

27. $5x + 7y + 3z^2 + 2z + 3x$

28. $x + y + z + x + z$

29. $5x^2y + 2xy + 3x + 2y + x^2$

30. $a^2b + 3ab + 2b^2 + ab + 2a^2b$

31. $3ax + 2x + 3y + 2ax + x$

32. $4xy + 3yz + 2xz + xz + yz$

33. $s^2 + 2t^2 + r^2 + 7s^2 + 2st$

34. $a^2 + b^2 + c^2 + ac + c^2$

35. $2a + 2b + 2c + 3ab + 3b$

36. $4y^2 + 2y + 3z^2 + 6y + 4z + z^2 + 4$

37. $x^2yz + xy^2z + 2xyz^2 + 3xy^2z + xyz^2 + x^2yz$

38. $abc + a^3bc + 2a^2bc + a^3bc + 5abc + 7a^2bc$

39. $10x^2yz + 3xyz + 8x^3yz + xyz + 5x^3yz + 3x^2yz$

40. $15xyz + 3wxy + 4wxz + 3wxy + 7wxz + 6xyz$

1.8 DIFFERENCES INVOLVING VARIABLES

We may subtract like terms in the same way that we added:

To subtract like terms, subtract their numerical coefficients.

For example,

$$7x - 3x \text{ is equivalent to } 4x,$$

and

$$3x^2 - x^2 \text{ is equivalent to } 2x^2.$$

In the case of $7x - 3y$, we can only indicate the subtraction, since these are unlike terms.

We can view an expression such as

$$6x^2 - 2x^2 + 3x^2 \quad \text{as} \quad (6x^2 - 2x^2) + 3x^2,$$

and hence as the sum

$$4x^2 + 3x^2 = 7x^2.$$

EXERCISES 1.8

■ *Simplify the following expressions.*

Sample problem

$(5y^3 - y^3) + 2y^3$

$4y^3 + 2y^3$ — Add or subtract coefficients of like terms.

Ans. $6y^3$

1. $5y^3 - 2y^3$ **2.** $6y^3 - 2y^3$ **3.** $7x^2 - x^2$

4. $4x^2 - x^2$ **5.** $(3x^2 - 2x^2) + 4x^2$ **6.** $(7x^2 + 2x^2) - 5x^2$

7. $(4a^2 + 3a^2) - a^2$ **8.** $(5b^2 - 2b^2) + 3b^2$ **9.** $(7a^2 - 3a^2) - a^2$

10. $(5b^2 - b^2) - 2b^2$ **11.** $(6a^3 - 2a^3) + 3a^3$ **12.** $(7b^3 + b^3) - 5b^3$

Sample Problem

$2x^2y + 7x^2y + 4xy^2 - xy^2$ — Add or subtract coefficients of like terms.

Ans. $9x^2y + 3xy^2$

13. $a^2 + 4a^2 + 3a$ **14.** $3b^3 - b^3 + 2b$ **15.** $6b^2 + 3b + b$

16. $3a^2 + 7a - 2a$ **17.** $2x + 7x - 3y$ **18.** $4x + 6y - 2y$

19. $4x^2y + 2x^2y + xy^2$ **20.** $5x^2y - 2x^2y - xy^2$ **21.** $4ab^2 + 3a^2b - 2a^2b$

22. $ab^2 + a^2b + 5a^2b$ **23.** $6a^2b - ab^2 + a^2b$ **24.** $3ab^2 + 2a^2b - ab^2$

25. $8x^2y + 3xy^2 - 2xy^2 - x^2y$ **26.** $5x^2yz + 2xy^2z - 3xyz^2 - 2xy^2z$

27. $11x^3yz + 5x^2yz - 3x^3yz + 2xyz$ **28.** $15ab^2c + 6a^2bc - 3ab^2c + 5abc^2$

29. $25xz^2 + 17x^2z - 8xz^2 + 5xz^2$ **30.** $23ab + 35ac - 17ab + 18ab$

1.9 PRODUCTS INVOLVING VARIABLES

We have adopted the use of exponents to indicate the number of times a given factor occurs in a product, e.g., $x^3 = (x)(x)(x)$. This notation not only has the advantage of brevity, but it also provides us with a simple means of multiplying expressions that contain similar factors. Consider the product $(x^2)(x^3)$, which in completely factored form appears as $(x)(x)(x)(x)(x)$. This, in turn, may be written x^5, since it contains x as a factor five times. Again, $(y^5)(y^2)$ is equivalent to $(y)(y)(y)(y)(y)(y)(y)$, which may be written as y^7. In both examples, the product of two expressions having the same base is obtained by adding the exponents of the powers to be multiplied. We can make a more general approach to this by considering the product $(a^m)(a^n)$. If this is written in completely factored form, it appears as

$$\underbrace{(aaa \cdot \cdots \cdot a)}_{m \text{ factors}}\underbrace{(aaa \cdot \cdots \cdot a)}_{n \text{ factors}} = \underbrace{aaa \cdot \cdots \cdot a,}_{m+n \text{ factors}}$$

which in exponent notation is written as a^{m+n}. Thus,

The product of two powers with the same base equals a power with the same base and with an exponent equal to the sum of the exponents of the powers.

In symbols this is expressed as

$$a^m \cdot a^n = a^{m+n}.$$

By applying both the commutative and associative properties for multiplication mentioned in Section 1.2, we may arrange the factors in a product in any order that we wish. For example, the product $(2x^2y)(5xy^2)$ can be written directly in completely factored form as $(2)(x)(x)(y)(5)(x)(y)(y)$, and by applying the associative and commutative laws, can be written as $(2)(5)(x)(x)(x)(y)(y)(y)$, which is equivalent to $10x^3y^3$. Wherever possible, as in the numerical coefficients, we perform the multiplication. Wherever it is not possible to multiply, as in the variable factors, we indicate the multiplication.

EXERCISES 1.9

■ *Find each of the following products by first expressing in a rearranged, completely factored form, then simplifying.*

Sample Problems

a. $x^2 \cdot x^3 \cdot x$

$xx \cdot xxx \cdot x$

Ans. x^6

b. $(3xy^2)(4x^3y^2)$

$3 \cdot 2 \cdot 2xxxxyyyy$

Ans. $12x^4y^4$

c. $(2xy)(4xy^2)(x^2y)$

$2 \cdot 2 \cdot 2xxxxyyyy$

Ans. $8x^4y^4$

1. $x^3 \cdot x^4$ **2.** $y^2 \cdot y^6$ **3.** $y \cdot y^4$

4. $x^3 \cdot x$ **5.** $(2x^2)(5x^3)$ **6.** $(3y)(4y^4)$

7. $(2a^2)(a^3)(2a)$ **8.** $(4a^2)(2a^2)(a^3)$ **9.** $(b^3)(3b)(4b)$

10. $(2b^4)(2b^3)(b)$ **11.** $(c^2)(3c^2)(4c^2)$ **12.** $(4c)(2c^4)(c^3)$

■ *Find the products* directly *by multiplying the numerical coefficients and indicating the product of the factors represented by variables.*

Sample Problems

a. $(4x^3)(3x^2)$ b. $(2xy^3)(5x^2y^2)(xy)$ c. $(x^3y^2)(4xy)(5xy^3)$

Ans. $12x^5$ *Ans.* $10x^4y^6$ *Ans.* $20x^5y^6$

13. $(2x)(4x^2)$

14. $(5y^2)(3y^2)$

15. $(y^4)(y^3)$

16. $(x)(3x^4)$

17. $(2a^2b)(7ab^3)$

18. $(a^3b^3)(b^2c)$

19. $(ab)(ab)(ab)$

20. $(a^2b)(a^3b^2)(b^3)$

21. $(a^2b)(b^2c)(a^3c^2)$

22. $(ac^2)(a^2b^3)(b^2c^2)$

23. $(2abc)(3b^2c^2)(ab^2c)$

24. $(4a^2bc)(2bc)(abc)$

Sample Problems

a. $a^3 + 3a^2(a)$ b. $(3x)(xy^3) - xy + 4x^2y^3$

Multiply.

$a^3 + 3a^3$ $3x^2y^3 - xy + 4x^2y^3$

Combine like terms.

Ans. $4a^3$ *Ans.* $7x^2y^3 - xy$

25. $4x^3 - x(3x^2)$

26. $3x^2(4x) - 2x^3$

27. $x^5 + 2x(x^4)$

28. $3b(a^2) + 3a^2(b) - a^2b$

29. $(x)(x^2) + (y^2)(2y) - x^3$

30. $x + x(xy) + x^2y$

31. $3a^2(b^2) - b^2$

32. $a^2 + 2a^2(3b)$

33. $2y(y^3) - 2y^4(3y)$

34. $6t^2(2t) + 4t(3)$

35. $2a(a^2) + 3a^2(a) - a^2$

36. $b^2(b^3) + 3b(b^2) - b^2(b)$

37. $4a(2b) - 2a(3b) + 3b$

38. $a(3b) + 2a(2b^2) - 3ab(b)$

39. $3a^2(ab) + a(a^2b) + 3a^3b$

40. $8a(ab^2) + 3ab(ab) - 2a^2b^2$

1.10 QUOTIENTS INVOLVING VARIABLES

From the meaning of a quotient (Section 1.2) and assuming that the denominator does not equal zero for any replacement of the variables *(as we shall in the remainder of this book)*, $\frac{x^5}{x^2} = x^3$ because x^3 is a number such that $(x^3)(x^2) = x^5$. The expressions

$$\frac{x^5}{x^2} \quad \text{and} \quad x^3$$

are equivalent. We may simplify quotients involving variable factors by expressing the dividend and divisor in completely factored form, and dividing out common factors. Using the previous example,

$$\frac{x^5}{x^2} = \frac{\overset{1}{\cancel{x}} \cdot \overset{1}{\cancel{x}} \cdot x \cdot x \cdot x}{\underset{1}{\cancel{x}} \cdot \underset{1}{\cancel{x}}} = x^3.$$

More generally,

The quotient of two powers with the same base equals a power with the same base with an exponent equal to the difference of the exponents of the powers.

If in the quotient $\frac{a^m}{a^n}$, n is less than m, we have

$$\frac{a^m}{a^n} = \frac{\overbrace{\overset{1}{\cancel{a}}\overset{1}{\cancel{a}}\overset{1}{\cancel{a}} \cdot \cdots \cdot \overset{1}{\cancel{a}}}^{n \text{ factors}}\overbrace{aa \cdot \cdots \cdot a}^{(m-n) \text{ factors}},}{\underbrace{\underset{1}{\cancel{a}}\underset{1}{\cancel{a}}\underset{1}{\cancel{a}} \cdot \cdots \cdot \underset{1}{\cancel{a}}}_{n \text{ factors}}}$$

and in exponential notation is written as

$$\frac{a^m}{a^n} = a^{m-n}.$$

Thus, as applied to the example above,

$$\frac{x^5}{x^2} = x^{5-2} = x^3.$$

In the case in which the dividend and divisor contain more than one variable and numerical coefficients as well, we have, for example,

$$\frac{12x^3y^4}{4x^2y} = \left(\frac{12}{4}\right)\left(\frac{x^3}{x^2}\right)\left(\frac{y^4}{y}\right) = 3x^{3-2}y^{4-1} = 3xy^3.$$

We sometimes show the process of finding quotients as follows:

$$\frac{\overset{3}{\cancel{12}}\,\overset{x}{\cancel{x^3}}\,\overset{y^3}{\cancel{y^4}}}{\underset{1}{\cancel{4}}\,\underset{1}{\cancel{x^2}}\,\underset{1}{\cancel{y}}} = 3xy^3.$$

Of course, if the exponents on the same variable in the dividend and divisor are equal, the quotient of these two powers is 1. Thus,

$$\frac{x^2}{x^2} = 1, \qquad \frac{y}{y} = 1, \qquad \text{and} \qquad \frac{z^5}{z^5} = 1.$$

EXERCISES 1.10

■ *Find each of the following quotients by first writing the expression in completely factored form, then simplify (if the expression is undefined, so state).*

Sample Problems

a. $\dfrac{24}{18}$

$$\frac{\overset{1}{\cancel{2}}\cdot 2\cdot 2\cdot \overset{1}{\cancel{3}}}{\underset{1}{\cancel{2}}\cdot 3\cdot \underset{1}{\cancel{3}}}$$

Ans. $\dfrac{4}{3}$

b. $\dfrac{12x^3}{4x}$

$$\frac{\overset{1}{\cancel{2}}\cdot\overset{1}{\cancel{2}}\cdot 3\overset{1}{\cancel{x}}xx}{\underset{1}{\cancel{2}}\cdot\underset{1}{\cancel{2}}\,\underset{1}{\cancel{x}}}$$

Ans. $3x^2$

c. $5a^3b^2c^3 \div bc^2$

$$\frac{5aaab\overset{1}{\cancel{b}}\overset{1}{\cancel{c}}\overset{1}{\cancel{c}}c}{\underset{1}{\cancel{b}}\underset{1}{\cancel{c}}\underset{1}{\cancel{c}}}$$

Ans. $5a^3bc$

1. $\dfrac{10}{2}$ **2.** $\dfrac{12}{4}$ **3.** $\dfrac{56}{7}$ **4.** $\dfrac{32}{8}$

5. $\dfrac{0}{6}$ **6.** $\dfrac{0}{8}$ **7.** $\dfrac{6}{0}$ **8.** $\dfrac{8}{0}$

9. $\dfrac{x^4}{x^3}$ **10.** $\dfrac{y^5}{y^2}$ **11.** $x^6 \div x^4$ **12.** $y^4 \div y$

13. $\dfrac{x^2y^4}{xy^2}$ **14.** $\dfrac{x^5y^3}{x^3y}$ **15.** $x^4y^2 \div xy$ **16.** $x^3y^5 \div x^2y^2$

17. $\dfrac{8x^2y^5}{4xy}$ **18.** $\dfrac{15x^4y^2}{3x^2y}$ **19.** $21x^2y^4 \div 7xy$ **20.** $12x^3y^3 \div 4xy^2$

■ *Find the quotients directly by finding the quotient of the numerical coefficients, and the quotient of the variable factors.*

Sample Problems

a. $\dfrac{\overset{2}{\cancel{4}}\overset{x}{\cancel{x^2}}\overset{y^2}{\cancel{y^3}}}{\underset{1}{\cancel{2}}\underset{1}{\cancel{x}}\underset{1}{\cancel{y}}}$ *Ans.* $2xy^2$

b. $\dfrac{\overset{1}{\cancel{5}}\overset{a}{\cancel{a^3}}b}{\underset{1}{\cancel{5}}\underset{1}{\cancel{a^2}}}$ *Ans.* ab

c. $\dfrac{\overset{3}{\cancel{12}}\overset{1}{\cancel{x^2}}\overset{y}{\cancel{y^2}}\overset{z}{\cancel{z^2}}}{\underset{1}{\cancel{4}}\underset{1}{\cancel{x^2}}\underset{1}{\cancel{y}}\underset{1}{\cancel{z}}}$ *Ans.* $3yz$

21. $\dfrac{a^3b^6}{b^2}$ **22.** $\dfrac{4b^2}{b^2}$ **23.** $\dfrac{6a^5b^4}{3a^5b}$ **24.** $\dfrac{3x^2y}{3x^2y}$

25. $\dfrac{0}{3x^2}$ **26.** $\dfrac{0}{x^2y}$ **27.** $\dfrac{a^2b^2c^2}{abc^2}$ **28.** $\dfrac{4abc^3}{2ac}$

29. $\dfrac{12a^7bc}{4a}$ **30.** $\dfrac{125x^3y^2z}{5x^3y^2z}$ **31.** $\dfrac{12a^2c^2}{0}$ **32.** $\dfrac{abc}{0}$

■ *Simplify.*

Sample Problem

$$\frac{2y^2 + y^2}{y} + 3y$$

Combine like terms above fraction bar; divide.

$$\frac{3\overset{y}{\cancel{y^2}}}{\underset{1}{\cancel{y}}} + 3y$$

$$3y + 3y$$

Combine like terms.

Ans. $6y$

33. $\dfrac{4x^2 + 2x^2}{x}$ **34.** $4x^2 + \dfrac{2x^2}{x}$ **35.** $\dfrac{3x^3}{x^2} - x$

36. $8x^4 + \dfrac{x^4}{x}$ **37.** $\dfrac{8x^4 + x^4}{x}$ **38.** $\dfrac{3x^2 + x^2}{x + 3x}$

39. $2x^2y + \dfrac{x^3y^2}{xy}$ **40.** $x^3y + \dfrac{2x^3y}{xy}$ **41.** $b^2 + \dfrac{b^2 + 5b^2}{2}$

42. $\dfrac{c^2 + 2c^2}{c} + 3c$ **43.** $\dfrac{0}{2} + c$ **44.** $\dfrac{0}{c} + c$

45. $\dfrac{x^2 + 2x^2}{x} + \dfrac{x^3}{x^2}$ **46.** $\dfrac{x + 5x}{3} + \dfrac{x^2 + 3x^2}{x}$ **47.** $\dfrac{y + 4y}{y} + \dfrac{6y^2 + 2y^2}{y}$

48. $\dfrac{2y^3 - y^3}{y} + \dfrac{3y^3 - y^3}{y^2}$ **49.** $\dfrac{2a^2 + a^2}{a - a} - 3a$ **50.** $4a^3 + \dfrac{3a^5 - a^5}{2a - 2a}$

CHAPTER REVIEW

1. List all prime numbers between 10 and 25.

2. Graph the first six natural numbers exactly divisible by four.

3. Express each word phrase in symbols.

a. The product of 6 and x.

b. Divide the sum of 4 and y by 6.

c. The product of y and the sum of 3 and x.

4. Write in exponential form.

a. $4aabbb$ b. $xyyzzz$ c. $3 \cdot 3ccd$

5. Write in completely factored form without exponents.

a. $6xy^2$ b. a^3b^2 c. $27cd^2$

6. Simplify.

a. $3 + 2(5)$ b. $8 + 0(4)$ c. $\frac{3^2 - 1}{4} + \frac{2^3 + 1}{3}$

7. If $a = 1$, $b = 0$, $c = 2$, find the value of each expression.

a. $a^2 + c$ b. $4a + 3b + c^2$ c. $\frac{c^2 - b}{2a}$

8. If $a = 2$ and $b = 3$, find the value of each expression.

a. $2(a + b)^2$ b. $2b^2 + 2a^2$ c. $(2b)^2 + (2a)^2$

■ *In Exercises 9–13, simplify the given expression.*

9. a. $3xy + 2y + 2xy + xy$ b. $6a^2 - a^2 - 3a$ c. $3r + 5s - r - s$

10. a. $(xy^2)(x^2y)$ b. $(3b^3)(2a)(2b)$ c. $(r^3)(s^2)(rs)$

11. a. $3x^2 - x^2(2x) + x^2$ b. $ab(b^2) - b^2$ c. $2r(rs^2) - r^2s^2$

12. a. $\frac{4a^2b}{2a}$ b. $\frac{3a^3b}{3a^3b}$ c. $\frac{12xy^3}{4y^2}$

13. a. $\frac{2x^2 + x^2}{x^2} - 2$ b. $\frac{4y - y}{3} + 6y$ c. $8a + \frac{6a^2 - a^2}{a}$

14. If x and y represent two numbers:

a. What is their sum? b. What is their product?

c. What is the quotient of x divided by y?

15. What are the parts of an algebraic expression separated by plus and minus signs called?

16. What is an algebraic expression consisting of two terms called?

17. In the expression $4x^3$, the number 4 is called the __?__ of x^3.

18. What is the numerical coefficient in the expression x^2?

19. What is the degree of the monomial $5x^3$?

20. What is the degree of the trinomial $2x^4 + x^2 + 6x$?

2.1 INTEGERS AND THEIR GRAPHICAL REPRESENTATION

On occasion, we use natural numbers to represent physical quantities such as money (5 dollars), temperature (20 degrees), and distance (10 miles). Since this representation does not differentiate between gains and losses, degrees above or below zero, or distances in opposite directions from a starting point, mathematicians have found it convenient to represent these ideas symbolically by the use of plus (+) or minus (−) signs. For example, we may represent:

A loss of five dollars as −\$5.
A gain of five dollars as +\$5.
Ten degrees below zero as $-10°$.
Ten degrees above zero as $+10°$.
Ten miles to the west of a starting point as −10 miles.
Ten miles to the east of a starting point as +10 miles.

Nonzero numbers whose numerals are preceded by a minus sign are called **negative numbers,** nonzero numbers whose numerals are preceded by a plus sign are called **positive numbers,** and together these kinds of numbers are called **signed numbers.** The signed numbers, together with the number 0 are called **integers.** When a numeral is written without a sign (e.g., 3, 5, 9), it is to be understood that a plus sign is intended. Furthermore, variables used in this chapter will represent integers unless otherwise specified.

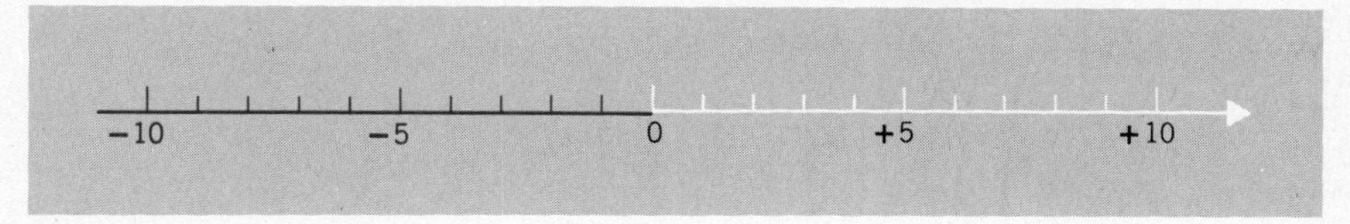

Figure 2.1

The line graph used to represent whole numbers can be extended to the left of the origin to include the graphs of negative numbers as shown in Figure 2.1. As before, the line graph is particularly useful in visualizing the relative order of two numbers.

The number whose graph lies on the left is less than the number whose graph lies on the right.

Thus, -4 is less than -2, -3 is less than 3, and -1 is less than 0. This notion is consistent with our physical experiences. A temperature of $-4°$ is lower or less than one of $-2°$, $-3°$ is less than $3°$, and $-1°$ is less than $0°$.

The **absolute value** of an integer is the value of the integer without regard to its sign and is designated symbolically by the use of two vertical bars.* Thus.

$$|-3| = 3, \quad |2| = 2, \quad |-5| = 5, \quad |0| = 0, \text{ etc.}$$

When we write $|-5| = 5$, we say we have *simplified the expression* $|-5|$.

EXERCISES 2.1

■ *Locate the graphs of the numbers on a line graph.*

Sample Problem

$-8, -6, -2, 1, 5$, and 9 (Graph on page 32.)

*Technically, absolute value is defined by $|x| = \begin{cases} x, \text{ if } x \text{ is greater than or equal to } 0 \\ -x, \text{ if } x \text{ is less than } 0. \end{cases}$

Line graph

Ans.

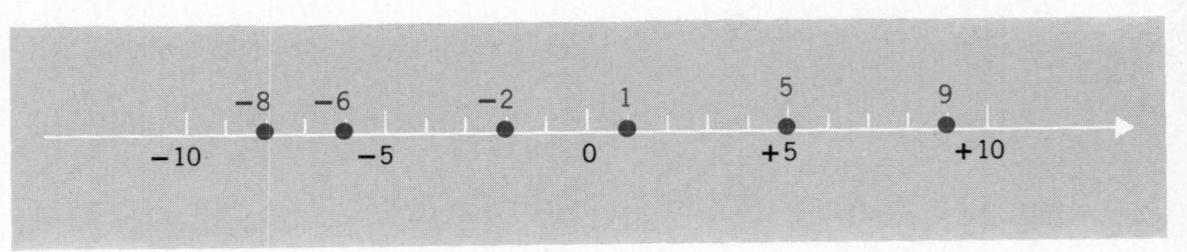

1. 0, −3, 5, −5, 2, and −1

2. −4, 3, −2, −1, 6, and 9

■ *Which number in each pair is the smaller? (Think of the number line).*

Sample Problems

a. 1, −3 b. −6, −9

Assign ODD 1–35

Ans.

a. −3 (the graph of −3 is to the left of the graph of 1).

b. −9 (the graph of −9 is to the left of the graph of −6).

3. 6, 8 **4.** 9, 4 **5.** 3, 0 **6.** 0, −2

7. 4, −2 **8.** −4, 2 **9.** −5, −20 **10.** 2, −20

11. −7, −9 **12.** −5, −2 **13.** −12, 5 **14.** −15, 1

15. −5, 0 **16.** 6, 0 **17.** −8, −2 **18.** −4, 4

■ *Locate the graphs of the numbers on a line graph.*

19. The natural numbers less than 10.

20. The integers between −3 and 7.

21. The even integers between −3 and 11.

22. The odd integers between −5 and 5.

23. The odd integers between −5 and −1.

24. The odd integers between −15 and 0.

■ *Simplify.*

Sample Problems

a. $|-7|$

b. $-|-5|$
$-(5)$

c. $|-2|^3$.
$(2)^3$

Ans. 7 *Ans.* −5 *Ans.* 8

25. $|5|$ **26.** $|-4|$ **27.** $|-8|$ **28.** $|6|$
29. $-|7|$ **30.** $-|-7|$ **31.** $-|-3|$ **32.** $-|3|$
33. $|-1|^3$ **34.** $|-3|^3$ **35.** $-|-2|^2$ **36.** $-|2|^2$

2.2 SUMS OF INTEGERS

To illustrate the meaning of the sum of two signed numbers, we may consider such numbers as representing gains and losses, (+) numbers denoting gains and (−) numbers denoting losses.

	+5 gain	−5 loss	−5 loss	+5 gain
	+3 gain	−3 loss	+3 gain	−3 loss
Sum:	+8 gain	−8 loss	−2 loss	+2 gain

These examples suggest a rule for the addition of integers.

To add two integers with
like signs: Add the absolute values of the numbers and prefix the common sign of the numbers to the sum.
unlike signs: Find the nonnegative difference of the absolute values of the numbers and prefix the sign of the number with the larger absolute value to the difference.
The sum of any integer *a* and 0 equals *a*.

In algebra, sums are usually written horizontally. In such an arrangement, the sum of +5 and +3 would appear as

$$+5 + (+3) = +8 \qquad \text{or} \qquad 5 + 3 = 8.$$

The sum of −5 and −3 would appear as

$$-5 + (-3) = -8 \qquad \text{or} \qquad -5 - 3 = -8,$$

where in the equation $-5 - 3 = -8$ the operation of addition is understood (−5 plus −3 equals −8). For the sum of −5 and +3, we would have

$$-5 + (+3) = -2 \qquad \text{or} \qquad -5 + 3 = -2,$$

and for the sum of +5 and −3,

$$+5 + (-3) = 2 \qquad \text{or} \qquad 5 - 3 = 2,$$

where again in the equation $5 - 3 = 2$ the operation of addition is understood (5 plus -3 equals 2).

In these examples, we observe two uses for the plus sign (+). It may be used as a sign of operation indicating the addition of two numbers, or it may be used as a sign of quality differentiating between a number and its negative. It is convenient to omit the addition (+) sign of operation and express $(-5) + (+3)$ as $-5 + 3$ and $(+5) + (-3)$ as $5 - 3$, etc., where all the signs are signs of quality. The operation is understood to be addition. This suggests the following rule.

In expressions involving only addition, parentheses which are preceded by a (+) sign may be dropped; each term within the parentheses retains its original sign.

We assume that the commutative and associative laws of addition hold with respect to integers. For example,

$$2 + (-3) = -3 + 2$$

and

$$[2 + (-3)] + 4 = 2 + [(-3) + 4].$$

EXERCISES 2.2

■ *Add.*

Sample Problems

a. $\begin{array}{r} -4 \\ \underline{-7} \end{array}$　　　　b. $\begin{array}{r} -5 \\ \underline{+2} \end{array}$

a. Since the signs are alike, add the absolute values

$$|-4| + |-7| = 4 + 7 = 11$$

and prefix the common sign "–" to the sum.

Ans. -11

b. Since the signs are unlike, find the nonnegative difference of the absolute values

$$|-5| - |+2| = 5 - 2 = 3$$

and prefix the "–" sign to 3 because -5 has a larger absolute value than $+2$.

Ans. -3

1. $\begin{array}{r} +5 \\ +4 \\ \hline \end{array}$	**2.** $\begin{array}{r} +6 \\ +2 \\ \hline \end{array}$	**3.** $\begin{array}{r} -4 \\ -6 \\ \hline \end{array}$	**4.** $\begin{array}{r} -8 \\ -5 \\ \hline \end{array}$
5. $\begin{array}{r} +6 \\ -1 \\ \hline \end{array}$	**6.** $\begin{array}{r} +8 \\ -5 \\ \hline \end{array}$	**7.** $\begin{array}{r} -4 \\ +3 \\ \hline \end{array}$	**8.** $\begin{array}{r} +5 \\ -9 \\ \hline \end{array}$
9. $\begin{array}{r} +7 \\ 0 \\ \hline \end{array}$	**10.** $\begin{array}{r} 0 \\ -8 \\ \hline \end{array}$	**11.** $\begin{array}{r} -6 \\ +6 \\ \hline \end{array}$	**12.** $\begin{array}{r} +8 \\ -8 \\ \hline \end{array}$

Sample Problems

a. $(-5) + (-7)$ b. $(+6) + (-2)$

a. The sum of the absolute values is

$$|-5| + |-7| = 5 + 7 = 12;$$

prefix this sum with the common sign "–".

Ans. -12

b. The nonnegative difference of the absolute values is

$$|+6| - |-2| = 6 - 2 = 4;$$

prefix this difference with "+" sign because $+6$ has the greater absolute value.

Ans. $+4$ or 4

13. $(+5) + (+3)$
14. $(+7) + (+5)$
15. $(-3) + (-8)$
16. $(-6) + (4)$
17. $(+8) + (-2)$
18. $(+6) + (-3)$
19. $(-5) + (+2)$
20. $(-12) + (+1)$
21. $(+5) + 0$
22. $0 + (-3)$
23. $(-6) + (+6)$
24. $(-4) + (+4)$
25. $(+4) + (+3)$
26. $(-4) + (-3)$
27. $(-2) + (0)$
28. $(-2) + (+4)$
29. $(+6) + (-3)$
30. $(-6) + (-3)$
31. $(-7) + (+9) + (+2)$
32. $(+8) + (-8) + (+4)$
33. $(-5) + (-2) + (+7)$
34. $(-6) + (-2) + (-3)$

35. $(+5) + (0) + (-8)$

36. $(-8) + (+8) + (0)$

37. $(+4) + (-2) + (-5) + (+3)$

38. $(+5) + (0) + (-9) + (-8)$

39. $(-6) + (0) + (+6) + (+3)$

40. $(-8) + (+4) + (+4) + (-2)$

2.3 SUMS INVOLVING VARIABLES

We learned in Chapter 1 that like algebraic terms may be combined by adding their numerical coefficients. Now, according to the laws of signs in Section 2.2, we can rewrite sums of terms with positive or negative coefficients. For example,

$$(+5x) + (+3x) = 5x + 3x = 8x,$$
$$(+5x) + (-3x) = 5x - 3x = 2x,$$
$$(-5x) + (-3x) = -5x - 3x = -8x,$$
$$(-5x) + (+3x) = -5x + 3x = -2x.$$

If a variable is preceded by a minus sign, the coefficient -1 is understood. Thus,

$$-x = -1 \cdot x,$$
$$-x^2 = -1 \cdot x^2, \text{ etc.}$$

EXERCISES 2.3

■ *Simplify.*

Sample Problems

a. $(+6x) + (-5x)$

$6x - 5x$

Ans. x

b. $(-4x^2) + (9x^2)$ — Remove ( ).

$-4x^2 + 9x^2$ — Add.

Ans. $5x^2$

1. $(2x) + (-3x)$

2. $(x) + (7x)$

3. $(-6y) + (4y)$

4. $(-3a) + (-4a)$

5. $(3x) + (-3x)$

6. $(-5a) + (-5a)$

7. $(6hk) + (-7hk)$

8. $(0) + (-3xz)$

9. $(-5cd) + (0)$

10. $(2x^2) + (5x^2)$

11. $(6xy) + (-5xy)$

12. $(-8abc) + (8abc)$

■ *Simplify.*

Sample Problems

a. $2x^2 - x + x^2 + 2x$

Ans. $3x^2 + x$

b. $(3x^3 - 2x^2 + x) + (x + 2x^2)$

Remove ( ).

$3x^3 - 2x^2 + x + x + 2x^2$

Combine like terms.

Ans. $3x^3 + 2x$

13. $3x^2 + 2x^2 - x$

14. $4y - 2y^2 - 6y$

15. $3a^2 - 4a - a$

16. $2x^2 - 9x^2 - 2x^2$

17. $x - x^2 + 4x$

18. $6y^2 - 3y - y$

19. $2x - x^2 - x^3 + 4x^2$

20. $y^3 - 2y^2 + y^2 - y$

21. $a^2 - 2a^3 + 4a^2 + 1$

22. $3a - 2b + c - b$

23. $(3x - y) + (3x - z)$

24. $2x^2 - 3y^2 + 2x^2 + y^2$

25. $3xy^2 - 2x^2y + xy^2 + 2x^2y$

26. $3ab - 4c + 3c - 2ab$

27. $(6x^2y - 5z) + (3 - 2z)$

28. $(6x^3 - 7xy) + (x^3 + xy)$

29. $7m^2 - 2m + 3 + 2m^2 - 3m$

30. $5h - 3k^2 + 2hk - 5h + 3k^2$

31. $(-2x^2 - 4x + 2) + (x^2 - 2)$

32. $3x^2y^2 - 2xy + x^2y^2 - 3xy$

33. $(ab^2 - 3a^2bt) + (abt^2 - a^2bt)$

34. $(2a^2b - abt) + (3ab^2 + 4abt)$

35. $3g^2 - 8ag + 6ag - 7g^2 + g^2 - 2g^2 + ag - 5ag$

36. $3 - a - a - 3 - 2 + a - a + 2 - 3 + 7 - 2a + 5$

37. $a^2bc - ab^2c - 3a^2bc + 2ab^2c + 3ab^2c - a^2bc$

38. $a + 2b + 3c - 3a - 2b - c + 4c - 2a - b$

39. $10ab - 14c + 21d - 3ab + 15c - 20d + 3c - 4d$

40. $-5x^2y + 3xy^2 + 4x^2y - 3xy^2 + x^2y - 2xy^2 + 6x^2y$

■ *Express the following word phrases by using algebraic expressions.*

41. Sum of -2 and x $x+(-2)$

42. Sum of $-x$ and 5

43. Sum of $2x$ and $-y$

44. Sum of $-x$ and $3y$

45. Sum of x, $-2y$, and $3z$

46. Sum of $5x$, $-2y$, and $-z$

47. Sum of y^2, $-3y$, and 2

48. sum of $3y^2$, $-6y$, and -4

2.4 DIFFERENCES OF INTEGERS

When discussing the difference of two natural numbers (Section 1.2), we observed that $5 - 3$ is a number, 2, which, when added to 3, gives 5; in general, $a - b$ is a number which, when added to b, gives a. The same idea holds for the difference of two integers. The following examples illustrate the process of subtraction:

$+5$ Subt. $\underline{+2}$	-5 Subt. $\underline{-2}$	$+2$ Subt. $\underline{+5}$
What number added to $+2$ gives $+5$? *Ans.* $+3$	What number added to -2 gives -5? *Ans.* -3	What number added to $+5$ gives $+2$? Ans. -3
-2 Subt. $\underline{-5}$	-5 Subt. $\underline{+2}$	$+2$ Subt. $\underline{-5}$
What number added to -5 gives -2? *Ans.* $+3$	What number added to $+2$ gives -5? *Ans.* -7	What number added to -5 gives $+2$? Ans. $+7$

These examples suggest a rule for the subtraction of integers.

To subtract an integer *b* from an integer *a*, change the sign of *b* and add the two according to the rules for adding integers.

In symbolic form, the rule is

$$a - b = a + (-b).$$

Using the rule for subtraction, we may simplify the above differences systematically as follows:

a.	$+5$ Subt. $\underline{+2}$	Change to and add	$+5$ $\underline{-2}$ $+3$	b.	-5 Subt. $\underline{-2}$	Change to and add	-5 $\underline{+2}$ -3
c.	$+2$ Subt. $\underline{+5}$	Change to and add	$+2$ $\underline{-5}$ -3	d.	-2 Subt. $\underline{-5}$	Change to and add	-2 $\underline{+5}$ $+3$
e.	-5 Subt. $\underline{+2}$	Change to and add	-5 $\underline{-2}$ -7	f.	$+2$ Subt. $\underline{-5}$	Change to and add	$+2$ $\underline{+5}$ $+7$

The same examples represented in horizontal form appear as:

a. $(+5) - (+2) =$
$(+5) + (-2) =$
$5 - 2 = +3$

b. $(-5) - (-2) =$
$(-5) + (+2) =$
$-5 + 2 = -3$

c. $(+2) - (+5) =$
$(+2) + (-5) =$
$2 - 5 = -3$

d. $(-2) - (-5) =$
$(-2) + (+5) =$
$-2 + 5 = +3$

e. $(-5) - (+2) =$
$(-5) + (-2) =$
$-5 - 2 = -7$

f. $(+2) - (-5) =$
$(+2) + (+5) =$
$+2 + 5 = +7$

The foregoing examples suggest the following rule.

In expressions involving only addition and subtraction, parentheses preceded by a (−) sign may be dropped, provided the sign of the term inside the parentheses is changed.

EXERCISES 2.4

■ *Subtract the bottom number from the top number.*

Sample Problem

a. $\begin{array}{r} +6 \\ \underline{+8} \end{array}$ Change to and add $\begin{array}{r} +6 \\ \underline{-8} \\ -2 \end{array}$

Ans. −2

b. $\begin{array}{r} -5 \\ \underline{-8} \end{array}$ Change to and add $\begin{array}{r} -5 \\ \underline{+8} \\ +3 \end{array}$

Ans. 3

1. $\begin{array}{r} +4 \\ \underline{+4} \end{array}$ **2.** $\begin{array}{r} +8 \\ \underline{+8} \end{array}$ **3.** $\begin{array}{r} +5 \\ \underline{+2} \end{array}$ **4.** $\begin{array}{r} +7 \\ \underline{+3} \end{array}$

5. $\begin{array}{r} +4 \\ \underline{+9} \end{array}$ **6.** $\begin{array}{r} +2 \\ \underline{+8} \end{array}$ **7.** $\begin{array}{r} -5 \\ \underline{0} \end{array}$ **8.** $\begin{array}{r} 0 \\ \underline{-5} \end{array}$

9. $\begin{array}{r} -9 \\ \underline{-2} \end{array}$ **10.** $\begin{array}{r} -4 \\ \underline{-1} \end{array}$ **11.** $\begin{array}{r} -6 \\ \underline{-8} \end{array}$ **12.** $\begin{array}{r} -1 \\ \underline{-4} \end{array}$

Sample Problems

a. $(+5) - (+2)$

$(+5) + (-2)$

Ans. 3

b. $(-6) - (-4)$

$(-6) + (+4)$

Ans. −2

Change signs and add.

13. $(+7) - (+2)$ **14.** $(+9) - (+1)$ **15.** $(+2) - (-5)$
16. $(+4) - (-9)$ **17.** $(-4) - (-6)$ **18.** $(-8) - (-9)$
19. $(-4) - (-2)$ **20.** $(-9) - (-1)$ **21.** $(4) - (2)$
22. $(8) - (4)$ **23.** $(3) - (7)$ **24.** $(1) - (9)$

Sample Problems

a. $(-5) + (-3) - (-4)$ b. $(6) - (5) + (-2)$

Remove ( ).

$-5 - 3 + 4$ $6 - 5 - 2$

Add.

Ans. -4 *Ans.* -1

25. $(-4) + (-5) - (+3)$ **26.** $(-2) - (-3) + (-4)$
27. $(+7) - (+2) + (-4)$ **28.** $(-6) + (+4) - (-2)$
29. $(+5) - (+2) - (-3)$ **30.** $(+6) - (-5) - (-1)$
31. $(4) - (2) + (-2)$ **32.** $(-2) + (3) + (5)$
33. $(-2) + (-2) - (-2) + (2)$ **34.** $(4) - (-4) + (4) - (4)$
35. $(-5) - (-2) + (3) + (1)$ **36.** $(8) - (5) + (-3) + (-1)$

2.5 DIFFERENCES INVOLVING VARIABLES

Similar to the process employed in Section 2.3 to rewrite sums of algebraic expressions, we can also rewrite differences in which the terms of the expression have positive or negative coefficients. For example,

$$\begin{aligned}(+7x) - (+2x) &= 7x - 2x = 5x,\\(+7x) - (-2x) &= 7x + 2x = 9x,\\(-7x) - (+2x) &= -7x - 2x = -9x,\\(-7x) - (-2x) &= -7x + 2x = -5x.\end{aligned}$$

Furthermore, if the expression inside the parentheses preceded by a "–" sign contains more than one term, the sign of *each* term is changed when the expression is written without parentheses. Thus,

$$-(a - b + c) = -a + b - c.$$

For example,

$$\begin{aligned}(+5x - 2y) - (+7x - 4y) &= 5x - 2y - 7x + 4y\\&= -2x + 2y\end{aligned}$$

and

$$(a^2 + 2b^2) - (4a^2 - b^2 + c^2) = a^2 + 2b^2 - 4a^2 + b^2 - c^2$$
$$= -3a^2 + 3b^2 - c^2.$$

EXERCISES 2.5

■ *Simplify.*

Sample Problems

a. $(6x) - (-2x)$

$6x + 2x$

Ans. $8x$

b. $(2y^2) - (4y^2) - (-5y^2)$

$2y^2 - 4y^2 + 5y^2$ Remove ( ).

Ans. $3y^2$ Add.

1. $(x) - (2x)$
2. $(3x) - (2x)$
3. $(-4x) - (-x)$
4. $(6x) - (5x)$
5. $(3y) - (-3y)$
6. $(-2y) - (2y)$
7. $(3xy) - (2xy)$
8. $(2xy) - (-xy)$
9. $(-2x^3y) - (3x^3y)$
10. $(-3xz) - (-2xz)$
11. $(3y^2) - (-y^2)$
12. $(-5y^2) - (y^2)$
13. $(3x) + (2x) - (-x)$
14. $(2y) - (-7y) - (3y)$
15. $(-6g) + (-3g) - (-7g)$
16. $(3rx) - (-2rx) + (-3rx)$
17. $(2ab^2) - (-ab^2) + (-3ab^2)$
18. $(6a) - (a) - (-2a)$

■ *Subtract the bottom polynomial from the top polynomial.*

Sample Problems

$$\begin{array}{r} 3x + 2y + z \\ x - 2y + z \\ \hline \end{array} \quad \text{Change to and add} \quad \begin{array}{r} 3x + 2y + z \\ -x + 2y - z \\ \hline 2x + 4y \end{array}$$

Ans. $2x + 4y$

19. $\begin{array}{r} 7a - 3b \\ 2a - 4b \\ \hline \end{array}$

20. $\begin{array}{r} 3xy - 2x + y \\ xy - 2x - y \\ \hline \end{array}$

21. $\begin{array}{r} x^2 - 4x + 3 \\ 2x^2 - 3x + 4 \\ \hline \end{array}$

22. $\begin{array}{r} 3a^2b \qquad\quad - 2 \\ -2a^2b + 2ab - 2 \\ \hline \end{array}$

23. $\begin{array}{l} 2x - 3y + 2z \\ 4x + 5y \\ \hline \end{array}$

24. $\begin{array}{r} -2x + 4 \\ 2x^2 + 3x + 1 \\ \hline \end{array}$

■ *Simplify.*

Sample Problem

$(3x^2 - 2y + z) - (-2x^2 + 3y - z)$ Remove ( ).

$3x^2 - 2y + z + 2x^2 - 3y + z$ Simplify.

Ans. $5x^2 - 5y + 2z$

25. $(3x^2 + 2x - 1) - (4x^2 - 2x + 3)$
26. $(7x^2 + 2x - 3) - (x^2 + 2x + 1)$
27. $(2y^2 - y + 1) - (3y^2 + 2y + 1)$
28. $(4x^2 - 2x - 1) - (3x^2 + x - 1)$
29. $(z^2 - 3z + 1) - (2z^2 + z + 1)$
30. $(y^2 - 3y + 4) - (y^2 + 2y - 3)$
31. $(2p^2 - 3p + 1) - (2p^2 - 3p + 1)$
32. $(y^2 - 3y + 1) - (2y^2 - 6y + 2)$
33. $(x^2y - xy + xy^2) - (2x^2y + 3xy)$
34. $(a^2b^2 + 2ab + 1) - (3 - ab)$
35. $(x^2y^2 - 2xy + 3) - (xy + 2)$
36. $(2g^2h + h - g) - (2g^2h + h)$
37. $(2xy^2 + 3xy - x) - (2xy + x)$
38. $(2ax^2 + 3ax + 4) - (2ax^2 - 3)$
39. $(x + y - z) + (x + y + 2z) - (x + y + z) + (3x - y + 2z)$
40. $(2x + y - z) + (x - 2y + z) - (x + y + 2z) - (x - 3y - 4z)$
41. $(a - b - c) + (a - b - c) - (a - b - c) + (a + b + c)$
42. $(2g + 3h - k) + (2g + h + k) - (2g + 2h + 2k) - (3g - h + k)$
43. $(2x + 2y - z) - (x + 2y - z) - (3x + 2y - z) + (x + 4y + 5z)$
44. $(a - b + c) - (2a + b - 2c) + (-a + b + c) - (a - 2b - 3c)$

Sample Problem

$3x - 2y - (4x + 2y) + x$ Remove ( ).

$3x - 2y - 4x - 2y + x$ Add.

Ans. $-4y$

45. $2x - y + (x + y)$
46. $3a - 2b + (2a + b)$
47. $2x + 3 - (x - 4)$
48. $(2x + 3) - x - 4$
49. $6a + 5b - (2a - 5b) - 2a$
50. $3c - 2d + 1 - (2 + 2c - d) + c - 1$

■ *Express the following word phrases by using algebraic expressions.*

51. x diminished by 4.
52. x diminished by -4.
53. y less x.
54. y less $-x$.
55. y less the quantity $(4 + x)$.
56. x less the quantity $(6 - 2x)$.

2.6 PRODUCTS OF INTEGERS

To find the product of two integers, we wish to adopt rules consistent with the rules for the multiplication of positive numbers and reflecting the properties of integers.

If we consider multiplication as a form of addition, that is, if we think of 3(2) as meaning the sum of three 2's $(2+2+2)$, then $3(-2)$ would represent the sum of three -2's, $(-2-2-2)$, which is -6. Thus it would appear that the product of a positive and a negative number should be a negative number.

To investigate the meaning of the product of two negative numbers, consider the sequence of products:

$$\begin{aligned} 4(-2) &= -8, \\ 3(-2) &= -6, \\ 2(-2) &= -4, \\ 1(-2) &= -2, \\ 0(-2) &= 0. \end{aligned}$$

If we continue the sequence on the left and give meaning to the product $-1(-2)$, it seems plausible to continue the sequence on the right with the number 2. That is, the sequence would continue:

$$\begin{aligned} -1(-2) &= 2, \\ -2(-2) &= 4, \\ -3(-2) &= 6, \text{ etc.} \end{aligned}$$

It appears (at least intuitively) that the product of two negative numbers should be a positive number. Therefore we shall adopt the rule.

To find the product of two integers, multiply the absolute values of the numbers. If the factors have like signs, the product is positive; if they have unlike signs, the product is negative. If one of the factors is 0, the product is zero.

Thus,

$$(+a)(+b) = +(ab), \qquad (+a)(-b) = -(ab),$$

$$(-a)(-b) = +(ab), \quad \text{and} \quad (-a)(+b) = -(ab).$$

Using this rule, we can determine the sign of the product of any number of factors. For example,

$$(-2)(3)(4) = (-6)(4) = -24,$$
$$(-2)(3)(-4) = (-6)(-4) = +24,$$
$$(-2)(-3)(-4) = (6)(-4) = -24.$$

We observe that if we have an *odd* number of negative factors, the product is negative; if we have an *even* number of negative factors, the product is positive.

Again, we assume that the commutative and associative laws of multiplication hold for signed numbers. For example,

$$(-2)(3) = 3(-2)$$

and

$$[(-2)(3)]\ (4) = (-2)\ [(3)(4)]).$$

The rules of signs for multiplying integers enable us to find products of monomials with positive or negative coefficients just as we found such products for monomials with natural number coefficients in Section 1.9. For example,

$$(-5x^2)(2x^3) = (-5)(2)(x^2)(x^3) = -10x^5$$

and

$$(-6xy^2)(-3x^3y) = (-6)(-3)(x)(x^3)(y^2)(y) = 18x^4y^3.$$

On page 36, we noted that

$$-x = -1 \cdot x,$$
$$-x^2 = -1 \cdot x^2, \text{ etc.}$$

Thus,

$$-3^2 = -1 \cdot 3^2 = -9.$$

On the other hand,

$$(-3)^2 = (-3)(-3) = 9.$$

Note that $-3^2 \neq (-3)^2$.

EXERCISES 2.6

■ *Multiply.*

Sample Problems

a. $+5$	b. -7	
$\underline{-2}$	$\underline{-3}$	
$-$	$+$	Determine sign of product.
10	21	Multiply numerical factors.
Ans. -10	*Ans.* $+21$ or 21	

1. $+5$ / $\underline{-3}$ **2.** -2 / $\underline{+4}$ **3.** -5 / $\underline{-6}$ **4.** $+4$ / $\underline{+8}$

5. $+5$ / $\underline{0}$ **6.** 0 / $\underline{-4}$ **7.** $+4$ / $\underline{-2}$ **8.** -3 / $\underline{-3}$

9. -4 / $\underline{+4}$ **10.** -2 / $\underline{-1}$ **11.** $+6$ / $\underline{-5}$ **12.** -6 / $\underline{+5}$

Sample Problems

a. $(-3)(-7)$	b. $(-4)(7)$	
$+$	$-$	Determine sign of product.
21	28	Multiply numerical factors.
Ans. $+21$ or 21	*Ans.* -28	

13. $(-6)(3)$ **14.** $(-5)(-4)$ **15.** $(2)(-8)$

16. $(6)(7)$ **17.** $(-4)(0)$ **18.** $(0)(5)$

19. $(2)(-3)(4)$ **20.** $(-2)(-3)(-4)$ **21.** $(-6)(-1)(2)$

22. $(4)(-1)(-7)$ **23.** $(-4)(0)(6)$ **24.** $(-5)(4)(0)$

Sample Problems

a. -4^2	b. $(-4)^2$	c. $-3 \cdot 2^2$
$-1 \cdot 4 \cdot 4$	$(-4)(-4)$	$-3 \cdot 2 \cdot 2$
Ans. -16	*Ans.* $+16$ or 16	*Ans.* -12

25. -2^2 **26.** $(-2)^2$ **27.** -2^3 **28.** $(-2)^3$

29. $-2 \cdot 3^2$ **30.** $(-2 \cdot 3)^2$ **31.** $(-1)^2 \cdot 3$ **32.** $(-1)^3 \cdot 3$

33. $-(-2)^2$ **34.** $-(-2)^3$ **35.** $(-1)(-4)^2$ **36.** $(-1)^2(-4)^3$

Sample Problems

a. $(-3x^2y)(2x)$ b. $(-2x^2y)(4x)(-3xy^2)$

Determine sign of product.

$-$ $+$

Multiply numerical factors.

-6 $+24$

Multiply variable factors.

Ans. $-6x^3y$ *Ans.* $+24x^4y^3$ or $24x^4y^3$

37. $3x(2xy)$ **38.** $2x(-2xy)$

39. $2x(-x^2y)$ **40.** $-xy(xy^2)$

41. $(-x)(x^2y)(-y)$ **42.** $(xy)(-x)(-y)$

43. $(-a)(-2a)(3a^2)$ **44.** $(3ab)(a^2b)(-3b)$

45. $(-abc)(-ab)(-bc)(b)$ **46.** $(ab)(b^2)(-ab)(c)$

47. $(-2)(3b)(-b^2)(b^3)$ **48.** $(-4)(-2a)(-a^2)(a^3)$

Sample Problems

a. $(-x)^2(-xy)$ b. $-x^2 \cdot x$

$(-x)(-x)(-xy)$ $-1 \cdot x \cdot x \cdot x$

Ans. $-x^3y$ *Ans.* $-x^3$

49. $(-x)^3$ **50.** $(-a)^2(a)$ **51.** $-(a)^3$

52. $(-x)^3(-xy)$ **53.** $-(-xy)^2(xy^2)$ **54.** $(-a)^2(ab)(-b^3)$

55. $(-3)^2(-x)^2(-y)^2$ **56.** $-2(-x)^3(y^2)$ **57.** $-2x(-x)^2(-y)^2$

58. $2x(-y)(-xy)^2$ **59.** $-2x(-3y)(-z)^2$ **60.** $(-3x^2)^2(y^2)^3$

■ *Express the following word phrases by using algebraic expressions and simplify.*

61. -4 multiplied by $3x$. **62.** $-5y$ multiplied by $-y$.

63. xy multiplied by $-y^2$. **64.** $-x^2$ multiplied by $-x^3y$.

65. Product of $-3xy$, x^2, and $-y^2$. **66.** Product of $4x$, $6x^2y$, and $-2xy^2$.

67. Product of -2, $(-x)^2$, and $-x^2$. **68.** Product of $3x^2$, $(3x)^2$, and $-3x^2$.

2.7 QUOTIENTS OF INTEGERS

The quotient of two integers is defined in the same way as the quotient of two natural numbers (Section 1.2); however, the sign of the quotient has to be consistent with the rule of signs for the multiplication of signed numbers. Recall that for natural numbers the quotient $\frac{a}{b}$ is the number q, such that $(b)(q) = a$, and let us examine the quotient of two signed numbers by considering a simple numerical case.

$$\frac{+6}{+3} = +2, \text{ because } (+3)(+2) = +6;$$

$$\frac{+6}{-3} = -2, \text{ because } (-3)(-2) = +6;$$

$$\frac{-6}{+3} = -2, \text{ because } (+3)(-2) = -6;$$

$$\frac{-6}{-3} = +2, \text{ because } (-3)(+2) = -6.$$

We may formalize these results in the following rule.

To find the quotient of two signed numbers, find the quotient of the absolute values of the numbers. If the dividend and divisor have like signs, the quotient is positive; if they have unlike signs, the quotient is negative.

As for whole numbers, a quotient equals zero if the dividend is zero and the divisor is not zero. Furthermore, as always, division by 0 is undefined. For example,

$$\frac{0}{3} = 0 \quad \text{and} \quad \frac{0}{-3} = 0,$$

however,

$$\frac{3}{0}, \quad \frac{-3}{0}, \quad \text{and} \quad \frac{0}{0} \quad \text{are undefined.}$$

EXERCISES 2.7

■ *Simplify.*

Sample Problems

a. $\frac{-12}{-2}$	b. $\frac{15}{-3}$	
$+$	$-$	Determine sign of quotient.
6	5	Divide absolute values of numerator and denominator.
Ans. $+6$	*Ans.* -5	

1. $\frac{-20}{-5}$ 2. $\frac{-24}{-8}$ 3. $\frac{-18}{9}$ 4. $\frac{16}{-4}$

5. $\frac{0}{5}$ 6. $\frac{0}{-2}$ 7. $\frac{6}{0}$ 8. $\frac{-3}{0}$

9. $-12 \div 2$ 10. $-20 \div (-4)$ 11. $12 \div (-3)$ 12. $-35 \div 5$

Sample Problem

a. $\frac{-24x^3y^3}{6xy^2}$	b. $\frac{-20x^2y^5}{-4xy^3}$	
$-$	$+$	Determine sign of quotient.
$\frac{\overset{4\ x^2y}{\cancel{24}\cancel{x^3}\cancel{y^3}}}{\underset{1\,1\,1}{\cancel{6}\cancel{x}\cancel{y^2}}}$	$\frac{\overset{5\ xy^2}{\cancel{20}\cancel{x^2}\cancel{y^5}}}{\underset{1\,1\,1}{\cancel{4}\cancel{x}\cancel{y^3}}}$	Divide out common factors.
Ans. $-4x^2y$	*Ans.* $5xy^2$	

13. $\frac{3x}{-x}$ 14. $\frac{-4x^2}{2x}$ 15. $\frac{-x^2y^3}{-xy}$ 16. $\frac{-16x^3}{-4x}$

17. $\frac{-6x}{6x}$ 18. $\frac{12y^3}{-12}$ 19. $\frac{16x^3}{-4x}$ 20. $\frac{-2xy}{y}$

21. $\frac{-6x^2y^3}{-xy}$ 22. $\frac{-x^3y^2}{-xy}$ 23. $\frac{6x^3}{-3x^2}$ 24. $\frac{3xy^3}{xy^2}$

25. $\frac{ax^2}{ax}$ 26. $\frac{8bc}{4c}$ 27. $\frac{-9xy^2}{3xy}$ 28. $\frac{-12c^2d^2}{12c^2d}$

29. $\frac{18x^3y}{-6xy}$ 30. $\frac{-36x^3y^3z^3}{18xy^2z^3}$ 31. $\frac{x^2y^2z^2}{x^2y^2z^2}$ 32. $\frac{26abc^2}{2ab}$

33. $\dfrac{-8x^2y^2}{-2xy}$ **34.** $\dfrac{-15a^3b^2}{-5a^3b^2}$ **35.** $\dfrac{30g^2h^3y^4}{15g^2h^2y^2}$ **36.** $\dfrac{24mn^2}{12mn}$

37. $\dfrac{-33x^2y}{11x^2y}$ **38.** $\dfrac{-26cd}{26c}$ **39.** $\dfrac{18xy^2z}{-xz}$ **40.** $\dfrac{-56x^2y^3z}{7x^2y^2z}$

■ *Simplify. (Review Section 1.4 for order of operations.)*

Sample Problem

$$y - \frac{3y + 5y}{4}$$

Combine like terms in numerator and divide.

$$y - \frac{\overset{2}{\cancel{8}}y}{\underset{1}{\cancel{4}}}$$

$$y - 2y$$

Combine like terms.

Ans. $-y$

41. $x - \dfrac{x + 2x}{x}$ **42.** $\dfrac{x^2 + 3x^2}{-2} - \dfrac{x^2}{x}$ **43.** $\dfrac{-3x + x}{-2} + x$

44. $x^2 + \dfrac{2x^2 - x^2}{x}$ **45.** $\dfrac{-xy}{x} + y$ **46.** $xy^2 + \dfrac{2x^3 - x^3}{x}$

47. $\dfrac{x^2y}{-y} + x^2$ **48.** $\dfrac{x^3y^3}{xy} - x^2$ **49.** $\dfrac{-xy^3}{y^3} - 2x$

50. $\dfrac{-y}{-y} - 6$ **51.** $\dfrac{x^2y^2}{-y} + \dfrac{-x^3y}{-x}$ **52.** $\dfrac{-y^2}{y} + \dfrac{-xy}{-x}$

53. $\dfrac{3x^2 - x^2}{-x^2} + \dfrac{4x^3 + 2x^3}{3x^3}$ **54.** $\dfrac{4y^3 + 8y^3}{6y^3 - 3y^3} - \dfrac{3x^3 + 7x^3}{5x^3 - 3x^3}$

55. $\dfrac{6xy^2 - 2xy^2}{2xy} + \dfrac{7x^2y + 8x^2y}{6x^2 - x^2}$ **56.** $\dfrac{14a^2b^2 - 2a^2b^2}{7ab - ab} - \dfrac{5ab^3 + 15ab^3}{(2b)^2}$

57. $\dfrac{24a(2b^3 - b^3)}{6b^2 - 2b^2} - \dfrac{15a^2b^2 + 3a^2b^2}{3a(4b - b)}$ **58.** $\left[\dfrac{(6a)^2 - (5a)^2}{10a + a}\right]^2 + \dfrac{7a^2b - a^2b}{4b - 10b}$

2.8 NUMERICAL EVALUATION

We may evaluate expressions using integers in the same manner that we evaluate expressions using natural numbers. (See Section 1.5.) It is helpful to use parentheses (as is done in the following sample problems) each time a substitution is made. Furthermore, the order for performing operations discussed in Section 1.4 is also valid for integers and should be followed carefully.

EXERCISES 2.8

■ *Find the value of each of the following expressions.*

Given $x = -2$.

Sample Problems

a. $2x^2 - 2x + 1$

$2(-2)^2 - 2(-2) + 1$

$2(4) + 4 + 1$

$8 + 4 + 1$

Ans. 13

b. $\dfrac{3x - 2}{4}$

$\dfrac{3(-2) - 2}{4}$

$\dfrac{-6 - 2}{4}$

$\dfrac{-8}{4}$

Ans. -2

1. $3x$ **2.** $-3x$ **3.** $4x^2$ **4.** $-4x^2$

5. $-x^2$ **6.** $(-x)^2$ **7.** $1 - 5x$ **8.** $2 + 3x$

9. $5x^3$ **10.** $-x^3$ **11.** x^4 **12.** x^5

13. $2x^2 + x$ **14.** $x^2 - x$ **15.** $-x^2 + 3x$ **16.** $4x^2 + 2x$

17. $\dfrac{2x + 2}{2} + 1$ **18.** $\dfrac{2x + 1}{3} - 2$ **19.** $\dfrac{x^2}{2} - \dfrac{2}{x}$ **20.** $\dfrac{6}{x} - \dfrac{4}{x^2}$

■ *Given* $x = 1$, $y = -2$.

Sample Problems

a. $2x - 3y$

$2(1) - 3(-2)$

$2 + 6$

Ans. 8

b. $2x^2 - 3xy + y^2$

$2(1)^2 - 3(1)(-2) + (-2)^2$

$2 + 6 + 4$

Ans. 12

21. xy **22.** $-2xy$ **23.** $x + y$

24. $x - y$ **25.** $2x + y$ **26.** x^2y

27. $-x^2y^2$
28. $3x^2y^3$
29. $-2x^3y^2$
30. $x^2 + y$
31. $2x^2 + y$
32. $x^2 - 2y^2$
33. $x^2 + xy + y^2$
34. $2x^2 - 3xy + y^2$
35. $-x^2 - y - y^2$
36. $2x^2 - (2y)^2$
37. $\frac{x - y}{3} + y^2$
38. $\frac{4x - 2y}{4} - 3y$

■ *Given* $a = 1$, $b = -2$, $c = -3$, $d = 0$.

39. $a + bc$
40. $a + b + c + d$
41. $2a + b - d$
42. $3a - 2b + 2c$
43. $-abc$
44. ab^2c
45. $a^2b^2c^2d^2$
46. $-abc^2$
47. $-3a^2bc$
48. $4abcd$
49. $a^2 + b^2$
50. $a^2 - b^2 - c$
51. $ab^2 - cd^2$
52. $a^2 + ac - b^2$
53. $2a^2 - 3bc + 2d^2$
54. $\frac{3ab}{c}$
55. $\frac{a - b}{-c}$
56. $\frac{a + cd - b}{bc - 3a}$
57. $\frac{bd}{ac} + \frac{bc^2}{a^2}$
58. $\frac{cb^2}{a} - \frac{cd}{b}$

CHAPTER REVIEW

1. Graph the integers $-6, -2, -1, 2, 5, 7$ on a line graph.

2. Graph the odd integers between -6 and 4 on a line graph.

3. Arrange the integers $2, -3, 5, -4, 0$ in order from smallest to largest.

4. What is the value of $|-3|$?

5. Add: a. $\begin{array}{r} +3 \\ \underline{-5} \end{array}$ b. $\begin{array}{r} -2 \\ \underline{-4} \end{array}$ c. $\begin{array}{r} -6 \\ \underline{7} \end{array}$

■ *Simplify.*

6. a. $(-2) + (-1)$ b. $(-3) + (0)$ c. $(-8) + (5)$

7. a. $5x + 7x$ b. $2x^2 + 5x^2 + 2x$ c. $3ab^2 - 2a^2b + ab^2 + 2a^2b$

8. Subtract the bottom number from the top number.

a. $\begin{array}{r} 6 \\ \underline{8} \end{array}$ b. $\begin{array}{r} -7 \\ \underline{-3} \end{array}$ c. $\begin{array}{r} 0 \\ \underline{-2} \end{array}$

■ *Simplify.*

9. a. $(-4) - (+3)$ b. $(3x) - (2x)$ c. $(-2x) - (-3x)$

10. a. $-2x + x + 3x$ b. $6x^2 + 2xy + 2x^2 - 3y^2 - 3xy$
c. $2xy^2 + 3xy + 3x^2y - 2x^2y + xy^2$

11. Subtract the bottom polynomial from the top polynomial.

a. $\begin{array}{r} 3x^2 - 2x + 1 \\ \underline{2x^2 - x + 2} \end{array}$ b. $\begin{array}{r} x^2 - 3x + 4 \\ \underline{2x - 9} \end{array}$ c. $\begin{array}{r} 2x^2 + 3y^2 - z^2 \\ \underline{3x^2 + z^2} \end{array}$

■ *Simplify.*

12. a. $(x + y + 2z) - (x - 2y + z)$ b. $(2a + 3b - 4c) - (a + b + c)$
c. $(x + y - 2z) + (2x + y - z) - (4x + 2y - 3z)$

13. a. $(3x)(4)$ b. $-3(-2x)$ c. $-(-2)^3$

14. a. $\frac{-15}{-3}$ b. $\frac{-8}{2}$ c. $\frac{-48}{-6}$

15. a. $\frac{xy}{-x}$ b. $\frac{-xyz}{xyz}$ c. $\frac{-4x}{-4}$

16. a. $\frac{3x - x}{2x} + 4$ b. $\frac{5x + 2x - 4x}{3x} - 1$ c. $\frac{3x^2 - 2x^2}{x} - \frac{2x^3 + 4x^3}{2x^2}$

17. a. $\frac{3x^2 - 5x^2}{x} + 6x$ b. $\frac{-x^3}{x^2} - \frac{12x}{3}$ c. $\frac{3x^2 - 4x^2 + x^2}{7} + 1$

18. If $x = -2$, $y = 3$, $z = 4$, find the value of each expression.

a. x^2 b. $\frac{2xy}{z}$ c. $\frac{3x + 2y}{z}$

19. If $x = -1$, $y = -2$, $z = 1$, find the value of each expression.

a. $\frac{-x^2y}{-z}$ b. $\frac{x^2 - z^2}{-2y}$ c. $\frac{x^2 - y}{z}$

20. Find the deviation (d) from the mean (M) of the measurement (m) if $d = m - M$, when

a. $m = 6$, $M = 8$ b. $m = -3$, $M = 4$ c. $m = 5$, $M = -2$

Cumulative Review

1. Graph the prime numbers between 17 and 25 on a line graph.
2. Graph the integers between −7 and 7 on a line graph.
3. Write in exponential form: $2 \cdot yyy$.
4. Write in completely factored form without exponents: $36xy^3$.
5. If $x = 1$, $y = 2$, $w = -2$, find the value of $x^2y^2w^2$.

■ *Simplify.*

6. $\frac{3 + 2 \cdot 6}{3} - 2^2 + (-2)^2$

7. $(-2 \cdot 3)^2 + (-2 \cdot 3^2) - (2 \cdot 3^2)$

8. $(2x)(x^2) - 3y^3 + x^3 + y^3$

9. $(3x)(x^2)(-x)$

10. $\frac{-3m^2n^2}{m^2n}$

11. $(a - 2b + c) + (2a - b + c) - (a + b + c)$

12. $\frac{3a^2b + 5a^2b}{4a^2} + b$

13. $(3x - x)^2 - (x - 3x)^2$

14. The number of feet (s) that an object falls in t seconds is given by the formula $s = 16t^2 + 24t$. How far will an object fall in 5 seconds?

15. If -3 is one of two factors of -12, what is the other factor?

16. If 2 and -3 are two factors of 30, what is the other factor?

17. $-a + b - c$ can be rewritten as $-($? $)$.

18. $-a - b + c$ can be rewritten as $-($? $)$.

19. For what value of y will the expression $\frac{3}{y}$ be undefined?

20. For what value of x will the expression $\frac{4 + x}{x}$ be undefined?

3.1 EQUATIONS AS SYMBOLIC SENTENCES

In this chapter, we wish to develop certain techniques to assist us in solving problems stated in words. These techniques involve symbolic representations of stated problems. For example, the stated problem,

"Find a number which, when added to 3, yields 7"

may be symbolized

$$3 + ? = 7; \qquad 3 + n = 7; \qquad 3 + x = 7; \text{ etc.},$$

where the symbols ?, n, or x are representations for the number we seek. We call such shorthand versions of stated problems **equations** or **symbolic sentences.** Equations such as $x + 3 = 7$ are more specifically identified as **first-degree equations** since the variable has an exponent of 1. The terms to the left of an equals sign are known as the **left-hand member** of the equation; those to the right comprise the **right-hand member.** Thus, in the equation $x + 3 = 7$, the left-hand member is $x + 3$ and the right-hand member is 7.

It is obvious that we do not need technical procedures to solve the equation $x + 3 = 7$, but simple examples are useful to illustrate concepts applicable to more complicated problems.

EXERCISES 3.1

■ *Write an equation that expresses each word sentence symbolically. In Exercises 1 to 12, use x as the variable.*

Sample Problem

Two times an integer added to three times itself is 25.

Ans. $3x + 2x = 25$ or $2x + 3x = 25$

1. The sum of 3 and an integer is 15.
2. The sum of a certain integer and 4 is 20.
3. Two times an integer added to four times the integer gives 42.
4. Four times an integer subtracted from six times the integer gives 10.
5. The sum of 8 and a certain integer is equal to twice the integer.
6. The sum of a certain integer and 4 is equal to five times the integer.
7. The sum of 3 and a certain integer is equal to twice the integer.
8. A certain integer increased by 24 is equal to three times the original integer.
9. If twice a certain integer is diminished by 5, the result is equal to the sum of 3 and the original integer.
10. If 5 is subtracted from three times a certain integer, the result is equal to the integer increased by 5.
11. Three times an integer exceeds the integer by 12.
12. If the sum of a certain integer and three times itself is divided by 2, the result is equal to the sum of the integer and 3.
13. The area (A) of a square is equal to the square of the length (s) of a side.
14. The perimeter (P) of a square is equal to four times the length (s) of a side.
15. The area (A) of a rectangle is equal to the product of its length (l) and its width (w).
16. The area (A) of a triangle is equal to one-half the product of its base (b) and altitude (h).
17. The volume (V) of a rectangular prism is found by multiplying its length (l) times its width (w) times its height (h).
18. The perimeter (P) of a rectangle is equal to the sum of twice its length (l) and twice its width (w).

3.2 SOLUTION OF EQUATIONS

Equations may be true or false as are word sentences. The equation $x + 3 = 7$ will be false if any number except 4 is substituted for the variable. The value of the variable for which the equation is true (4 in this example) is called a **solution** or **root** of the equation. We may determine whether a given number is or is not a solution of a given equation by substituting the number in place of the variable and determining the truth or falsity of the result. A first-degree equation has only one solution.

EXERCISES 3.2

■ *Determine whether each equation is true for the indicated value of the variable; that is, determine whether the given number is, or is not, the solution of the given equation.*

Sample Problems

a. $3x + 4 = 10$, for $x = 2$

$$3(2) + 4 = 10$$
$$10 = 10$$

Ans. 2 is the solution.

b. $y - 4 = 3y + 1$, for $y = -2$

$$(-2) - 4 = 3(-2) + 1$$
$$-6 = -5$$

Ans. -2 is not the solution.

1. $x + 9 = 12$, for $x = 3$
2. $2x + 4 = -2$, for $x = -3$
3. $0 = 4 + y$, for $y = -4$
4. $y - 3 = 0$, for $y = -3$
5. $3a + 8 = 14 + a$, for $a = 3$
6. $a - 3 = 3a$, for $a = -2$
7. $8x - 3 = -2x + 4$, for $x = 3$
8. $4 - y = 2y$, for $y = 2$
9. $0 = 6r - 24$, for $r = 4$
10. $0 = 2r + 12$, for $r = -6$
11. $\dfrac{x}{-5} = -3$, for $x = 15$
12. $\dfrac{2x}{3} = x - 2$, for $x = 6$
13. $\dfrac{x}{4} - 3 = x + 2$, for $x = 8$
14. $\dfrac{2x}{3} - 5 = \dfrac{3x}{4} - 6$, for $x = 12$
15. $\dfrac{x + 3}{4} = 2$, for $x = 5$
16. $\dfrac{x - 2}{5} = x - 10$, for $x = 12$
17. $2x - b = x + b$, for $x = 2b$
18. $x + a = 3x + 5a$, for $x = -2a$
19. $2y - a = -5a$, for $y = -2a$
20. $\dfrac{x}{2} - b = 5b + x$, for $x = -12b$
21. $3x + b = 5x - 5b$, for $x = 3b$
22. $x - 3a = 4x + a$, for $x = -a$
23. $\dfrac{y - 3c}{c} = -8c$, for $y = -5c$
24. $\dfrac{3y + 2d}{-2} = 3y + 4d$, for $y = -2d$

3.3 SOLUTION OF EQUATIONS USING ADDITION AND SUBTRACTION PROPERTIES

Equivalent equations are equations that have identical solutions. Thus,

$$3x + 3 = x + 13, \quad 3x = x + 10, \quad 2x = 10, \quad \text{and} \quad x = 5$$

are equivalent equations, because 5 is the only solution of each of them. In solving any equation, we transform a given equation whose solution may not be obvious to an equivalent equation whose solution is readily discernible.

Since an equation is simply a statement that the left-hand and right-hand members are different names for the same number, same quantities added to or subtracted from each member will produce another equality. Thus, if

$$x + 3 = 7,$$

then

$$x + 3 - 3 = 7 - 3,$$

or

$$x = 4.$$

We formalize these properties in the following assumption or **axiom** as such statements are sometimes called.

If the same quantity is added to or subtracted from equal quantities, the resulting quantities are equal.

In symbols this axiom is expressed as follows:

$$\text{If} \quad a = b, \quad \text{then} \quad a + c = b + c.$$

The addition or subtraction axiom can be applied to transform a given equation to an equivalent equation of the form $x = a$, from which the solution can be obtained by inspection. For an equation such as

$$2x + 1 = x - 2,$$

we wish to obtain an equivalent equation in which all terms containing x are in one member and all terms not containing x are in the other in order to have x by itself as one member. If we first add -1 to (or subtract 1 from) each member, we obtain

$$2x + 1 - 1 = x - 2 - 1,$$
$$2x = x - 3.$$

If we now add $-x$ to (or subtract x from) each member, we obtain

$$2x - x = x - 3 - x,$$
$$x = -3,$$

from which the solution -3 is obvious. (Note that we are using color to indicate quantities to be added to or subtracted from each member of the equation.)

Since each equation obtained in the process is equivalent to the original equation, -3 is also a solution of $2x + 1 = x - 2$. The solution can be checked by substitution. Thus,

$$2(-3) + 1 = (-3) - 2,$$
$$-5 = -5.$$

The *symmetric property of equality* is also helpful in the solution of equations. This property asserts:

$$\text{If} \quad a = b, \quad \text{then} \quad b = a.$$

This enables us to interchange the members of an equation whenever we please without having to be concerned with any changes of sign. Thus,

$$\text{if} \quad 4 = x + 2, \quad \text{then} \quad x + 2 = 4;$$
$$\text{if} \quad x + 3 = 2x - 5, \quad \text{then} \quad 2x - 5 = x + 3;$$
$$\text{if} \quad d = rt, \quad \text{then} \quad rt = d; \text{ etc.}$$

EXERCISES 3.3

■ *Write an equation equivalent to the given equation.*

Sample Problem

$x - 4 = 7$, by adding 4 to each member.

$x - 4 + 4 = 7 + 4$

Ans. $x = 11$

1. $x + 2 = 9$, by adding -2 to each member.

2. $x - 6 = 8$, by adding 6 to each member.

3. $4 - y = 0$, by adding y to each member.

4. $8 = 5 + y$, by adding -5 to each member.

5. $3z = 2z + 8$, by adding $-2z$ to each member.

6. $z - 9 = 16$, by adding 9 to each member.

■ *Write an equation equivalent to the given equation.*

Sample Problem

$4x - 2 - 3x = 4 + 6$, by (a) combining like terms and (b) adding 2 to each member.

(a) $x - 2 = 10$

(b) $x - 2 + 2 = 10 + 2$

Ans. $x = 12$

7. $2x - x + 5 = 8$, by (a) combining like terms and (b) adding -5 to each member.

8. $3y - 4 + 2 - 2y = 8$, by (a) combining like terms and (b) adding 2 to each member.

9. $2z = 9 + z$, by (a) adding $-z$ to each member and (b) combining like terms.

10. $3x + 4 - 2x = 8 + 3$, by (a) combining like terms and (b) adding -4 to each member.

11. $3y - 5 - 2y = 7$, by (a) combining like terms and (b) adding 5 to each member.

12. $3z - z = z + 1$, by (a) combining like terms and (b) adding $-z$ to each member.

■ *Solve each equation.*

Sample Problem

$x + 7 = 12$

Add -7 to each member.

$x + 7 - 7 = 12 - 7$

Combine like terms.

*Ans.** $x = 5$ *Check.* $(5) + 7 = 12$

13. $x + 3 = 8$ **14.** $5 + y = 9$ **15.** $z - 3 = 8$

16. $x - 4 = 0$ **17.** $2 + z = 7$ **18.** $-3 + y = 12$

*The solution of the original equation is the number 5; however, the answer is often displayed in the form of the trival equation $x = 5$.

Sample Problem

$3x - x = 5 + x$

Combine like terms.

$2x = 5 + x$

Add $-x$ to each member.

$2x - x = 5 + x - x$

Combine like terms.

Ans. $x = 5$

19. $4 + y - 7 = 6$

20. $5 + x - 2 = 8$

21. $2x - x = 4$

22. $3y - 2y = 6$

23. $4x = 3x + 5$

24. $5x = 4x - 6$

25. $6x = 7 + 5x$

26. $8x = 3 + 7x$

27. $3x + 5 = 2x + 3$

28. $7y - 2 = 6y + 4$

29. $6a - a = 4a + 2$

30. $4a + a = 4a - 3$

31. $6 = 4x - 3x + 2$

32. $8 = 5x - 4x - 5$

33. $2 + 3b = 4b - 1$

34. $6b - 4 = 7b + 2$

35. $4x - 2 - 3x + 5 = 0$

36. $7 - 2x + 3x + 1 = 0$

37. $0 = 4x + 5 - 3x$

38. $0 = 5x - 4x - 6$

39. $2r + 3r = 4r + 1$

40. $6s - 4s = s - 2$

41. $8 + 3t - t = 6 + t$

42. $u + 2u + 3u = 5u - 1$

43. $5z + 3 + 6z - 1 = 5z + 3 + 2 + 5z$

44. $6t + 7 - 3t - 2 = 4t + 6 - 2t - 2$

45. $2(4t - t) + 6 = 2(2t + t) + 8 - t$

46. $-3(x - 3x) + 5 = -4(3x - x) + 7 + 13x$

47. $\frac{4x - 2x}{2} + 3(x + 2x) = 2(3x + x) + x$

48. $5(2x + x) - \frac{3(2x + x)}{9} = 2(3x + 4x) + x$

49. $\frac{3(6y - y)}{5 - 2} + \frac{8y - 2y}{3} = 2(5y - 2y) + 4$

50. $\frac{3(5y - y)}{4 + 2} - \frac{4(2y - y)}{2} = 8 + y$

3.4 SOLUTION OF EQUATIONS USING THE DIVISION PROPERTY

For reasons similar to those used in justifying the addition and subtraction axiom, we may divided each member of an equality by the same nonzero number and produce an equivalent equation. More formally, we have the following rule.

If equal quantities are divided by the same (nonzero) quantity, the quotients are equal.

This rule can be represented in symbols as follows:

$$\text{If} \quad a = b, \quad \text{then} \quad \frac{a}{c} = \frac{b}{c} \ (c \neq 0).$$

In solving equations, we use this axiom to produce equivalent equations in which the variable has a coefficient of 1.

EXERCISES 3.4

■ *Write an equation equivalent to the given equation.*

Sample Problem

$-4x = 12$, by dividing each member by -4.

$$\frac{\overset{1}{\cancel{-4}}x}{\underset{1}{\cancel{-4}}} = \frac{\overset{-3}{\cancel{12}}}{\underset{1}{\cancel{-4}}}$$

Ans. $x = -3$

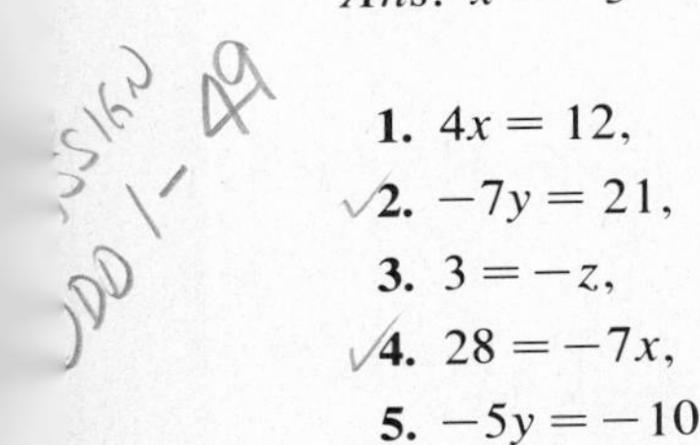

1. $4x = 12$, by dividing each member by 4.
2. $-7y = 21$, by dividing each member by -7.
3. $3 = -z$, by dividing each member by -1.
4. $28 = -7x$, by dividing each member by -7.
5. $-5y = -10$, by dividing each member by -5.
6. $3z = -18$, by dividing each member by 3.

■ *Write an equation equivalent to each given equation.*

7. $3x - x = 8$, by (a) combining like terms and (b) dividing each member by 2.
8. $5y - y = 7 + 9$, by (a) combining like terms and (b) dividing each member by 4.

9. $24 - 9 = z - 4z$, by (*a*) combining like terms and (*b*) dividing each member by -3.

10. $7 + 5 = x - 5x$, by (*a*) combining like terms and (*b*) dividing each member by -4.

11. $4y + 5y = 4 + 14$, by (*a*) combining like terms and (*b*) dividing each member by 9.

12. $4z + z = 16 - 1$, by (*a*) combining like terms and (*b*) dividing each member by 5.

■ *Solve each equation.*

Sample Problem

$3y + 2y = 15$ Combine like terms.

$5y = 15$ Divide each member by 5.

Ans. $y = 3$

Check. $3(3) + 2(3) = 15$

13. $5x = 15$ **14.** $6y = 24$ **15.** $12 = 3y$

16. $35 = 7x$ **17.** $-12 = 3y$ **18.** $-25 = 5z$

19. $-x = 6$ **20.** $-y = 3$ **21.** $8z - 2z = 12$

22. $7x - 2x = 20$ **23.** $2y + 4y = -6$ **24.** $2z + z = -18$

Sample Problem

$2x + 5 = 5x - 4$ Add $+4$ and $-2x$ to each member.

$2x + 5 + 4 - 2x = 5x - 4 + 4 - 2x$ Combine like terms.

$9 = 3x$ Divide each member by 3.

$3 = x$ Exchange members.

Ans. $x = 3$

25. $5 = 6p - 13$ **26.** $8 = 3y + 2$ **27.** $13 = 7x + 6$

28. $7 = 2p - 9$ **29.** $-5 = -2 - 3x$ **30.** $-20 = 1 - 7t$

31. $-6t = 3t$ **32.** $7r = 5r$ **33.** $2x - 2 = 6$

34. $4z - 3 = 9$

35. $-3a + 4 = 2a + 24$

36. $2x = -4x + 6$

37. $7x = 14 + 5x$

38. $5y + 3 = 13 - 5y$

39. $3d + 2 - 4d = 6$

40. $3x - 4 = 4x + 2$

41. $3 = 6x - 3 - 3x$

42. $30 = 6r - 24 + 3r$

43. $2t - 8 = 0$

44. $0 = 3t + 21$

45. $6(2y + y) - 2(3y - y) = 2y + 24$

46. $6(y - 3y) + 2(6y - y) + 8 = 3(2y - y) + 23$

47. $\frac{3(6y - 2y)}{4} + 2(3y + y) - 6 = 2(y + 2y) + 9$

48. $\frac{2(3t + 2t)}{7 - 2} - \frac{4(t + 2t)}{7 - 5} = 2(t + 3t) + \frac{4(2t - 6t)}{5 - 3}$

49. $\frac{3x(5 - 2)}{4 - 1} - 3(2x - 4x) = x(2^2 - 1) + 24$

50. $\frac{4(6z + 5z)}{5 - 3} - \frac{z(17 - 2)}{3 + 2} = 18(2z + z) - 70$

3.5 SOLUTION OF EQUATIONS USING THE MULTIPLICATION PROPERTY

For reasons identical with those discussed in the preceding two sections, we have the following rule.

If equal quantities are multiplied by the same quantity, their products are equal.

This rule can be represented in symbols by:

$$\text{If} \quad a = b, \quad \text{then} \quad a \cdot c = b \cdot c.$$

In solving equations we use this axiom to produce equivalent equations that are free of fractions.

Exercises 3.5

■ *Write an equation equivalent to the given equation.*

Sample Problem

$\frac{x}{3} = -4,$ by multiplying each member by 3.

$$\overset{1}{\cancel{3}} \cdot \frac{x}{\underset{1}{\cancel{3}}} = 3(-4)$$

Ans. $x = -12$ *Check.* $\frac{-12}{3} = -4$

1. $\frac{x}{2} = 6,$ by multiplying each member by 2.
2. $\frac{y}{3} = -8,$ by multiplying each member by 3.
3. $4 = \frac{z}{-5},$ by multiplying each member by -5.
4. $-3 = \frac{x}{-4},$ by multiplying each member by -4.
5. $\frac{y}{3} = 6,$ by multiplying each member by 3.
6. $\frac{z}{-2} = -4,$ by multiplying each member by -2.

■ *Solve each equation.*

Sample Problems

a. $\frac{x}{2} = 15$ b. $\frac{3x}{5} = 9$

Multiply each member in Problem a by 2 and in Problem b by 5.

$$\overset{1}{\cancel{2}}\left(\frac{x}{\underset{1}{\cancel{2}}}\right) = 2(15) \qquad \overset{1}{\cancel{5}}\left(\frac{3x}{\underset{1}{\cancel{5}}}\right) = 5(9)$$

$$3x = 45$$

In Problem b, divide each member by 3.

Ans. $x = 30$ *Ans.* $x = 15$

7. $\frac{x}{2} = 8$ **8.** $\frac{x}{3} = 5$ **9.** $7 = \frac{y}{6}$ **10.** $9 = \frac{y}{4}$

11. $\frac{z}{3} = -4$ **12.** $\frac{z}{7} = -4$ **13.** $-6 = \frac{x}{3}$ **14.** $-5 = \frac{x}{9}$

15. $\frac{y}{-2} = 5$ **16.** $\frac{y}{-4} = 6$ **17.** $\frac{z}{4} = 0$ **18.** $\frac{z}{5} = 0$

19. $\frac{2a}{5} = 8$ **20.** $\frac{4a}{5} = -12$ **21.** $8 = \frac{2b}{3}$ **22.** $4 = \frac{2b}{5}$

23. $\frac{2c}{4} = -10$ **24.** $\frac{4c}{2} = -8$ **25.** $-8 = \frac{4a}{5}$ **26.** $-10 = \frac{5a}{3}$

27. $\frac{-b}{3} = 12$ **28.** $\frac{-b}{5} = 6$ **29.** $\frac{-2c}{3} = -10$ **30.** $\frac{-3c}{4} = -12$

31. $\frac{2a}{3} = -8$ **32.** $\frac{4a}{5} = -8$ **33.** $\frac{-b}{2} = 16$ **34.** $\frac{-b}{6} = 5$

35. $\frac{2c}{3} = -2$ **36.** $\frac{3c}{4} = -3$ **37.** $15 = \frac{3a}{5}$ **38.** $12 = \frac{4a}{3}$

39. $\frac{4t - t}{6} = 5$ **40.** $\frac{7t - t}{18} = -2$

41. $\frac{2(4t + 6t)}{40} = \frac{7 + 2}{3}$ **42.** $\frac{3(t - 3t)}{9} = \frac{2 - 18}{4}$

43. $\frac{3^2 + 4^2}{5} = \frac{2(3x + 5x)}{48}$ **44.** $\frac{7^2 - 5^2}{-6} = \frac{2y + 4}{9}$

3.6 FURTHER SOLUTIONS OF EQUATIONS

We are now in possession of all the techniques necessary to solve most first-degree equations. There is no specific order in which the axioms are to be applied. Any one or more of the following steps may be appropriate.

1. Combine like terms in each member of an equation.
2. By application of the addition or subtraction axiom, write the equation with all terms containing the unknown in one member and all terms not containing the unknown in the other.
3. Combine like terms in each member.
4. Use the multiplication axiom to remove fractions.
5. Use the division axiom to obtain a coefficient of 1 for the variable.

EXERCISES 3.6

■ *Solve.*

Sample Problem

$$\frac{4x - 2x}{3} = 2$$

Combine like terms, $4x - 2x$.

$$\frac{2x}{3} = 2$$

Multiply each member by 3

$$\overset{1}{\cancel{3}}\left(\frac{2x}{\underset{1}{\cancel{3}}}\right) = 3(2)$$

Divide each member by 2

$$\frac{\overset{1}{\cancel{2}}x}{\underset{1}{\cancel{2}}} = \frac{6}{2}$$

Ans. $x = 3$

1. $8 = x + 3$
2. $x + 4 = 5$
3. $9 = 3y$
4. $5y - 36 = 2y$
5. $3z - 5z = 6$
6. $4z - 7z = 12$
7. $a = 8 - a$
8. $2a = 6 - a$
9. $\frac{b}{2} = 6$
10. $7 = \frac{b}{5}$
11. $7 - 2c = c + 1$
12. $c + 2 = 6 - 3c$
13. $x - 3 = 2x$
14. $12 + 4x = 6x$
15. $-6 = \frac{3y}{5}$
16. $12 = \frac{2y}{3}$
17. $4 + 3z = z$
18. $z + 11 = 2z$
19. $\frac{4x - 2x}{2} = 8 - 5$
20. $\frac{x - 3x}{4} = 9 - 2$
21. $4x - 3 = 2x + 5$
22. $6x - 5 = 2x + 7$
23. $2x - 3 + 2x = 4 - x + 8$
24. $2y - 3 + 3y = 4y + 2$
25. $6z + 5 - 7z = 10 - 2z + 3$
26. $6a - 4 + 2 = 3a + 1$
27. $5y + 3 - y = 10 + y + 2$
28. $3x + 4 - 5x + 2 = 0$
29. $5x + 7 - 2x - 16 = 0$
30. $0 = 7 - 2x + 3 - 3x$

31. $0 = 3x + 5 - 7x + 3$

32. $-2x - 5x = 2x + 16 - x$

33. $\frac{3x + 2x}{2} = \frac{10 + 4}{2} - 2$

34. $\frac{5x - 3x}{4} = \frac{8 + 2}{5} - 2$

35. $\frac{4x + x}{3} = 10$

36. $0 = \frac{5x}{2} + 10$

37. $0 = 6 - \frac{2y}{3}$

38. $4 = 1 - \frac{3x}{7}$

39. $6 - x = 6 + 2x$

40. $8 + x = 8 - 5x$

41. $3x - 14 = 5x - 4x + 2$

42. $3y(7 - 2) + 17 = 16y + y - 1$

43. $\frac{(3z - z)}{6} + 3 = 2 + 9$

44. $\frac{5(4x - x)}{3} + x(3^2 - 1) = x - 36$

45. $\frac{8(2t + 5t)}{3^2 - 2} + \frac{2(3t + t)}{2^2} = 8(2t + t) + 28$

46. $\frac{6(5u - u)}{2^3} - \frac{5(u + 5u)}{3} = 3(2u - u) + 30$

3.7 LITERAL EQUATIONS—FORMULAS

It is often necessary to solve equations or formulas in which there is more than one variable. We may solve for any specified variable in terms of the others. The procedures used are identical to those developed in the preceding sections.

EXERCISES 3.7

■ *Solve for x.*

Sample Problem

$$5ax - 2c = 2ax$$

Add $-2ax$ and $+2c$ to each member.

$$5ax - 2c - 2ax + 2c = 2ax - 2ax + 2c$$

Combine like terms.

$$3ax = 2c$$

Divide each member by $3a$.

Ans. $x = \frac{2c}{3a}$

1. $x - a = 0$
2. $x - 3a = 0$
3. $3x - 3a = x - a$
4. $2x + 3a = 9a - x$
5. $5a - 2x = 2a - x$
6. $5a + x = 2x - a$

■ *Solve for y.*

7. $ay - b = 0$
8. $2ay + 2b = 4b + ay$
9. $3a - 3by = -9a + by$
10. $2aby + 6a = aby$
11. $5aby - 3b = 7b - aby$
12. $3ay - 4ab + ay = 0$

■ *Solve for x or y.*

13. $cx - a^2 = 0$
14. $dy - a^2 + 3dy = 0$
15. $bx + 2b = 5b - bx + b$
16. $ax - 2ar^2 = ar^2$
17. $\frac{a}{b}x - c = 0$
18. $0 = \frac{b}{c}y + a$

■ *Solve each of the following formulas for the symbol in color.*

Sample Problem

$$c = 2\pi r$$

Exchange members.

$$2\pi r = c$$

Divide each member by 2π.

$$\frac{\overset{1}{\cancel{2}}\overset{1}{\cancel{\pi}}r}{\underset{1}{\cancel{2}}\underset{1}{\cancel{\pi}}} = \frac{c}{2\pi}$$

Ans. $r = \frac{c}{2\pi}$

19. $d = rt$
20. $v = k + gt$
21. $v = lwh$
22. $f = ma$
23. $c = \pi d$
24. $I = prt$
25. $d = rt$
26. $v = lwh$
27. $v = lwh$
28. $f = ma$
29. $I = prt$
30. $I = prt$
31. $v = k + gt$
32. $s = \frac{at^2}{2}$
33. $F = \frac{kmM}{d^2}$
34. $F = \frac{kmM}{d^2}$

■ *Solve for x.*

35. $a(bx - 2bx) = a^2b$
36. $k(3cx + cx) = 8c^2k^2$
37. $\frac{c^2x + c^2x}{2c} = c^3 - cx$
38. $\frac{b^2x - 3b^2x}{b} = 6b^2 - 4bx$
39. $\frac{acx + 3acx}{2a} + \frac{4bcx + 6bcx}{20b} = c(4x - 2x) + c^2$

40. $\dfrac{2a^2bx + a^2bx}{ab} + \dfrac{6ab^2x - ab^2x}{12b^2} = a(4x - x) + 10a^2$

3.8 APPLICATIONS

In this section, we shall first be concerned with representing word sentences symbolically, that is, in finding equations that represent stated problems. Then we shall solve a variety of word problems.

To aid in finding a correct symbolic representation of a stated problem, the following steps are suggested:

1. Represent the quantity you wish to find both as a symbol and as a word phrase.
2. Where applicable, draw a sketch and indicate all known quantities thereon.
3. Write an equation which represents symbolically a word sentence relating the known and unknown quantities. The equation may be obtained from:
   a. The problem itself, which may state the relationship explicitly. For example, "8 is the sum of what number and 5?" can be written as $x + 5 = 8$.
   b. From formulas or relationships which are a part of our general mathematical background, such as $I = prt$, $d = rt$, etc.

It may prove helpful to write an actual word sentence before attempting to write out the symbolic sentence.

EXERCISES 3.8

■ *In Exercises 1–16:*

a. *Write a simple statement of what is to be found and represent the unknown quantities by appropriate symbols. Use a single variable.*
b. *Write an equation relating unknown quantities with known quantities. (Do not solve.)*

Sample Problem

The sum of an integer and 15 is 72. Find the integer.

Ans. a. Let $x =$ the integer b. $x + 15 = 72$

1. The sum of an integer and 21 is 59. Find the integer.

2. Six more than a certain integer is 22. Find the integer.

3. If a certain integer is added to 15, the sum is 53. Find the integer.
4. If four times an integer is increased by one, the sum is 29. Find the integer.

Sample Problem

Jack earns \$130 per week of which \$37 is deducted before he is paid. What is his weekly take-home pay?

Ans. a. Let $x =$ weekly take-home pay. b. $x + 37 = 130$, or
$x = 130 - 37$, or
$130 - x = 37$

5. If the flying time from Los Angeles to San Francisco is 47 minutes, how many minutes from San Francisco is a plane that is 19 minutes from Los Angeles?
6. A man weighing 78 kilograms steps on a scale while carrying a briefcase. If the scale reads 88 kilograms, what is the weight of the briefcase?
7. How many minutes can be devoted to the entertainment portion of a half-hour television show if 9 minutes are taken up by commercials?
8. A car 72 meters from a fallen tree skidded to a stop 5 meters from the tree. How far did the car skid?
9. A 340-centimeter board is cut into four pieces of equal length and a 24-centimeter piece remains. How long is each of the four pieces?
10. If a 36-ounce solution fills four glass containers of the same size with 4 ounces of the solution left over, how many ounces will each container hold?

Sample Problem

The sum of two consecutive even integers is 46. What are the integers?

Ans. a. Let $x =$ the smaller integer.
Then $x + 2 =$ the next consecutive even integer.

b. $x + (x + 2) = 46$

11. The sum of two consecutive even integers is 26. Find the integers.
12. The sum of three consecutive integers is 57. Find the integers.
13. The sum of two consecutive odd integers is 28. Find the integers.

14. The sum of three consecutive even integers equals four times the smallest integer. What are the integers?

15. A board 112 centimeters long is cut into two pieces so that one piece is three times as long as the other. How long are the two pieces?

16. A board 24 feet long is cut into three pieces of which the second is three times as long as the first, and the third is 4 feet longer than the first. How long are the three pieces?

■ *Solve completely.*

Sample Problem

The sum of a certain number and 12 is equal to three times the number. What is the number?

Specify what the variable represents.

Let $x =$ the number.

Write an equation relating known and unknown quantities.

$x + 12 = 3x$

Solve the equation.

$12 = 2x$

$6 = x$

Ans. The required number is 6.

Check. Is the sum of 6 and 12 equivalent to the product of 3 and 6? Yes; the answer checks. Note that it is not sufficient that 6 satisfy the equation, since the equation itself may be in error.

17. If 5 is added to twice a certain number, the result is 19. What is the number?

18. What number subtracted from three times itself gives 14?

19. The sum of two consecutive even integers is 86. Find the integers.

20. Find three consecutive even integers whose sum is 84.

21. Find three consecutive integers whose sum is -33.

22. Find three consecutive odd integers whose sum is -21.

23. A man drove from town A to town B and then returned to town A. Leaving town A again to return to town B, the man found that, after 5 kilometers on the road, he had traveled a total of 19 kilometers since he first left town A. How far is it from town A to town B?

24. Two ships leave port at the same time, traveling in the same direction. In one hour, one ship sails three times as far as the other. If the ships are then 14 kilometers apart, how far has the slower ship sailed?

25. Two keypunch operators compared their output over a period of time and found that one had punched 24 more cards than the other, but after combining their cards neither could remember how many she had, herself, punched. If together they had produced 212 cards, how many cards had each operator punched?

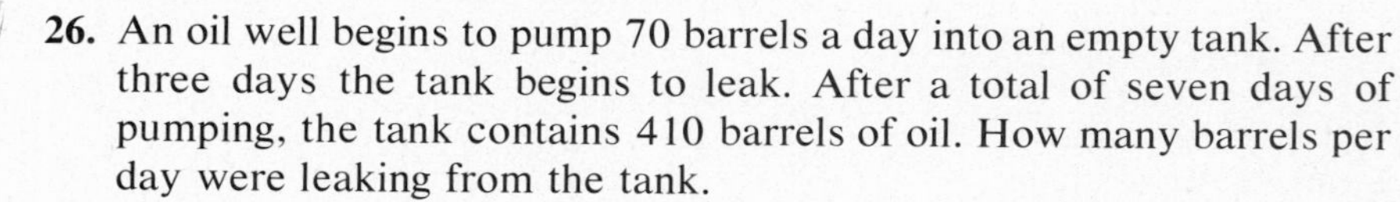

26. An oil well begins to pump 70 barrels a day into an empty tank. After three days the tank begins to leak. After a total of seven days of pumping, the tank contains 410 barrels of oil. How many barrels per day were leaking from the tank.

27. At a recent election, the winning candidate received 50 votes more than his opponent. If there were 4376 votes cast in all, how many votes did each candidate receive?

28. There were 12,822 votes cast in a recent election. The winning candidate received 132 votes more than his opponent. How many votes did each candidate receive?

3.9 APPLICATIONS FROM GEOMETRY

In this section we consider formulas from geometry and evaluate these formulas for given measures of geometric quantities.

The following relationships, which should be familiar to you from your studies in arithmetic, are stated here for reference.

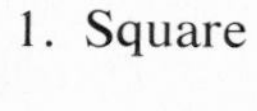

1. Square

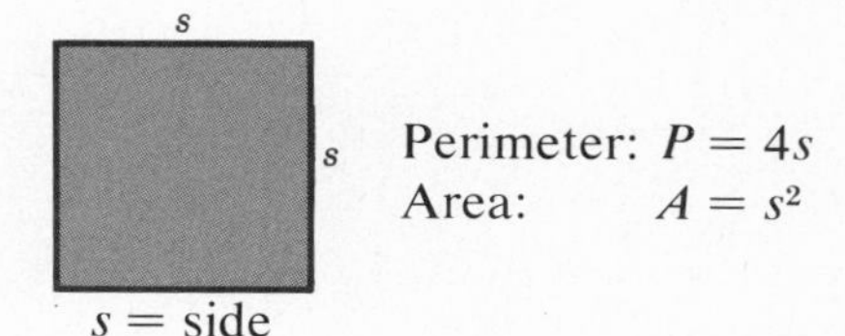

Perimeter: $P = 4s$
Area: $A = s^2$

2. Rectangle

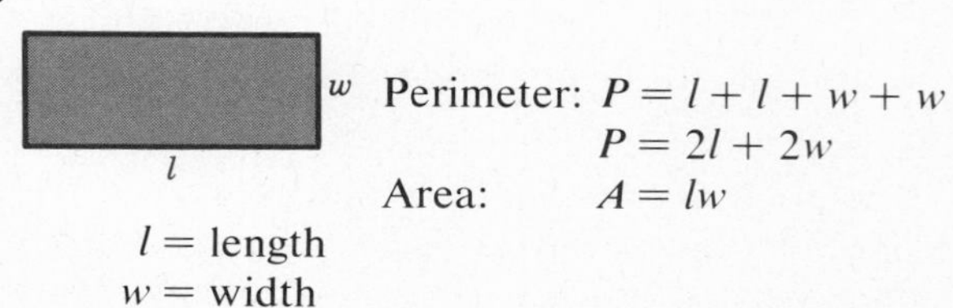

Perimeter: $P = l + l + w + w$
$P = 2l + 2w$
Area: $A = lw$

3. Triangle

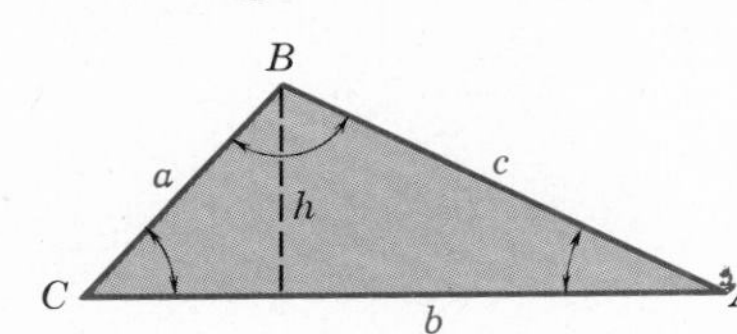

Perimeter: $P = a + b + c$

Area: $A = \dfrac{hb}{2}$

Sum of interior angles:
$\angle A + \angle B + \angle C = 180°$

$b =$ base
$h =$ height or altitude

(a) Isosceles triangle

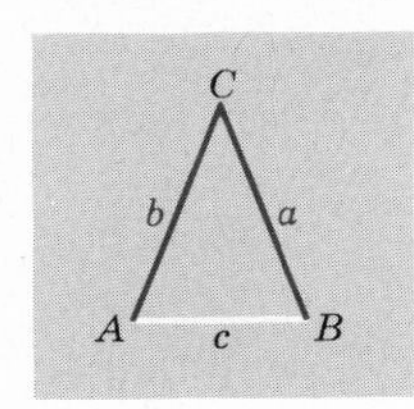

2 equal sides: $a = b$
2 equal angles: $\angle A = \angle B$

(b) Equilateral triangle

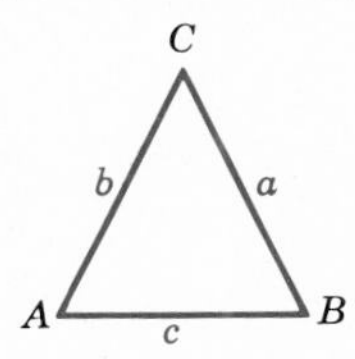

3 equal sides: $a = b = c$
3 equal angles: $\angle A = \angle B = \angle C$

(c) Right triangle

$a =$ leg, $b =$ leg, $c =$ hypotenuse;
angle $C = 90°$

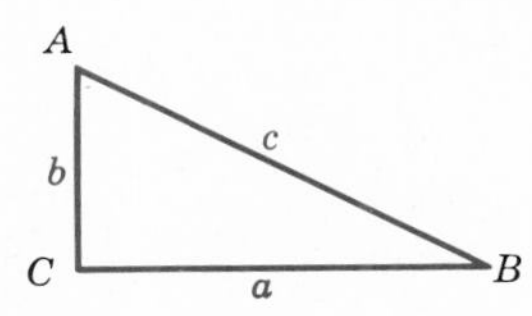

4. Angles

Acute angle: 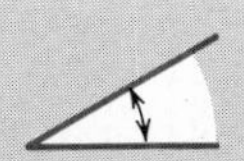 less than 90°

Right angle:  equals 90°

Obtuse angle: 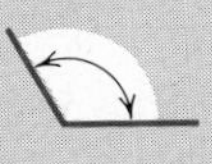 Greater than 90° and less than 180°

Straight angle: 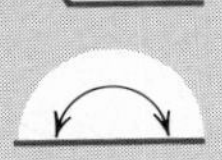 equals 180°

Supplementary angles: two angles whose sum is 180°
Complementary angles: two angles whose sum is 90°

5. Circle

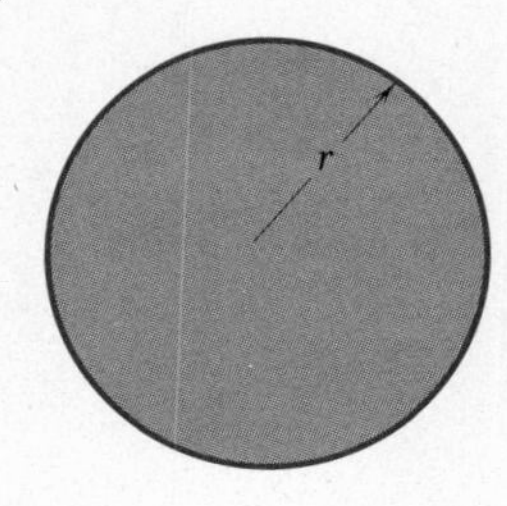

Diameter: $d = 2r$
Circumference: $C = 2\pi r$ or πd
Area: $A = \pi r^2$
(π is approximately equal to 3.14)

EXERCISES 3.9

■ *Solve.*

Sample Problem

Find the area of a circle with radius equal to 2 centimeters.

Use the appropriate formula relating the area and radius of a circle.

$A = \pi r^2$

Substitute 2 for r and 3.14 for π, and solve for A.

$A = \pi(2)^2$

$A = 3.14(4)$

Ans. $A = 12.56$ square centimeters.

1. What is the area of a triangle with a height of 6 inches and a base of 3 inches?

2. What is the area of a circle with a radius of length 5 inches (use $\pi = 3.14$)?

3. What is the area of a rectangle with a length of 12 feet and a width of 10 feet?

4. What is the perimeter of a triangle if the sides are of length, 3, 4, and 5 inches?

■ *Find the area of the shaded portion of each figure. All curves shown are parts of circles, and all horizontal and vertical lines meet at right angles.*

Sample Problem

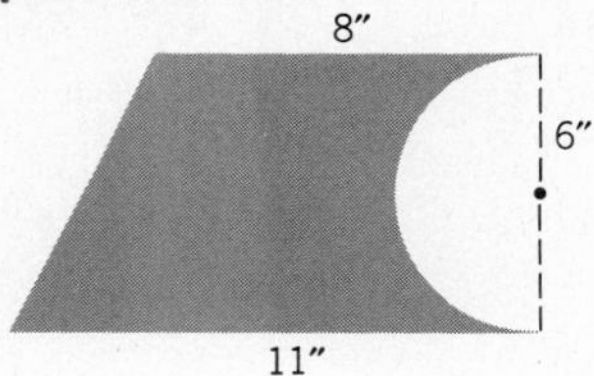

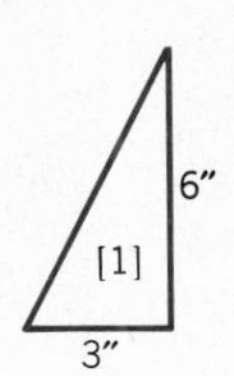

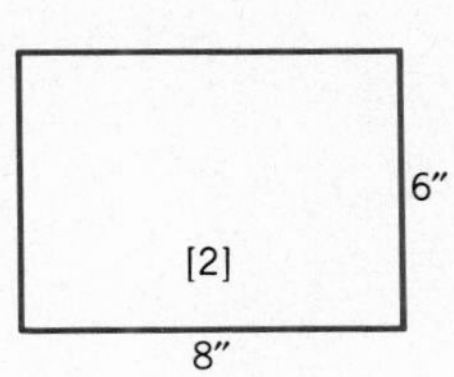

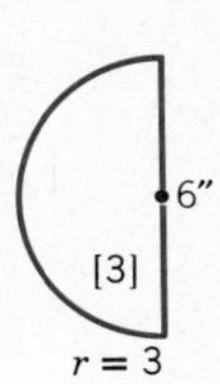

Consider the area of the three separate regions shown.

$$A_{[1]} = \frac{1}{2}(3)(6) = 9;$$
$$A_{[2]} = (6)(8) = 48;$$
$$A_{[3]} = \left(\frac{1}{2}\right) 3.14(3)^2 = 14.13$$

The area of the shaded region is given by

$$A = A_{[1]} + A_{[2]} - A_{[3]}$$
$$= 9 + 48 - 14.13 = 42.87$$

Ans. 42.87 square inches.

5.

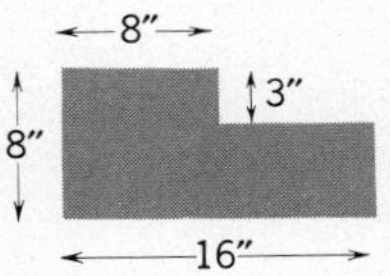

6.

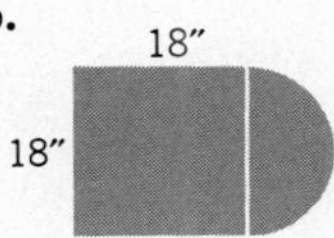

7.

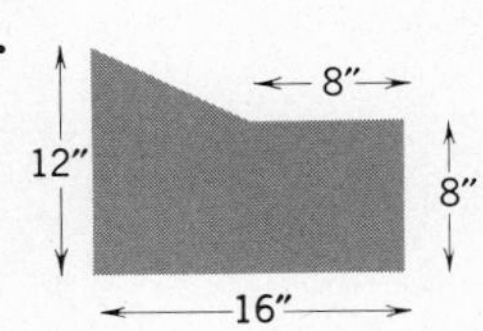

8.

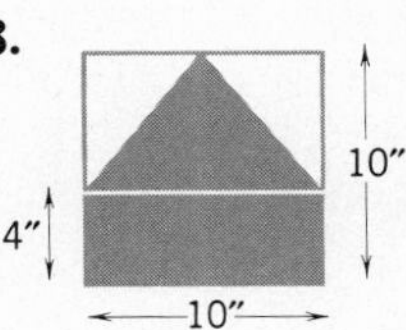

9.

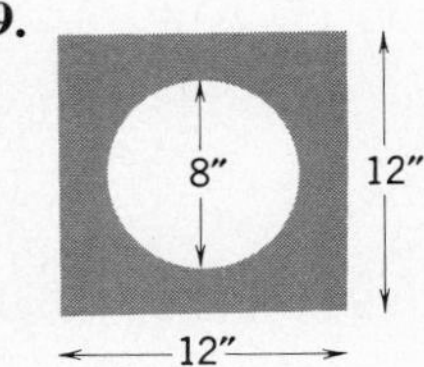

10.

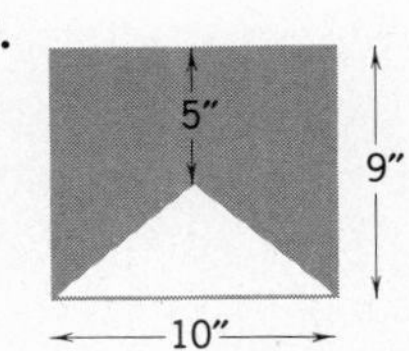

11.

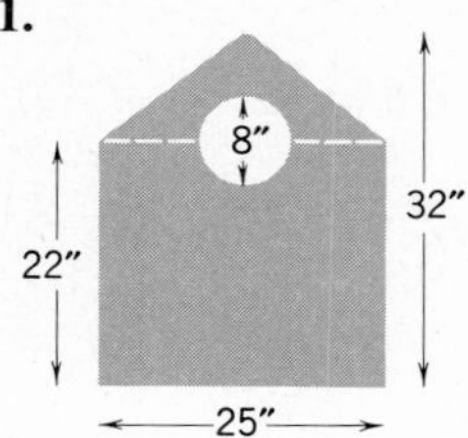

12.

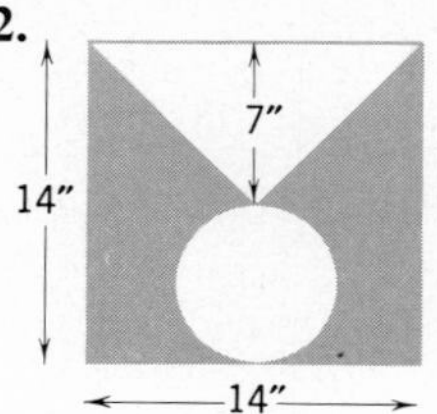

Sample Problem

The perimeter of a rectangle is 120 feet. The length is 10 feet greater than the width. What are the dimensions of the rectangle?

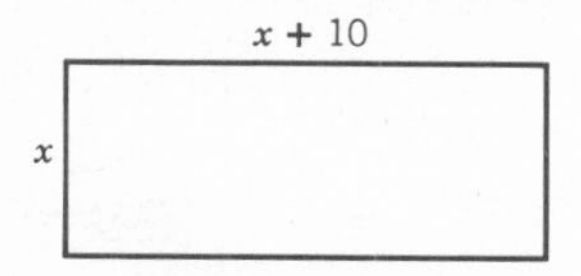

Sketch a figure and label.

Specify what the variable represents.

Let $x =$ the width.
Then $x + 10 =$ the length.

Write an equation relating the known and unknown quantities.

$$x + (x + 10) + x + (x + 10) = 120$$

Solve the equation.

$$4x + 20 = 120$$
$$4x = 100$$
$$x = 25$$
$$x + 10 = 35$$

List both answers.

Ans. width: 25 feet
length: 35 feet

Check. Does $25 + 35 + 25 + 35 = 120$?
Yes; the answer checks.

13. Where should a 39-inch board be cut so that one part will be 5 inches longer than the other?

14. A rectangle is 26 meters long and has an area of 169 square meters. Find the width of the rectangle.

15. A triangle whose base is 16 centimeters has an area of 144 square centimeters. Find the altitude of the triangle.

16. A man wishes to enclose a rectangular garden 18 meters longer than it is wide with 180 meters of wire fencing. What should be the dimensions of the garden?

17. A rectangle is 10 meters longer than it is wide and its perimeter is 164 meters. What are its dimensions?

18. A tennis court for singles is 24 feet longer than twice its width and its perimeter is 210 feet. Find its dimensions.

19. The perimeter of an isosceles triangle is 56 meters. The two equal sides are each 4 meters longer than the base. Find the length of each side.

20. One angle of a triangle is 10° larger than another, and the third angle is 29° larger than the smallest. How large is each angle?

21. One angle of a triangle is twice as large as another, and the third angle is 10° less than the larger of the other two. How large is each angle?

22. Two angles of a triangle are equal and the third is 20° less than the sum of the equal angles. How large is each angle?

23. The perimeter of a rectangle is 56 feet. What are its dimensions if the length is three times the width?

24. The length of a rectangle is five times its width. What are the dimensions of the rectangle if the perimeter is 36 feet?

25. How long is the side of a square whose perimeter is 24 feet?

26. The perimeter of a triangle is 104 inches. The second side is twice the first side, and the third side is 4 inches more than the second. How long is each side?

27. Two angles of a triangle are 40° and 70°. How large is the third angle?

28. Two angles of a triangle are equal, and the third angle is 40°. How large is each of the equal angles?

29. One angle of a triangle is 10° more than another, and the third is 20° more than the smallest. How large is each angle?

30. The largest angle of a triangle is three times the smallest, and the third angle is 30° smaller than the largest. How large is each angle?

CHAPTER REVIEW

1. Write an equation expressing the following:

a. A number added to 3 equals two less than twice the same number.
b. A number subtracted from 12 equals twice the same number.
c. Three times a number divided by 4 equals 21 less 6.

■ *Solve.*

2. a. $2 + x = 8$ b. $5y = 2 + 4y$ c. $\frac{2a + 4a}{3} = 5a + 3$

3. a. $4x + 3x = 35$ b. $4x - 4 = 2x - 4$ c. $8z + 6z = 2z - 12$

4. a. $\frac{2a}{3} = -12$ b. $\frac{b + 4b}{3} = 15$ c. $\frac{6x - 2x}{3} = -4$

5. a. $\frac{-9a - a}{2} = 10$ b. $\frac{3x + 5x}{2} = 6 + 3x$ c. $\frac{8x - 4x}{2} = \frac{8 + 10}{3}$

6. Show by direct substitution that the solutions you obtained in Exercises 2a, 2b, and 2c are correct.

7. Show by direct substitution that the solutions you obtained in Exercises 3a, 3b, and 3c are correct.

8. Solve each of the following formulas for the symbol in color.

   a. $f = ma$ b. $v = k + gt$ c. $M = \frac{a + b}{2}$.

9. If an odd integer is represented by x, how may the next consecutive odd integer be represented in terms of x?

10. If an even integer is represented by x, how may the next consecutive even integer be represented in terms of x?

11. If an integer is represented by x, how may the next four consecutive integers be represented in terms of x?

12. If the width of a rectangular garden is represented by x, how may the length which is three times this width be represented in terms of x?

13. If a man's height is represented by x, how may the height of a second man be represented who is 7 centimeters taller? 7 centimeters shorter?

14. If a man's weight is represented by x, how may the weight of a second man be represented who weighs 18 kilograms more? 12 kilograms less?

15. The sum of four consecutive integers is 54. Find the integers.

16. The sum of three consecutive odd integers is five times the smaller integer. Find the integers.

17. A delivery man delivered 431 papers to a location where two boys were to pick them up for house delivery. The man could not remember how many customers each boy had on his route but did remember that one boy had 27 more customers than the other. How should he have divided the papers?

18. A 32-foot board is cut into three pieces, so that one of the pieces is 3 feet longer than a second piece, and the third is 5 feet longer than the second. How long is each piece?

19. It takes 144 meters of wire to enclose a rectangular garden of width 34 meters. What is the length of the garden?

20. One angle of a triangle is 15° larger than a second angle, and the third

angle is 30° larger than the smaller of the other two. How large is each angle?

Cumulative Review

1. Simplify: $\frac{6^2 - 4^2}{2} - \frac{2^2 + 1}{5}$.
2. Which of the following are natural numbers?

$$3^2, \quad \frac{2^3}{3}, \quad \frac{4^2}{2}, \quad \frac{4 + 2^3}{12}, \quad 5^2 - 3^2.$$

3. In the expression $3b^4$, the number 3 is called the __?__ of b^4.
4. In the expression $3b^4$, the number 4 is called an __?__.
5. If two numbers have like signs, the sign of their product is __?__.
6. Which is greater, $|-5|$ or $|3|$?
7. If a is greater than b, and b is greater than c, then c is __?__ a.
8. Simplify: a. $(x^3)(x^2)$. b. $(x^3)^2$.
9. What is the value of $(-3)^4$? of -3^4?
10. The velocity v (in feet per second) of a falling body is related to time t (in seconds) by $v = 32t$. What is the velocity of the body after 6 seconds?
11. How long would it take a falling body to attain a velocity of 128 feet per second? ($v = 32t$)
12. The distance traveled in time t by a car moving with a constant velocity v is given by $d = vt$. How long will it take the car to travel 312 kilometers at 52 kilometers per hour?
13. The product of three factors is $48xy^2$. If two of the factors are -3 and $4x$, what is the third factor?
14. In the simplification of arithmetic or algebraic expressions, multiplication operations are always performed __?__ addition operations.
15. Like terms may be combined by adding the numerical __?__ of the variable factors of the terms.
16. Two linear equations in one unknown which have the same solution are called __?__ equations.
17. Find the number b that satisfies the equation $7b - 3 = 6 - 2b$.
18. Show by direct substitution that -2 is a solution of $-2a - 3 = 7 + 3a$.
19. Solve the equation $3y - b = y + 2$ for y.
20. The sum of three consecutive even integers is -48. What are the integers?

4.1 THE DISTRIBUTIVE LAW

If a sum is to be multiplied by another number, we can either multiply each term of the sum by the number before we add or we can first add the terms and then multiply. For example,

$$\begin{aligned} 6(3+2) &= 6 \cdot 3 + 6 \cdot 2 \\ &= 18 + 12 \\ &= 30, \end{aligned}$$

or

$$\begin{aligned} 6(3+2) &= 6(5) \\ &= 30. \end{aligned}$$

In either case the result is the same.

This property, which was first introduced in Section 1.7, is formalized as the **distributive law,** and may be represented symbolically by either

$$a(b+c) = ab + ac \quad \text{or} \quad (b+c)a = ab + ac.$$

By applying the distributive law to algebraic expressions containing parentheses, we can obtain equivalent expressions without parentheses.

EXERCISES 4.1

■ *Remove parentheses and simplify.*

Sample Problems

a. $2x(x-3)$ b. $-y(y^2+3y-4)$ c. $xy(x-2y+1)$

Ans. $2x^2-6x$ *Ans.* $-y^3-3y^2+4y$ *Ans.* x^2y-2xy^2+xy

1. $3(x-4)$ **2.** $4(x+1)$ **3.** $5(2y-2)$
4. $2(3y+6)$ **5.** $-2(x+4)$ **6.** $-3(x-7)$
7. $2a(5a+3)$ **8.** $3a(3a-1)$ **9.** $-b(b-2)$
10. $-3b(2b+1)$ **11.** $xy(x+y)$ **12.** $xy(x-2y)$
13. $-x^2(2x+3y)$ **14.** $-y^2(x-2y^2)$ **15.** $x(x^2-2x+1)$
16. $y(y^2+3y+4)$ **17.** $-y(y^2-y+2)$ **18.** $-x(x^2+3x-2)$
19. $4x^3(x^2-3x+4)$ **20.** $2y^3(y^2+y-2)$ **21.** $-y^3(y^3-y+1)$
22. $-x^3(y^3+2y^2-1)$ **23.** $-xy(x^2+xy+y^2)$ **24.** $-xy(2x^2-xy+3y^2)$

Sample Problems

a. $c(y-3)+2cy$ b. $a(3-a)-2(a+a^2)$

Remove parentheses.

$cy-3c+2cy$ $3a-a^2-2a-2a^2$

Combine like terms.

Ans. $3cy-3c$ *Ans.* $a-3a^2$

25. $-a(x+1)+ax$ **26.** $by-b(1-y)$
27. $a(x+1)+x(a+1)$ **28.** $2a(x+3)-3a(x-3)$
29. $a(x+y)-2(ax+y)$ **30.** $3(x^2+2x-1)-2(x^2+x-2)$
31. $ax(x^2+2x-3)-a(x^3+2x^2)$ **32.** $3(x-2y)-2(x+3y)+2x$
33. $2x(3-x)+2(x^2-2x+1)-2$ **34.** $3y(2y-5)+2y^2-5(y^2+2y)$
35. $3(y^2-2y+1)+3(1+2y-y^2)$
36. $3(ax^2+ax-a)-2a(x^2+x-1)$
37. $3x^2(a-b+c)-2x(ax-bx+cx)$
38. $2(x+3y)-2x(1+y)+2y(x-2)$
39. $-3ab(x+y-2)-2a(bx-by+2b)+b(ax+1)$
40. $3ab^2(2+3a)-2ab(3ab+2b)-2b^2(a^2-2a)$

Sample Problems

a. $+(3a-2b)$
$+1(3a-2b)$

b. $-(2a-3b)$
$-1(2a-3b)$

Ans. $3a-2b$ *Ans.* $-2a+3b$

41. $-(a+c)$

42. $-(2-x)$

43. $+(a-2b+c)$

44. $+(2a+b-c)$

45. $-(3x+2y-z)$

46. $-(2r-s-t)$

47. $-(1-3x+x^2)$

48. $-(3+3x-2x^2)$

49. $-(a-b)+(a-b)$

50. $(a-2b)-(a-b)$

51. $-(x-y)-(x+y)$

52. $-(x^2-x)+(x^2+x)$

4.2 FACTORING MONOMIALS FROM POLYNOMIALS

From the symmetric property of equality, we know that if

$$a(b+c) = ab + ac, \quad \text{then} \quad ab + ac = a(b+c).$$

Thus, if there is a monomial factor common to all terms in a polynomial, we may write the polynomial as the product of the common factor and another polynomial. For instance, since each term in $x^2 + 3x$ contains x as a factor, we may write the expression as the product $x(x+3)$. The process of rewriting a polynomial in this way is called **factoring,** and the number x is said to be factored "from" or "out of" the polynomial $x^2 + 3x$.

To factor a monomial from a polynomial:

1. Write a set of parentheses preceded by the monomial common to each term in the polynomial.
2. Divide the monomial factor into each term in the polynomial and write the quotient in the parentheses.

Thus, to factor the common monomial $2x$ from the polynomial

$$4x^3 - 6x^2 + 2x,$$

we write

$$2x(\qquad\qquad),$$

and, upon dividing each term in the polynomial by $2x$, we obtain

$$2x(2x^2 - 3x + 1).$$

Generally polynomials of this type can be factored by inspection. The result can be checked by multiplying the factors obtained and verifying that the product is the original polynomial.

In this book, we shall restrict such factors to monomials consisting of numerical coefficients that are integers, and to integral powers of the variables. Such monomials can be determined by inspection. The choice of sign for the monomial factor is a matter of convenience, and we can use the sign most suitable to our purpose. Thus,

$$-3x^2 - 6x$$

may be factored either as

$$-3x(x + 2) \quad \text{or as} \quad 3x(-x - 2).$$

The first form is usually more convenient.

EXERCISES 4.2

■ *Factor.*

Sample Problems

a. $4x + 4y$	b. $3xy - 6y$	c. $2x^3 - 4x^2 + 8x$
$4(\quad)$	$3y(\quad)$	$2x(\quad)$
Ans. $4(x + y)$	*Ans.* $3y(x - 2)$	*Ans.* $2x(x^2 - 2x + 4)$

1. $3x + 6$ **2.** $4x - 8$ **3.** $2x - 6y$

4. $10x + 5y$ **5.** $2y^2 - 2y$ **6.** $3y^2 + 6y$

7. $ay^2 + y$ **8.** $by^2 - b$ **9.** $9ay^2 + 6y$

10. $4bx^2 - 12x$ **11.** $3y^2 - 3y + 3$ **12.** $2y^2 - 4y - 2$

13. $ax + ay - az$ **14.** $2bx - 6by + 4bz$ **15.** $x^2 - 3x + xy$

16. $y^2 + 2y - 4xy$ **17.** $4y^3 - 2y^2 + 2y$ **18.** $6x^3 + 6x^2 - 9x$

19. $6ax^2y - 18axy^2 + 24axy$ **20.** $3a^2x^2y - 12ax^2y^2 + 9ax^2y$

Sample Problems

a. $-3x^2 - 3xy$	b. $-a - b + c$	c. $-x^2 - x + 1$
$-3x(\quad)$	$-1(\quad)$	$-1(\quad)$
Ans. $-3x(x + y)$	*Ans.* $-1(a + b - c)$	*Ans.* $-1(x^2 + x - 1)$
	or $-(a + b - c)$	or $-(x^2 + x - 1)$

21. $-a^2 - ab$ **22.** $-a^2 - a$ **23.** $-x - x^2$

24. $-ab - ac$ **25.** $-abc - ab - bc$ **26.** $-b^2 - bc - ab$

27. $-6y^3 - 3y^2 - 3y$ **28.** $-2x^2 - 4x - 2$ **29.** $-x + x^2 - x^3$

30. $-3x^2 + 3xy - 3x$ **31.** $-xy^5 - xy^4 + xy^2$ **32.** $-x^2y + xy^2 - 3xy$

■ *Factor the right-hand member of each of the following equations.*

Sample Problems

a. $A = P + PRT$
$A = P(\qquad)$

Ans. $A = P(1 + RT)$

b. $S = 4kR^2 - 4kr^2$
$S = 4k(\qquad)$

Ans. $S = 4k(R^2 - r^2)$

33. $d = k + kat$

34. $A = kR^2 + 2kr^2$

35. $S = kr^2h + kr^2$

36. $R = r + rat$

37. $V = 2ga^2D - 2ga^2d$

38. $L = 2an + n^2 - nd$

39. $A = ar^2 + br^2 + cr^2$

40. $S = 2kr^2 + 2krh$

4.3 BINOMIAL PRODUCTS I

The distributive law can be used to multiply two binomials. Although there is little necessity for multiplying binomials in arithmetic as shown in the first of the following examples, the principle also applies to expressions containing variables.

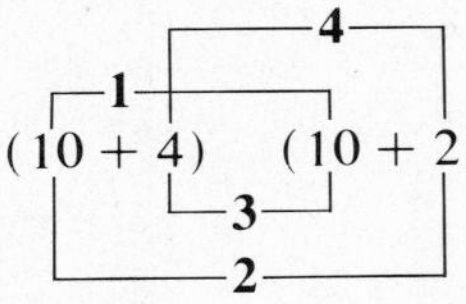

$$\begin{aligned}(10 + 4)(10 + 2) &= 10 \cdot 10 + 10 \cdot 2 + 4 \cdot 10 + 4 \cdot 2 \\ &= 100 + 20 + 40 + 8 \\ &= 168.\end{aligned}$$

(Brackets in the diagram label the products 1, 2, 3, 4.)

Now consider this procedure for an expression containing variables. For example,

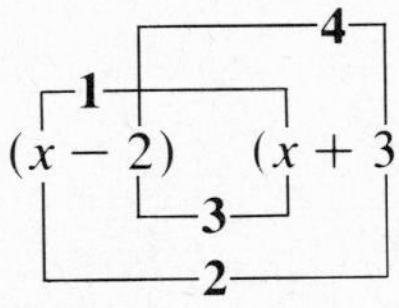

$$\begin{aligned}(x - 2)(x + 3) &= x^2 + 3x - 2x - 6 \\ &= x^2 + x - 6.\end{aligned}$$

(Brackets in the diagram label the products 1, 2, 3, 4.)

With practice, the second and third products can be added mentally.

EXERCISES 4.3

■ *Write as a polynomial.*

Sample Problems

a.

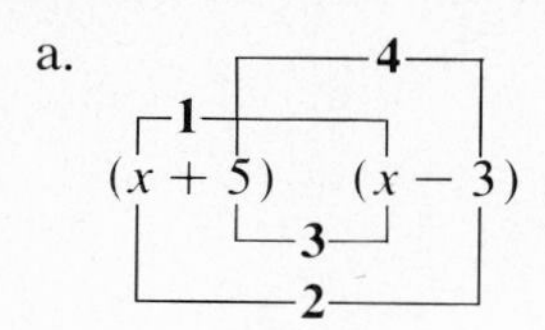

b.

4
1
$(x - 3)$ $(x + 3)$
3
2

Apply distributive law.

a. $x^2 - 3x + 5x - 15$ b. $x^2 + 3x - 3x - 9$

Simplify.

Ans. $x^2 + 2x - 15$ *Ans.* $x^2 - 9$

1. $(x + 3)(x + 4)$ **2.** $(x - 2)(x - 3)$ **3.** $(y - 3)(y + 1)$
4. $(y + 4)(y - 2)$ **5.** $(a + 5)(a + 2)$ **6.** $(a - 5)(a - 2)$
7. $(b - 4)(b + 2)$ **8.** $(b + 5)(b - 3)$ **9.** $(x + 1)(x + 8)$
10. $(x - 2)(x - 9)$ **11.** $(y - 1)(y - 7)$ **12.** $(y + 1)(y - 6)$
13. $(a + 4)(a + 4)$ **14.** $(a - 3)(a - 3)$ **15.** $(b - 5)(b + 5)$
16. $(b + 7)(b - 7)$ **17.** $(x + 1)(x + 1)$ **18.** $(x - 9)(x - 9)$
19. $(y - 1)(y + 1)$ **20.** $(y + 4)(y - 4)$ **21.** $(2 + x)(2 - x)$
22. $(5 - x)(5 - x)$ **23.** $(6 + y)(6 - y)$ **24.** $(9 - y)(9 + y)$

Sample Problems

a. $(x + 6)^2$ b. $(y - 3)^2$ Rewrite expression.

$(x + 6)(x + 6)$ $(y - 3)(y - 3)$

Apply distributive law.

$x^2 + 6x + 6x + 36$ $y^2 - 3y - 3y + 9$

Simplify.

Ans. $x^2 + 12x + 36$* *Ans.* $y^2 - 6y + 9$**

25. $(x + 4)^2$ **26.** $(y - 5)^2$ **27.** $(x - 7)^2$ **28.** $(y + 1)^2$
29. $(x - 1)^2$ **30.** $(y + 8)^2$ **31.** $(x + 2)^2$ **32.** $(y - 10)^2$

Sample Problems

a. $(x - 2b)(x + 3b)$ b. $(y - a)(y + a)$

Apply distributive law.

$x^2 + 3bx - 2bx - 6b^2$ $y^2 + ay - ay - a^2$

Simplify.

Ans. $x^2 + bx - 6b^2$ *Ans.* $y^2 - a^2$

*Note that $(x + 6)^2 \neq x^2 + 6^2$ or $x^2 + 36$.
**Note that $(y - 3)^2 \neq y^2 - 3^2$ or $y^2 - 9$.

33. $(x-3b)(x-b)$ **34.** $(x-a)(x-2a)$ **35.** $(x+2y)(x-y)$
36. $(x+b)(x+2b)$ **37.** $(x+2a)(x+2a)$ **38.** $(x+3b)(x+3b)$
39. $(a-b)^2$ **40.** $(x-y)^2$ **41.** $(y-6a)(y+6a)$
42. $(x+3z)(x-3z)$ **43.** $(x-t)(x+t)$ **44.** $(y-c)(y+c)$

Sample Problems

a. $3(x-2)(x+3)$ b. $a(a+1)(a+3)$

Multiply binomial factors.

$3(x^2+3x-2x-6)$ $a(a^2+3a+a+3)$

Simplify.

$3(x^2+x-6)$ $a(a^2+4a+3)$

Multiply by monomial.

Ans. $3x^2+3x-18$ *Ans.* a^3+4a^2+3a

45. $2(x+1)(x+2)$ **46.** $4(x-3)(x+2)$ **47.** $6(y+5)(y+5)$
48. $3(y-7)(y+1)$ **49.** $6(x-1)^2$ **50.** $3(y+3)^2$
51. $a(a-1)(a+5)$ **52.** $b(b-2)(b+7)$ **53.** $a(a-2)(a+2)$
54. $b(b+3)(b-3)$ **55.** $x(y-3)^2$ **56.** $x^2(y-4)^2$

4.4 FACTORING TRINOMIALS I

In the preceding section, we were concerned with finding the product of two binomials. In the present section, we propose to reverse this process, that is, given the product of two binomials, to find the binomials. The process involved is another example of factoring. As before, we shall confine our attention to factors containing only integral numerical coefficients. Such factors do not always exist, but we shall study cases where they do.

We observe that in multiplying two binomials,

$$(x+b)(x+a) = x^2+ax+bx+ab = x^2+(a+b)x+ab,$$

(diagram labels: 1 = $x \cdot x$, 2 = $x \cdot a$, 3 = $b \cdot x$, 4 = $b \cdot a$)

the first term in the trinomial is product 1; the last term in the trinomial is product 4; the middle term in the trinomial is the sum of the products 2 and 3. This then, is the process we wish to reverse.

We illustrate the factoring of a trinomial by example.

$x^2 - 3x - 10$

1. Write two first-degree factors whose product is the first term in the trinomial.

$(x \quad\quad)(x \quad\quad)$

2. Write two factors whose product is the last term of the trinomial. Consider all combinations.

```
(x    5)(x    2)
(x   10)(x    1)
 |     └3┘    |
 └─────2──────┘
```

3. Select the combination(s) which yield(s) the middle term, $-3x$, of the trinomial upon the addition of products 2 and 3.

$(x \quad 5)(x \quad 2)$

4. Insert proper signs.
   a. If the third term of the trinomial is (+), the signs on the last terms in the factors will be alike and will be the same as the sign of the second term of the trinomial.
   b. If the third term of the trinomial is (−), the signs on the last term in the factors will be opposite. The signs must be such as to yield the correct term in the trinomial.

$(x - 5)(x + 2)$

5. Check the answer by multiplying the binomials.

$x^2 - 3x - 10$

Skill at factoring is usually the result of extensive practice. You will find that the more the process is applied, the clearer it will become. The second and third steps in the process outlined above should be done mentally if possible and the answer written directly.

It is easier to factor a trinomial completely if any monomial factor common to each term of the trinomial is factored first. For example,

$$12x^2 + 36x + 24$$

might be factored as

$$(12x + 24)(x + 1), \quad (12x + 12)(x + 2), \quad (6x + 12)(2x + 2),$$
$$(2x + 4)(6x + 6), \quad (4x + 8)(3x + 3), \quad \text{or} \quad (3x + 6)(4x + 4).$$

A monomial factor may then be factored from several of these binomial factors. However, first factoring the common factor 12 from the original expression yields

$$12(x^2 + 3x + 2),$$

from which we have

$$12(x + 2)(x + 1).$$

which is then said to be in **completely factored form.**

EXERCISES 4.4

■ *Factor.*

Sample Problem

$x^2 - 6x - 16$

Write first-degree factors of x^2.

$(x \quad)(x \quad)$

Write all possible factors of 16.

$(x \quad 16)(x \quad 1)$
$(x \quad 8)(x \quad 2)$
$(x \quad 4)(x \quad 4)$

(inner products: 3; outer products: 2)

Select the combination(s) that will yield $6x$ upon the addition of products 2 and 3.

$(x \quad 8)(x \quad 2)$

Insert proper signs.

Ans. $(x - 8)(x + 2)$

Check by multiplying.

Check. $x^2 - 6x - 16$

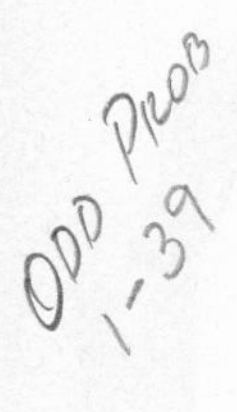

1. $x^2 + 5x + 6$
2. $x^2 + 9x + 20$
3. $y^2 - 8y + 15$
4. $y^2 - 3y + 2$
5. $x^2 - x - 12$
6. $x^2 - 3x - 10$
7. $y^2 + y - 20$
8. $y^2 + 2y - 8$
9. $a^2 + 2a - 35$
10. $a^2 + 8a - 20$
11. $b^2 - 19b - 20$
12. $b^2 - 4b - 12$
13. $x^2 + 11x + 30$
14. $x^2 + 20x + 100$
15. $y^2 - 14y + 13$
16. $y^2 - 16y + 63$
17. $a^2 - 5a - 50$
18. $a^2 - a - 72$
19. $b^2 - 4b - 45$
20. $b^2 - 12b - 45$
21. $x^2 - 46x + 45$

22. $x^2 + 14x + 45$ **23.** $y^2 - 44y - 45$ **24.** $y^2 - 14y - 51$

Sample Problem

$x^2 - 12xy + 32y^2$

$(x \quad)(x \quad)$ Write first-degree factors of x^2.

Write all first-degree factors of $32y^2$.

$(x \quad 32y)(x \quad y)$
$(x \quad 16y)(x \quad 2y)$
$(x \quad 8y)(x \quad 4y)$

(inner products marked 3, outer products marked 2)

Select combination(s) that yield $12xy$ upon addition of products 2 and 3.

$(x \quad 8y)(x \quad 4y)$

Insert proper signs.

Ans. $(x - 8y)(x - 4y)$

Check by multiplying.

Check. $x^2 - 12xy + 32y^2$

25. $x^2 + 4ax + 4a^2$ **26.** $x^2 - xy - 2y^2$ **27.** $a^2 - 3ab + 2b^2$
28. $r^2 + 4rx + 3x^2$ **29.** $s^2 + as - 6a^2$ **30.** $x^2 + 15xy + 36y^2$
31. $a^2b^2 - ab - 2$ **32.** $x^2y^2 + xy - 2$

■ *Factor completely.*

Sample Problems

a. $3x^2 + 12x + 12$
$3(x^2 + 4x + 4)$

Ans. $3(x + 2)(x + 2)$

Check. $3x^2 + 12x + 12$

b. $2b^3 - 8b^2 - 10b$
$2b(b^2 - 4b - 5)$

Ans. $2b(b - 5)(b + 1)$

Check by multiplying.

Check. $2b^3 - 8b^2 - 10b$

33. $2x^2 + 10x + 12$ **34.** $3a^2 - 3a - 18$
35. $y^3 - 2y^2 - 3y$ **36.** $b^3 + 2b^2 + b$
37. $5c^2 - 25c + 30$ **38.** $2a^2 - 38a - 40$
39. $4a^2b + 12ab - 72b$ **40.** $3x^2y - 6xy - 105y$

Sample Problem

$6 - 5x - x^2$

Write all possible factors of 6 and first-degree factors of x^2.

$(6 \quad x)(1 \quad x)$
$(3 \quad x)(2 \quad x)$

(products: 3 inner, 2 outer)

Select combination(s) that yield $5x$ upon addition of products 2 and 3. In this case, both combinations might work.

$(6 \quad x)(1 \quad x)$

Inserting proper signs eliminates $(3 \quad x)(2 \quad x)$ since there exists no combination of signs that yields both the middle term and the last term of the trinominal.

Ans. $(6 + x)(1 - x)$

Check by multiplying.

Check. $6 - 5x - x^2$

41. $21 - 4x - x^2$ **42.** $6 - x - x^2$ **43.** $10 + 7y + y^2$
44. $30 - 11y + y^2$ **45.** $63 - 2y - y^2$ **46.** $18 + 7y - y^2$
47. $32 - 12z + z^2$ **48.** $24 + 10z - z^2$ **49.** $8 - 9x + x^2$
50. $81 + 18x + x^2$ **51.** $56 - 15y + y^2$ **52.** $54 + 3y - y^2$

4.5 BINOMIAL PRODUCTS II

In this section, we apply the procedure developed in Section 4.3 to multiply binomial factors whose first terms have numerical coefficients other than 1.

EXERCISES 4.5

■ *Write as a polynomial.*

Sample Problems

a. $(2x - 3)(x + 1)$ b. $(3x - y)(3x + y)$

Multiply as indicated.

$2x^2 + 2x - 3x - 3$ $\qquad$ $9x^2 + 3xy - 3xy - y^2$

Combine like terms.

Ans. $2x^2 - x - 3$ $\qquad$ *Ans.* $9x^2 - y^2$

1. $(2x + 1)(x + 3)$ **2.** $(4x - 2)(x - 1)$ **3.** $(3y - 2)(y + 1)$
4. $(y + 5)(4y - 2)$ **5.** $(3y + 1)(2y + 3)$ **6.** $(4y - 1)(3y + 2)$
7. $(5x - 2)(4x + 3)$ **8.** $(6x - 1)(2x + 5)$ **9.** $(2x + 3)(2x - 3)$
10. $(3x - 1)(3x + 1)$ **11.** $(4y - 3)(4y + 3)$ **12.** $(6y - 5)(6y + 5)$
13. $(2x + 1)^2$ **14.** $(3x + 1)^2$ **15.** $(5x + 2)^2$
16. $(2x - 3)^2$ **17.** $(4y + 5)^2$ **18.** $(3y + 7)^2$
19. $(2x - a)(x + 2a)$ **20.** $(2x - a)(x - a)$ **21.** $(x + a)(3x + a)$
22. $(2x - a)(2x + a)$ **23.** $(3x - a)(2x + a)$ **24.** $(2x - a)(3x - a)$
25. $(3x - 2y)(3x + 2y)$ **26.** $(2x - 5y)(2x + 5y)$ **27.** $(4x + 7y)(4x - 7y)$
28. $(5x + 9y)(5x - 9y)$ **29.** $(x - 2y)^2$ **30.** $(2x - 3y)^2$
31. $(3x - y)^2$ **32.** $(3x - 2y)^2$ **33.** $(8x + 3y)^2$
34. $(2x + y)^2$ **35.** $(2x + 3y)^2$ **36.** $(3x - 4y)^2$

Sample Problems

a. $3(2x - 1)(x + 2)$ $\qquad$ b. $x(x + 2)(3x - 5)$

Multiply binomials.

$3(2x^2 + 4x - x - 2)$ $\qquad$ $x(3x^2 - 5x + 6x - 10)$

Simplify.

$3(2x^2 + 3x - 2)$ $\qquad$ $x(3x^2 + x - 10)$

Multiply by the monomial.

Ans. $6x^2 + 9x - 6$ $\qquad$ *Ans.* $3x^3 + x^2 - 10x$

37. $2(3x + 1)(x - 3)$ **38.** $4(x - 2)(2x - 3)$ **39.** $3(2y + 1)(2y - 1)$
40. $6(3y + 2)(3y - 2)$ **41.** $3(2x - 5)^2$ **42.** $3(x + 1)^2$
43. $x(x - 2)(2x + 5)$ **44.** $y(y + 2)(y - 1)$ **45.** $x(2x - 1)^2$
46. $y(y + 1)^2$ **47.** $r(3r - 1)(3r + 1)$ **48.** $s(2s - 3)(2s + 3)$

4.6 FACTORING TRINOMIALS II

The factoring procedure developed in Section 4.4 is applicable to trinomials which have a coefficient other than 1 on their second-degree term. We illustrate by example:

$8x^2 + 2x - 15$

1. Consider all combinations of first-degree factors of the first term in the trinomial.

$$\begin{array}{llll} (8x &)(x &) \\ (4x &)(2x &) \end{array}$$

2. Consider all combinations of factors of the last term of the trinomial. These factors must be arranged in all possible combinations with the first terms.

$$\begin{array}{lrlr} (8x & 1)(x & & 15) \\ (8x & 15)(x & & 1) \\ (8x & 3)(x & & 5) \\ (8x & 5)(x & & 3) \\ (4x & 1)(2x & & 15) \\ (4x & 15)(2x & & 1) \\ (4x & 3)(2x & & 5) \\ (4x & 5)(2x & & 3) \end{array}$$

└3┘

└———2———┘

3. Select the combination(s) which will yield the second term of the trinomial upon addition of products 2 and 3.

$(4x \quad 5)(2x \quad 3)$

4. Insert proper signs.

$(4x - 5)(2x + 3)$

5. Check by multiplying.

Check. $8x^2 + 2x - 15$

With practice you will be able to check the combinations mentally and will not need to write out all of the possibilities.

EXERCISES 4.6

■ *Factor.*

Sample Problem

$9x^2 - 8 - 21x$

Write in decreasing powers of x.

$9x^2 - 21x - 8$

Write all possible first-degree factors of $9x^2$.

$$\begin{array}{llll} (9x &)(x &) \\ (3x &)(3x &) \end{array}$$

Write all possible factors of 8. List factors in all possible combinations with the first terms.

(9x 8)(x 1)
(9x 1)(x 8)
(9x 2)(x 4)
(9x 4)(x 2)
(3x 8)(3x 1)
(3x 2)(3x 4)
3
2

Select the combination(s) that yield 21x upon addition of products 2 and 3.

(3x 8)(3x 1)

Insert proper signs.

Ans. $(3x - 8)(3x + 1)$

Check by multiplying.

Check. $9x^2 - 21x - 8$

1. $3a^2 + 4a + 1$ **2.** $2r^2 + 3r + 1$ **3.** $2x^2 - 3x + 1$
4. $2y^2 + 5y + 3$ **5.** $9b^2 - 6b + 1$ **6.** $4a^2 + 4a + 1$
7. $2x^2 - 3 + x$ **8.** $2x^2 + 3 + 7x$ **9.** $2x^2 - 3 - x$
10. $2x^2 - 7x + 3$ **11.** $6a^2 - 1 - a$ **12.** $1 + 6a^2 + 5a$
13. $4y^2 - 4y + 1$ **14.** $4y^2 - 5y + 1$ **15.** $4y^2 - 3y - 1$
16. $4y^2 + 3y - 1$ **17.** $4a^2 - 11a + 6$ **18.** $23a + 4a^2 - 6$
19. $4a^2 + a - 5$ **20.** $16x^2 - 5 - 16x$ **21.** $16x^2 - 2x - 5$
22. $16x^2 - 38x - 5$ **23.** $16x^2 - 11x - 5$ **24.** $16x^2 + 79x - 5$
25. $9x^2 - 21x - 8$ **26.** $64x^2 + 64x + 15$ **27.** $16y + 4y^2 + 15$
28. $10y - 8 + 25y^2$ **29.** $2t^2 - 5st - 3s^2$ **30.** $2a^2 + 5ab - 3b^2$
31. $3x^2 + 2a^2 - 7ax$ **32.** $9y^2 - 3yz - 2z^2$ **33.** $5by + 4y^2 + b^2$
34. $9a^2 + 9ab - 4b^2$ **35.** $4a^2 + 16ab + 15b^2$ **36.** $9x^2 + 3xy - 2y^2$

Sample Problems

a. $6x^2 + 15x - 9$
$3(2x^2 + 5x - 3)$

b. $4a^2b + 10ab + 6b$
$2b(2a^2 + 5a + 3)$

Ans. $3(2x - 1)(x + 3)$ *Ans.* $2b(2a + 3)(a + 1)$

Check by multiplying.

Check. $6x^2 + 15x - 9$ *Check.* $4a^2b + 10ab + 6b$

37. $6x^2 + 8x + 2$

38. $6x^3 + 21x^2 + 9x$

39. $8y^2 - 6y - 2$

40. $18x^2 - 3x - 3$

41. $18x^2 - 9x - 27$

42. $4x^3 - 10x^2 - 6x$

43. $27y^3 - 9y^2 - 6y$

44. $4y^3 + 10y^2 + 6y$

45. $12ab^2 + 15a^2b + 3a^3$

46. $27a^2b + 27ab^2 - 12b^3$

47. $50xy^3 + 20x^2y^2 - 16x^3y$

48. $4a^2bx^2 - 2abx - 12b$

4.7 FACTORING THE DIFFERENCE OF TWO SQUARES

Some polynomials occur frequently enough to justify an effort to recognize these special forms, which in turn will enable us to write their factored form directly. We have in previous sections observed that

$$(a + b)(a - b) = a^2 - b^2.$$

In this section we are interested in viewing this relationship from right to left; that is, from a polynomial form to its factored form. In words:

The difference of two squares, $a^2 - b^2$, equals the product of the sum $a + b$ and the difference $a - b$.

For example,

$$x^2 - 9 = x^2 - 3^2 = (x + 3)(x - 3),$$

and

$$4y^2 - 25x^2 = (2y)^2 - (5x)^2 = (2y + 5x)(2y - 5x).$$

EXERCISES 4.7

■ *Factor.*

Sample Problem

$x^2 - 16$ — Write the factored form directly.

Ans. $(x + 4)(x - 4)$ — Check by multiplying.

Check. $x^2 - 16$

1. $x^2 - 9$ **2.** $y^2 - 25$ **3.** $x^2 - 1$ **4.** $x^2 - 81$

5. $x^2 - z^2$ **6.** $x^2 - 9y^2$ **7.** $x^2y^2 - 16$ **8.** $x^2y^2 - 36$

9. $a^2x^2 - 49b^2$ **10.** $x^2 - 100a^2b^2$ **11.** $36 - x^2$ **12.** $b^2 - y^2$

Sample Problems

a. $9x^2 - 4$
$(3x)^2 - 2^2$
Ans. $(3x - 2)(3x + 2)$

b. $16a^2 - b^2y^2$
$(4a)^2 - (by)^2$
Ans. $(4a - by)(4a + by)$

Check by multiplying.

Check. $9x^2 - 4$

Check. $16a^2 - b^2y^2$

13. $4b^2 - 9$ **14.** $9b^2 - 1$ **15.** $25x^2 - 16$

16. $4y^2 - 25$ **17.** $9 - 4x^2$ **18.** $25 - 9y^2$

19. $81 - 4x^2$ **20.** $9 - 64y^2$ **21.** $4a^2 - 121b^2$

22. $64x^2 - 9y^2$ **23.** $25y^2 - 49x^2$ **24.** $100x^2 - 81y^2$

25. $49a^2x^2 - 144b^2y^2$ **26.** $49a^2x^2 - 36b^2y^2$ **27.** $4x^2y^2 - 81$

28. $121 - 49x^2y^2$ **29.** $36a^2b^2 - 1$ **30.** $1 - 100a^2b^2$

Sample Problems

a. $x^3 - x^5$
$x^3(1 - x^2)$
Ans. $x^3(1 - x)(1 + x)$

b. $a^2x^2y - 16y$
$y(a^2x^2 - 16)$
$y[(ax)^2 - 4^2]$
Ans. $y(ax - 4)(ax + 4)$

Check by multiplying.

Check. $x^3 - x^5$

Check. $a^2x^2y - 16y$

31. $5x^2 - 5$ **32.** $2x^2 - 8$ **33.** $3x^3 - 3x$

34. $3a^2 - 75$ **35.** $2x^2 - 8y^2$ **36.** $3xy^2 - 12xb^2$

37. $3a^2b^2 - 12c^2d^2$ **38.** $8x^2y^2z^2 - 18$

4.8 EQUATIONS INVOLVING PARENTHESES

It is frequently necessary to solve equations in which the variable is included as a part of an expression enclosed in parentheses. These equations can be solved in the usual manner after they have been simplified by applying the distributive law to remove the parentheses.

EXERCISES 4.8

■ *Solve.*

Sample Problem

$$4(5-y)+3(2y-1)=3$$

Apply the distributive law.

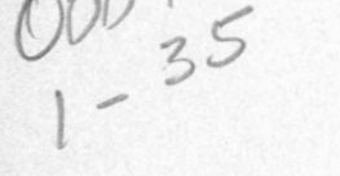

$$20-4y+6y-3=3$$

Solve for y.

$$2y+17=3$$
$$2y=-14$$

Ans. $y=-7$

1. $3(x-5)=6$
2. $5(3x-2)=35$
3. $2(4y+5)=2$
4. $6(3y-4)=-60$
5. $28=4(1+2x)$
6. $0=7(8-2x)$
7. $-3(2x+1)-4=-1$
8. $-5(2x-3)+2=47$
9. $7(y-3)=2y-31$
10. $4(5-y)=10-6y$
11. $3y+35=4(2y+5)$
12. $5x-64=-2(3x-1)$
13. $-x-(8+x)=2$
14. $3(7+2x)=30+7(x-1)$
15. $5x-(x+2)=7+(x+3)$
16. $4(y-1)=5(y-2)$
17. $-2y+5(y+1)=25+7y$
18. $25+5y=-2(y-4)-18$
19. $(a-1)-(a+2)=a+(a-3)$
20. $3(2a-1)+2(a+5)=15$
21. $(b+5)-(b-1)=3b$
22. $5a-4(1-a)=11-6(a-5)$
23. $b=4(b+6)+3b$
24. $b=(b+1)-(b-5)$

Sample Problem

$$(x+5)(x+3)-x=x^2+1$$

Apply the distributive law.

$$x^2+8x+15-x=x^2+1$$

Solve for x.

$$7x=-14$$

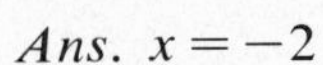

Ans. $x=-2$

25. $(x-1)(x+2)=x^2+1$
26. $(x-4)(x-1)=9+x^2$
27. $(x+2)(x-2)=x^2-4x$
28. $(y+3)(y+1)=y^2-5$
29. $(y+2)(y+4)=y(y+8)$
30. $(y-4)(y+4)=y^2+4y$

31. $(a + 3)(a - 2) + 1 = a^2$

32. $(a - 1)(a - 1) = a^2 - 11$

33. $(a - 2)(a + 2) + 3 = a^2 - a$

34. $2(a - 1)(a + 1) = 2a^2 + 2a$

35. $(a - 3)^2 = a^2 - 15$

36. $(a + 2)^2 = a^2 + 5a - 2$

4.9 APPLICATIONS

Parentheses are useful in representing products in which the variable is contained in one or more terms in any factor. As an example, consider an integer x and the next consecutive integer $x + 1$. Five times the smaller integer is represented as $5x$ and five times the larger integer is represented as $5(x + 1)$. As another example, the expression $2x - 1$ represents a number for a given value of x. Three times this number is represented as $3(2x - 1)$.

EXERCISES 4.9

■ *In each of the following statements, assign a variable to one of the unknowns and represent all other unknowns in terms of this variable.*

Sample Problem

One integer is three more than another. If x represents the smaller integer, represent in terms of x:

a. The larger integer.

b. Five times the smaller integer.

c. Five times the larger integer.

Ans. $x + 3$

Ans. $5x$

Ans. $5(x + 3)$

1. One integer is four more than a second integer. If x represents the smaller integer, represent in terms of x:

a. The larger integer.

b. Five times the smaller integer.

c. Five times the larger integer.

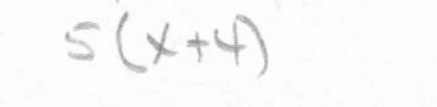

2. One integer is six less than a second integer. If n represents the larger integer, represent in terms of n:

a. The smaller integer.

b. Two times the smaller integer.

c. Two times the larger integer.

3. One integer is six more than another. If n represents the smaller integer, represent in terms of n, three times the larger integer.

4. One integer is five less than another. If n represents the larger integer, represent in terms of n, six times the smaller integer.

Sample Problem

The sum of two integers is 13. If x represents the smaller integer, represent in terms of x:

a. The larger integer.	b. Five times the smaller integer.	c. Five times the larger integer.
Ans. $13 - x$	*Ans.* $5x$	*Ans.* $5(13 - x)$

5. The sum of two integers is 27. If n represents the smaller integer, represent in terms of n:

a. The larger integer.
b. Three times the smaller integer.
c. Three times the larger integer.

6. The sum of two integers is 39. If n represents the larger integer, represent in terms of n:

a. The smaller integer.
b. Six times the smaller integer.
c. Four times the larger integer.

7. The difference of two integers is 16. If n represents the smaller integer, represent in terms of n:

a. The larger integer.
b. Five times the smaller integer.
c. Two times the larger integer.

8. The difference of two integers is 21. If n represents the larger integer, represent in terms of n:

a. The smaller integer.
b. Two times the larger integer.
c. Three times the smaller integer.

9. A 10-meter board is divided into two pieces. If x represents the shorter piece, represent in terms of x:

a. The longer piece.
b. Two times the shorter piece.
c. Five times the longer piece.

10. A 24-meter board is divided into two parts. If y represents the longer piece, represent in terms of y:

a. The shorter piece.
b. Three times the longer piece.
c. Ten times the shorter piece.

Sample Problem

Represent (in cents) the *value* of each of the following collections of coins.

a. two dimes	b. x dimes	c. $(x + 3)$ dimes
Ans. 10(2) or 20 cents	*Ans.* $10x$ cents	*Ans.* $10(x + 3)$ cents
d. Three nickels	e. n nickels	f. $(n - 2)$ nickels
Ans. 5(3) or 15 cents	*Ans.* $5n$ cents	*Ans.* $5(n - 2)$ cents

11. In a collection of coins there are four more dimes than quarters. If x represents the number of quarters, represent in terms of x:

a. The number of dimes.
b. The value (in cents) of the quarters.
c. The value (in cents) of the dimes.

12. In a collection of coins there are five fewer pennies than dimes. If n represents the dimes, represent in terms of n:

a. The number of pennies.
b. The value (in cents) of the dimes.
c. The value (in cents) of the pennies.

13. In a collection of coins there are three fewer dimes than quarters and two more pennies than quarters. If n represents the number of quarters, represent in terms of n:

a. The number of dimes.
b. The number of pennies.
c. The value (in cents) of the quarters.
d. The value (in cents) of the dimes.
e. The value (in cents) of the pennies.

14. In a collection of coins, there are four more dimes than pennies and two more nickels than dimes. If x represents the number of pennies, represent in terms of x:

a. The number of dimes.
b. The number of nickels.
c. The value (in cents) of the pennies.
d. The value (in cents) of the dimes.
e. The value (in cents) of the nickels.

15. In a collection of coins, there are three more dimes than nickels. If n represents the number of nickels, represent in terms of n, the total value of the collection in cents.

16. In a collection of coins there are twice as many dimes as nickels and three more quarters than dimes. If n represents the number of nickels, represent in terms of n, the total value of the collection in cents.

Sample Problem

One truck has a capacity 3 tons greater than another truck. If x represents the capacity of the smaller truck, represent in terms of x:

a. The capacity of the larger truck.

Ans. $x + 3$

b. The total tons hauled by the smaller truck in three trips.

Ans. $3x$

c. The total tons hauled by the larger truck in five trips.

Ans. $5(x + 3)$

17. One truck has a capacity of 4 tons less than another truck. If x represents the capacity of the larger truck, represent in terms of x:

a. The capacity of the smaller truck.
b. The total tons hauled by the larger truck in seven trips.
c. The total tons hauled by the smaller truck in six trips.

18. One truck can carry 7 tons, and another truck can carry 10 tons. Let x represent the number of trips made by the larger truck. If the smaller truck makes four more trips than the larger truck, represent in terms of x:

a. The number of trips made by the smaller truck.
b. The total tons hauled by the larger truck.
c. The total tons hauled by the smaller truck.

■ *Solve each of the following problems completely.*

19. One integer is two less than a second integer. The larger plus four times the smaller equals 17. Find the integers.

20. One integer is three more than a second integer. Four times the second plus twice the first equals 42. Find the integer.

21. The gasoline tank of one truck holds 4 gallons more than the tank of another truck. If it takes 144 gallons of gasoline to fill each tank three times, how many gallons does each tank hold?

22. In a mathematics textbook containing 368 pages, there were 28 pages more of exercises than there were of explanatory material. How many pages of each were there in the book?

23. A board is 20 feet long. Where should it be cut so that one piece will be 4 feet shorter than the other? (Use a sketch.)

24. Where should a 30-foot cable be cut so that twice the longer piece is equal to three times the shorter?(Use a sketch.)

Sample Problem

A collection of coins consisting of dimes and quarters has a value of \$5.80. There are sixteen more dimes than quarters. How many dimes and quarters are in the collection?

Represent the unknown quantities in terms of x.

Let $x =$ the number of quarters
$x + 16 =$ the number of dimes

Write an equation relating the value of the quarters and dimes to the value of the entire collection.

$$\begin{bmatrix}\text{value of}\\ \text{quarters}\\ \text{in cents}\end{bmatrix} + \begin{bmatrix}\text{value of}\\ \text{dimes}\\ \text{in cents}\end{bmatrix} = \begin{bmatrix}\text{value of}\\ \text{collection}\\ \text{in cents}\end{bmatrix}$$

$$25x + 10(x + 16) = 580$$

Solve for x.

$$\begin{aligned} 25x + 10x + 160 &= 580 \\ 35x &= 420 \\ x &= 12 \\ x + 16 &= 28 \end{aligned}$$

Ans. There are 12 quarters and 28 dimes in the collection.

25. A man had \$1.80 in change. The change was entirely in the form of dimes and nickels. If he had three more dimes than nickels, how many of each coin did he have?

26. A man had \$1.45 in change. The money consisted of quarters and dimes only. If he had four fewer quarters than he had dimes, how many of each coin did he have?

27. A man had \$1.14 in change consisting of pennies, nickels, and dimes. He had six more nickels than pennies and six less dimes than pennies. How many of each coin did he have?

28. A man had \$1.47 in change consisting of pennies, nickels, and quarters. He had three more pennies than quarters and one more nickel than pennies. How many of each coin did he have?

Sample Problem

How much candy worth 80 cents a kilogram (Kg) must a grocer blend with 60 kg of candy worth $1 a kilogram to make a mixture worth 90 cents a kilogram?

Represent the unknown quantities in terms of x.

Let $x =$ kilograms of 80-cent candy,
then $x + 60 =$ kilograms of 90-cent candy

Write an equation relating the values of the amounts of 80-cent and $1 candy with the value of the amount of 90-cent candy.

$$\begin{bmatrix}\text{value of}\\ x\text{ kilograms of}\\ \text{80-cent candy}\end{bmatrix} + \begin{bmatrix}\text{value of}\\ \text{60 kilograms of}\\ \text{\$1 candy}\end{bmatrix} = \begin{bmatrix}\text{value of}\\ (x+60)\text{ kilograms}\\ \text{of 90-cent candy}\end{bmatrix}$$

$$80x + 60(100) = 90(60 + x)$$

Solve for x.

$$80x + 6000 = 5400 + 90x$$

$$600 = 10x$$

$$x = 60$$

Ans. The grocer should use 60 kilograms of the 80-cent candy.

29. How many grams of metal worth 50 cents a gram should be mixed with 20 grams of metal worth 32 cents a gram to produce an alloy worth 40 cents a gram?

30. How many pounds of dog food worth 5 cents a pound should a pet store owner mix with 15 pounds of dog food worth 8 cents a pound to produce a mixture worth 6 cents a pound?

31. Fine powder is worth 30 cents a kilogram, and coarse powder is worth 12 cents a kilogram. How many kilograms of the fine powder should be mixed with 50 kilograms of the coarse powder for the mixture to sell for 20 cents a kilogram?

32. A man uses 60 kgs of fine powder worth 30 cents a kilogram and a coarse powder worth 25 cents a kilogram to make a mixture that he wishes to sell for 28 cents a kilogram. How many kilograms of the coarse powder does he use?

33. One thousand tickets were sold at a football game. Adults paid \$1.80 each for their tickets, and children paid 80 cents each. If the total receipts for the game were \$1200, how many tickets of each kind were sold?

34. Three hundred tickets were sold at a baseball game. Adults paid 90 cents each for their tickets, and children paid 40 cents each. If the total receipts for the game were \$200, how many tickets of each kind were sold?

35. Two trucks are carrying material to a road construction job. One truck can carry 4 tons more per trip than the other. If the smaller truck makes five trips and the larger truck makes seven trips, they can deliver a total of 112 tons of material. What is the capacity of each truck?

36. A 10-ton truck and a 12-ton truck are carrying material to a road construction job. If the smaller truck makes three trips more than the larger truck, how many trips does each make if together they deliver 140 tons?

CHAPTER REVIEW

■ *Write as a polynomial.*

1. a. $3x(x^2 + x)$ b. $2xy(y - x)$ c. $-(x^2 - y + 1)$
2. a. $a(2 - a)$ b. $-b(a - b)$ c. $3b(a + b + c)$

■ *Factor.*

3. a. $3a^2 - 6a^2b$ b. $2x^3 + 4x^2 + 6x$ c. $-y^2 - y^3$
4. a. $a^2 + a^2b$ b. $4b - 4$ c. $b - b^2 - b^3$

■ *Write as a polynomial.*

5. a. $(x - 2)(x + 3)$ b. $(2a - 3)(3a - 4)$ c. $(2a - 3)^2$
6. a. $(x + a)(x - 2a)$ b. $(2x - b)(x + b)$ c. $(2b + 1)^2$

■ *Factor.*

7. a. $x^2 - 4x - 21$ b. $10a^2 + 17a + 3$ c. $4x^2 - 9$
8. a. $a^2 - 10a + 21$ b. $3b^2 + 4b + 1$ c. $2b^2 + 3b - 2$
9. a. $2x^2 + 14x + 24$ b. $3y^2 + 24y - 60$ c. $4x^3 - 4x$
10. a. $x^2 - 3ax + 2a^2$ b. $x^2 - a^2$ c. $4b^2 + 6bc - 4c^2$

11. Solve.
    a. $3(x - 5) = 45$
    b. $6(4 - b) + 8 = 4(3 + b)$
    c. $7 - (b - 2) = 3(11 + b)$
12. Factor the right-hand member of the formula $A = 2krh + 2kr^2$, and evaluate for $k = 3.14$, $r = 7$, and $h = 10$.
13. The sum of two numbers is 24. If one of the numbers is represented by x, how can the second number be represented in terms of x?
14. How can the value (in cents) of x dimes be represented in terms of x?
15. How can the value (in cents) of $(x + 3)$ quarters be represented in terms of x?
16. How can the value (in cents) of $(y-2)$ nickels be represented in terms of y?
17. If oranges cost 30 cents per dozen, how can the cost (in cents) of $(x + 4)$ dozen oranges be represented in terms of x?
18. One number is six more than a second number. Ten times the smaller minus four times the larger equals six. Find the numbers.
19. One number is 10 more than a second number. Eight times the smaller number added to three times the larger number equals 134. Find the numbers.
20. A man had $2.65 in change, consisting of eight more nickels than dimes. How many of each coin did he have?

Cumulative Review

1. Graph on a line graph all natural numbers between 6 and 26 which are exactly divisible by 4.
2. Write $81x^3y^3$ in completely factored form.
3. If $a = 2$, $b = 3$, $c = -2$, find the value of $a^2c^2 - 8abc$.

■ *Simplify.*

4. $x^2 - 2x + 3 - 4x + 2 - 2x^2$.
5. $(x^2 - 3) - (2x^2 + x)$.
6. $x(x - 3) - 2x(x + 4) + 3x^2$.
7. $(x - 3)^2 - 2(x + 1)^2 + 2x^2 - 1$
8. Solve for x: $x - b = 2x + 3b$.
9. Show by substitution that the answer you obtained in Exercise 8 is correct.

10. Write an algebraic expression for the cost of n articles at c cents each.
11. The temperature dropped $y°$ from a maximum value of 80°. Express the new temperature in terms of y.
12. The temperature dropped 6° from a maximum of $y°$. Express the new temperature in terms of y.
13. The minimum temperature of the day was y degrees; the maximum temperature was x degrees. Express the increase of temperature during the day in terms of x and y.
14. The sum of two numbers is 47. If x represents the smaller number, what expression in terms of x, represents the larger number?
15. Find two consecutive odd integers whose sum is 56.
16. Find three consecutive integers whose sum is 111.
17. One integer is five less than a second integer. The larger plus four times the smaller equals 40. Find the integers.
18. A man had $1.45 in change consisting of nickels, dimes, and quarters. He had one more dime than quarters and three more nickels than quarters. How many of each coin did he have?
19. A 35-centimeter stick is cut into three pieces so that the two end pieces are each equal in length to one-third of the middle piece. How long is each piece? (Hint: the length of the middle piece is three times the length of each end piece.)
20. A boy earns 60 cents per hour, and his older brother earns 90 cents per hour. One week the older boy worked 10 hours longer than his brother. How long did each boy work if their total income was $39?

5.1 FRACTIONS AND THEIR GRAPHICAL REPRESENTATION

In arithmetic, the indicated quotient of two numbers is called a **fraction.** In algebra, we define a fraction to be the indicated quotient of two algebraic expressions. For example,

$$\frac{x}{3}, \quad \frac{x+y}{1}, \quad \text{and} \quad \frac{a^2-b}{a+b}$$

are fractions.

We may graph arithmetic fractions on a line graph in the same manner in which we graph integers. By dividing a unit on a number line into a number of equal divisions corresponding to the denominator of the fraction, we can locate the point corresponding to the fraction itself by counting off the number of divisions corresponding to the numerator. For example, the graphs of $-3/2$, $-1/4$, and $3/4$ are shown on the line graph in Figure 5.1.

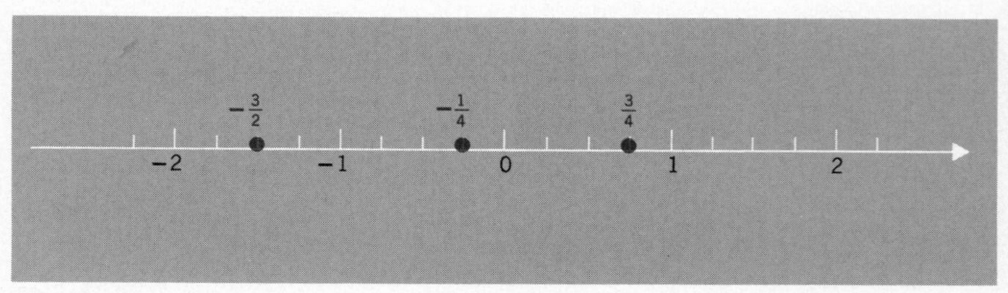

Figure 5.1

There are three signs associated with a fraction; the sign of the numerator, the sign of the denominator, and the sign of the fraction itself. We can divide all possible combinations of these signs into two categories—one where no negative signs or two negative signs are involved, and one where one or three negative signs are involved.

Zero or two (−) signs:

$$+\frac{+a}{+b} = +\left(+\frac{a}{b}\right) = \frac{a}{b}$$

$$+\frac{-a}{-b} = +\left(+\frac{a}{b}\right) = \frac{a}{b}$$

$$-\frac{-a}{+b} = -\left(-\frac{a}{b}\right) = \frac{a}{b}$$

$$-\frac{+a}{-b} = -\left(-\frac{a}{b}\right) = \frac{a}{b}$$

One or three (−) signs:

$$+\frac{-a}{+b} = +\left(-\frac{a}{b}\right) = -\frac{a}{b}$$

$$+\frac{+a}{-b} = +\left(-\frac{a}{b}\right) = -\frac{a}{b}$$

$$-\frac{+a}{+b} = -\left(+\frac{a}{b}\right) = -\frac{a}{b}$$

$$-\frac{-a}{-b} = -\left(+\frac{a}{b}\right) = -\frac{a}{b}$$

The foregoing chart may be summarized by the following rule.

Any two of the three signs of a fraction may be changed without changing the value of the fraction.

In this book, the two forms $\frac{a}{b}$ and $\frac{-a}{b}$, which have positive signs on the denominator and on the fraction itself, will be considered *standard forms* for fractions. Observe that

$$\frac{\mathbf{a}}{\mathbf{b}} = \frac{-\mathbf{a}}{-\mathbf{b}} = -\frac{-\mathbf{a}}{\mathbf{b}} = -\frac{\mathbf{a}}{-\mathbf{b}},$$

and

$$\frac{-\mathbf{a}}{\mathbf{b}} = \frac{\mathbf{a}}{-\mathbf{b}} = -\frac{\mathbf{a}}{\mathbf{b}} = -\frac{-\mathbf{a}}{-\mathbf{b}}.$$

If the numerator contains more than one term, there are alternative standard forms for a fraction. For example,

$$-\frac{x-3}{4} = \frac{-(x-3)}{4}$$

$$= \frac{-x+3}{4}$$

$$= \frac{3-x}{4},$$

and any of the three forms on the right-hand side of the equal sign may be taken as standard form.

EXERCISES 5.1

■ *Represent each quotient in fractional form.*

1. $4 \div 5$ **2.** $7 \div 2$ **3.** $2x \div y$ **4.** $x \div 3y$

5. $5 \div (x - y)$ **6.** $(2x + y) \div 3$ **7.** $5x \div (x - 5)$ **8.** $(y + 2) \div 7y$

■ *Graph each set of numbers on a line graph. Use a separate graph for each exercise.*

9. $\frac{1}{4}, \frac{3}{4}$ **10.** $\frac{1}{3}, \frac{2}{3}$ **11.** $\frac{1}{2}, \frac{5}{2}$ **12.** $-\frac{1}{4}, -\frac{3}{4}$

13. $-\frac{5}{6}, \frac{1}{6}$ **14.** $-\frac{1}{2}, \frac{1}{2}$ **15.** $-\frac{5}{2}, \frac{5}{4}$ **16.** $3, \frac{3}{4}, \frac{3}{2}$

17. $-3, -\frac{3}{4}, \frac{3}{2}$ **18.** $-\frac{2}{3}, \frac{1}{3}, 0$ **19.** $\frac{2}{5}, \frac{3}{5}, \frac{4}{5}$ **20.** $3, -\frac{5}{3}, 0$

■ *Rewrite each fraction in standard form.*

Sample Problems

a. $-\frac{1}{2}$ b. $\frac{-x}{-y}$ c. $\frac{x-1}{-3}$

Ans. $\frac{-1}{2}$ *Ans.* $\frac{x}{y}$ *Ans.* $\frac{-(x-1)}{3}$, $\frac{-x+1}{3}$ or $\frac{1-x}{3}$

21. $\frac{-3}{-4}$ **22.** $-\frac{-1}{2}$ **23.** $-\frac{2}{-3}$ **24.** $-\frac{-1}{-3}$

25. $\frac{3}{-5}$ **26.** $-\frac{-2}{5}$ **27.** $-\frac{-a}{-b}$ **28.** $-\frac{-a}{b}$

29. $\frac{a}{-b}$ **30.** $-\frac{a}{-b}$ **31.** $\frac{-x}{y}$ **32.** $-\frac{3y}{x}$

33. $-\frac{7x}{-8y}$ **34.** $-\frac{2c}{-1}$ **35.** $-\frac{-c}{-1}$ **36.** $\frac{c}{-1}$

37. $-\frac{x+2}{4}$ **38.** $\frac{x+3}{-3}$ **39.** $\frac{-5}{-(x+2)}$ **40.** $-\frac{x+2}{x-3}$

5.2 REDUCING FRACTIONS TO LOWER TERMS

In algebra, as in arithmetic, to reduce a fraction to lower terms, we use the following fundamental principle.

If both the numerator and the denominator of a given fraction are divided by the same nonzero number, the resulting fraction is equivalent to the given fraction.

In symbols this appears as

$$\frac{\mathbf{a}}{\mathbf{b}} = \frac{\frac{\mathbf{a}}{\mathbf{c}}}{\frac{\mathbf{b}}{\mathbf{c}}}, \quad \mathbf{c} \neq \mathbf{0}.$$

This fundamental principle is particularly useful in the following form:

$$\frac{\mathbf{a} \cdot \mathbf{c}}{\mathbf{b} \cdot \mathbf{c}} = \frac{\frac{\mathbf{a} \cdot \mathbf{c}}{\mathbf{c}}}{\frac{\mathbf{b} \cdot \mathbf{c}}{\mathbf{c}}} = \frac{\mathbf{a}}{\mathbf{b}}, \quad \mathbf{c} \neq \mathbf{0}.$$

In Section 1.10, we divided the numerator and the denominator of a fraction by the same number by writing the numerator and the denominator in completely factored form and dividing out common factors. For example.

$$\frac{a^5}{a^2} = \frac{\not{a}\not{a}aaa}{\not{a}\not{a}} = a^3.$$

In cases such as this where no quotient is indicated above or below the factors "divided out," the quotient 1 is to be understood.

Observe that from the above rule,

$$\frac{1 \cdot 3}{4 \cdot 3} = \frac{1}{4}.$$

On the other hand,

$$\frac{1 + 3}{4 + 3} = \frac{4}{7}.$$

These examples illustrate that while we can divide out common factors, we cannot divide out common terms. As another example, note that

$$\frac{\not{2} \cdot 4}{\not{2}} = 4$$

while

$$\frac{2+4}{2} = \frac{6}{2} = 3.$$

As a last example of this common error of dividing out common terms, observe that

$$\frac{\not{2} \cdot (3+4)}{\not{2}} = 3 + 4 = 7$$

while

$$\frac{2 \cdot 3 + 4}{2} = \frac{6+4}{2} = 5$$

We observed on page 25, that in the general case, if m is greater than n, then

$$\frac{a^m}{a^n} = a^{m-n}.$$

The quotient a^{m-n} is obtained by subtracting the exponent in the denominator from the exponent in the numerator. If the greater exponent is in the denominator, that is, if n is greater than m, then

$$\frac{a^m}{a^n} = \frac{\overbrace{\not{a}\,\not{a}\,\not{a}\,\not{a}\,\not{a} \cdot \cdots \cdot \not{a}}^{m \text{ factors}}}{\underbrace{\not{a}\,\not{a}\,\not{a}\,\not{a}\,\not{a} \cdot \cdots \cdot \not{a}\; a \cdot \cdots \cdot a}_{n \text{ factors}}},$$

or

$$\frac{a^m}{a^n} = \frac{1}{a^{n-m}}.$$

EXERCISES 5.2

■ *Reduce each fraction to an equivalent fraction in lowest terms by first completely factoring the numerator and the denominator and then dividing each by their common factors. Express your answer in standard form.*

Sample Problems

a. $\dfrac{6}{9}$ b. $\dfrac{10y^2}{4y}$ c. $\dfrac{xy^2}{-x^2y}$ d. $\dfrac{-6a^3b}{15ab^2}$

$\dfrac{\cancel{3} \cdot 2}{\cancel{3} \cdot 3}$ $\dfrac{5 \cdot \cancel{2}\cancel{y}y}{2 \cdot \cancel{2}\cancel{y}}$ $\dfrac{\cancel{x}\cancel{y}y}{-1x\cancel{x}\cancel{y}}$ $\dfrac{-2 \cdot \cancel{3}\cancel{a}aa\cancel{b}}{\cancel{3} \cdot 5\cancel{a}\cancel{b}b}$

Ans. $\dfrac{2}{3}$ *Ans.* $\dfrac{5y}{2}$ *Ans.* $\dfrac{-y}{x}$ *Ans.* $\dfrac{-2a^2}{5b}$

1. $\dfrac{4}{6}$ **2.** $\dfrac{9}{12}$ **3.** $-\dfrac{18}{21}$ **4.** $-\dfrac{12}{30}$

5. $\dfrac{32}{20}$ **6.** $\dfrac{40}{24}$ **7.** $\dfrac{2x^3}{8x^4}$ **8.** $\dfrac{8x^2}{12x^5}$

9. $\dfrac{-6y}{9y^5}$ **10.** $\dfrac{-8y}{18y^3}$ **11.** $\dfrac{14}{-28x}$ **12.** $\dfrac{48}{-9x^3}$

13. $\dfrac{x^3y}{x}$ **14.** $\dfrac{xy^4}{y^2}$ **15.** $\dfrac{-x^2}{x^3y^2}$ **16.** $\dfrac{-y}{xy^3}$

17. $\dfrac{4ax^2}{2a}$ **18.** $\dfrac{6bx}{3b}$ **19.** $\dfrac{12bx^4}{8bx^2}$ **20.** $\dfrac{12ax^5}{20ax}$

21. $\dfrac{5ab^2c^3}{4abc}$ **22.** $\dfrac{12a^2b^3c}{10ab^2c}$ **23.** $\dfrac{26a^3b^2c}{6ab^2c^3}$ **24.** $\dfrac{24abc}{6a^2b^2c^2}$

Sample Problems

a. $\dfrac{(x-2)(x+3)}{(x-5)(x+3)}$ a. $\dfrac{6x-4y}{12x-8y}$ c. $\dfrac{x^2+x-12}{x^2+2x-15}$

$\dfrac{(x-2)\cancel{(x+3)}}{(x-5)\cancel{(x+3)}}$ $\dfrac{\cancel{2}\cancel{(3x-2y)}}{2 \cdot \cancel{2}\cancel{(3x-2y)}}$ $\dfrac{(x+4)\cancel{(x-3)}}{(x+5)\cancel{(x-3)}}$

Ans. $\dfrac{x-2}{x-5}$ *Ans.* $\dfrac{1}{2}$ *Ans.* $\dfrac{x+4}{x+5}$

25. $\dfrac{3(a+b)}{4(a+b)}$ **26.** $\dfrac{4(a+2b)}{6(a+2b)}$ **27.** $\dfrac{12(x-y)}{-3}$ **28.** $\dfrac{15(a+b)}{-5}$

29. $\dfrac{(a-b)}{(a-b)}$ **30.** $\dfrac{(2x-y)}{(2x-y)}$ **31.** $\dfrac{2x+2y}{-(x+y)}$ **32.** $\dfrac{3x-3y}{-(3x+3y)}$

33. $\dfrac{2x-2a}{(x-a)^2}$ **34.** $\dfrac{3x-3a}{2(x-a)^2}$ **35.** $\dfrac{-4x}{4x^2+16x}$ **36.** $\dfrac{-3x}{6x^2+9x}$

37. $\dfrac{x+1}{x^2+2x+1}$ **38.** $\dfrac{x-4}{x^2-3x-4}$ **39.** $\dfrac{a-b}{a^2-2ab+b^2}$ **40.** $\dfrac{a-b}{a^2-b^2}$

41. $\dfrac{(a-b)^2}{a^2-b^2}$ **42.** $\dfrac{(x-2y)^2}{x^2-4y^2}$ **43.** $\dfrac{a^2-3a}{a^2-2a-3}$ **44.** $\dfrac{a^2-a}{a^2+a-2}$

45. $\dfrac{x^2+x-6}{x^2-9}$ **46.** $\dfrac{x^2+5x+6}{x^2-4}$ **47.** $\dfrac{a^2+6a+9}{a^2+2a-3}$ **48.** $\dfrac{x^2+5x+6}{x^2+6x+9}$

■ *In Exercises 49–60, state whether the second expression is equivalent to the first expression.*

Sample Problems

a. $\dfrac{x-1}{x^2-1}, \dfrac{1}{x+1}$ b. $\dfrac{3a+a^2}{3a}, \quad a^2$

Simplify fractions.

$\dfrac{\cancel{(x-1)}}{\cancel{(x-1)}(x+1)}, \dfrac{1}{x+1}$ $\dfrac{\cancel{a}(3+a)}{3\cancel{a}}, \quad a^2$

Compare fractions.

$\dfrac{1}{x+1}, \dfrac{1}{x+1}$ $\dfrac{3+a}{3}, \quad a^2$

Ans. Yes, equivalent. *Ans.* No, not equivalent.

49. $\dfrac{x+y}{y}, x$ **50.** $\dfrac{-2(x+y)}{-2x}, -2y$ **51.** $\dfrac{2x+2}{2}, x$

52. $\dfrac{x}{x+xy}, \dfrac{1}{1+y}$ **53.** $\dfrac{x-1}{x+1}, -1$ **54.** $\dfrac{y^2}{y^3-y^2}, \dfrac{1}{y-1}$

55. $\dfrac{2b+2a}{2a}, 2b$ **56.** $\dfrac{y^2-1}{y-1}, y$ **57.** $\dfrac{y}{ay-by}, \dfrac{1}{a-b}$

58. $\dfrac{a^3}{a^4-a^3}, \dfrac{1}{a-1}$ **59.** $\dfrac{2+y}{4}, \dfrac{y}{2}$ **60.** $\dfrac{6+2x}{12}, \dfrac{x}{2}$

5.3 QUOTIENTS OF POLYNOMIALS

In Section 5.2, we reduced fractions to lower terms by expressing the numerator and denominator in completely factored form and dividing out common factors. The same procedure may be used to divide a polynomial by a monomial, provided the monomial occurs as a factor in each term of the polynomial. For example,

$$\frac{2x^3+4x^2+2x}{2x} = \frac{\cancel{2x}(x^2+2x+1)}{\cancel{2x}}$$
$$= x^2+2x+1.$$

The same division can be accomplished in another manner. Because the sum of several fractions with common denominators may be expressed as a single fraction,

$$\frac{a}{d} + \frac{b}{d} + \frac{c}{d} = \frac{a+b+c}{d},$$

by the symmetric property of equality we have

$$\frac{a+b+c}{d} = \frac{a}{d} + \frac{b}{d} + \frac{c}{d}.$$

Thus, a fraction whose numerator is a polynomial may be expressed as the sum of a number of fractions whose numerators are the terms of the polynomial and whose denominators are the same as that of the original fraction. In the example above, we have

$$\frac{2x^3 + 4x^2 + 2x}{2x} = \frac{2x^3}{2x} + \frac{4x^2}{2x} + \frac{2x}{2x},$$

and the right-hand member may be simplified term by term to produce

$$x^2 + 2x + 1.$$

This particular approach has the advantage of being applicable to fractions where the numerator does not contain the denominator as a factor.

For example,

$$\begin{aligned} \frac{3x^2 + 2x + 1}{x} &= \frac{3x^2}{x} + \frac{2x}{x} + \frac{1}{x} \\ &= 3x + 2 + \frac{1}{x}. \end{aligned}$$

In Section 5.2 we divided one polynomial by another polynomial in which the divisor was a factor of the dividend by first representing the division in fractional form and then dividing out common factors. A polynomial may also be divided by another polynomial in a manner similar to that used in long division in arithmetic as shown in the following example.

$21\overline{)672}$ $x + 3\overline{)x^2 + x - 6}$

Divide 2 into 6.   Divide x into x^2; the quotient is x.

$$\begin{array}{r} 3 \\ 21\overline{)672} \end{array} \qquad \begin{array}{r} x \\ x + 3\overline{)x^2 + x - 6} \end{array}$$

Subtract product of 3 and 21 from 67.   Subtract product of x and $x + 3$ from $x^2 + x$; the difference is $-2x$.

$$
\begin{array}{r}
3 \\
21\overline{)672} \\
\underline{63} \\
4
\end{array}
\qquad\qquad
\begin{array}{r}
x\phantom{{}+x-6} \\
x+3\overline{)x^2+x-6} \\
\underline{x^2+3x}\phantom{{}-6} \\
-2x\phantom{{}-6}
\end{array}
$$

"Bring down" 2.

"Bring down" -6.

$$
\begin{array}{r}
3 \\
21\overline{)672} \\
\underline{63} \\
42
\end{array}
\qquad\qquad
\begin{array}{r}
x\phantom{{}+x-6} \\
x+3\overline{)x^2+x-6} \\
\underline{x^2+3x}\phantom{{}-6} \\
-2x-6
\end{array}
$$

Divide 2 into 4.

Divide x into $-2x$; the quotient is -2.

$$
\begin{array}{r}
32 \\
21\overline{)672} \\
\underline{63} \\
42
\end{array}
\qquad\qquad
\begin{array}{r}
x-2\phantom{{}-6} \\
x+3\overline{)x^2+x-6} \\
\underline{x^2+3x}\phantom{{}-6} \\
-2x-6
\end{array}
$$

Subtract product of 2 and 21 from 42.

Subtract product of -2 and $x+3$ from $-2x-6$; the difference is 0.

$$
\begin{array}{r}
32 \\
21\overline{)672} \\
\underline{63} \\
42 \\
\underline{42} \\
0
\end{array}
\qquad\qquad
\begin{array}{r}
x-2\phantom{{}-6} \\
x+3\overline{)x^2+x-6} \\
\underline{x^2+3x}\phantom{{}-6} \\
-2x-6 \\
\underline{-2x-6} \\
0
\end{array}
$$

The result may be checked by multiplying the quotient and the divisor to ensure that the product is the dividend. As always, the division is not valid if the divisor is 0. Thus, in the example above, where the divisor is $x+3$, we must restrict x from having a value of -3.

This process is most useful when the divisor is not a factor of the dividend. If such is the case, the division process will produce a remainder which can be expressed in terms of a fraction. For example, in dividing

$$(x^2+x+1) \quad \text{by} \quad (x+2),$$

we have

$$
\begin{array}{r}
x-1\phantom{{}+1} \\
x+2\overline{)x^2+x+1} \\
\underline{x^2+2x}\phantom{{}+1} \\
-x+1 \\
\underline{-x-2} \\
3
\end{array}.
$$

The result can then be expressed as

$$x - 1 + \frac{3}{x + 2}.$$

EXERCISES 5.3

■ *Rewrite each quotient in two ways as shown in the sample problem.*

Sample Problem

$$\frac{6x - 8}{2} \qquad\qquad \frac{6x - 8}{2}$$

$$\frac{\cancel{2}(3x - 4)}{\cancel{2}} \qquad\qquad \frac{\overset{3}{\cancel{6}}x}{\cancel{2}} - \frac{\overset{4}{\cancel{8}}}{\cancel{2}}$$

Ans. $3x - 4$ *Ans.* $3x - 4$

1. $\dfrac{8x - 4}{4}$
2. $\dfrac{6x + 3}{3}$
3. $\dfrac{y^2 + 2y}{y}$
4. $\dfrac{y^2 - 4y}{y}$
5. $\dfrac{3x^2 + 9x}{3x}$
6. $\dfrac{4x^2 - 2x}{2x}$
7. $\dfrac{3y^3 - 2y^2 + y}{y}$
8. $\dfrac{y^4 + 2y^3 + y^2}{y^2}$
9. $\dfrac{6y^3 - 3y^2 + 9y}{3y}$
10. $\dfrac{12y^3 + 4y^2 - 8y}{4y}$
11. $\dfrac{4xy^2 - x^2y + xy}{xy}$
12. $\dfrac{9x^2y^2 + 3xy^2 - 3x^2y}{3xy}$

■ *Rewrite each quotient as shown in the sample problems.*

Sample Problems

a. $\dfrac{4x^2 + 2x + 1}{2}$

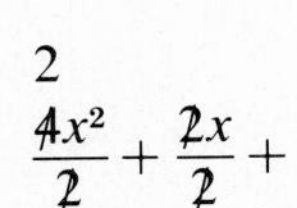

$$\frac{\overset{2}{\cancel{4}}x^2}{\cancel{2}} + \frac{\cancel{2}x}{\cancel{2}} + \frac{1}{2}$$

Ans. $2x^2 + x + \dfrac{1}{2}$

b. $\dfrac{2x^3 - x^2 + 4}{-x}$

$$\frac{2\overset{x^2}{\cancel{x^3}}}{-\cancel{x}} + \frac{-\overset{x}{\cancel{x^2}}}{-\cancel{x}} + \frac{4}{-x}$$

Ans. $-2x^2 + x + \dfrac{-4}{x}$

13. $\dfrac{9x^2 + 6x - 1}{3}$
14. $\dfrac{8x^2 + 4x - 1}{4}$
15. $\dfrac{y^2 + 2y - 1}{y}$

16. $\dfrac{6y^2 + 4y - 3}{2y}$ 17. $\dfrac{9x^4 - 6x^2 - 2}{3x^2}$ 18. $\dfrac{6x^4 - 6x^2 - 4}{6x^2}$

19. $\dfrac{y^3 - 3y^2 + 2y - 1}{y}$ 20. $\dfrac{2y^3 + 8y^2 + 2y - 1}{2y}$ 21. $\dfrac{xy^2 + xy + x}{xy}$

22. $\dfrac{x^3y^2 + x^2y^3 + xy}{xy^2}$ 23. $\dfrac{2x^2y^2 - 4xy^2 + 6xy}{2xy^2}$ 24. $\dfrac{8x^3y + 4x^2y - 4xy}{4x^2y}$

Sample Problems

a. $(x^2 - 3x - 4) \div (x + 1)$

$$\begin{array}{r l}
 & x - 4 \\
x + 1 & \overline{)\, x^2 - 3x - 4} \\
 & \underline{x^2 + x} \\
 & -4x - 4 \\
 & \underline{-4x - 4} \\
 & 0
\end{array}$$

Ans. $x - 4$

b. $(x^2 - 1) \div (x + 1)$

$$\begin{array}{r l}
 & x - 1 \\
x + 1 & \overline{)\, x^2 + 0x - 1} \\
 & \underline{x^2 + x} \\
 & -x - 1 \\
 & \underline{-x - 1} \\
 & 0
\end{array}$$

Ans. $x - 1$

25. $(x^2 + 5x - 6) \div (x - 1)$

26. $(x^2 + x - 6) \div (x - 2)$

27. $(x^2 + 6x + 5) \div (x + 5)$

28. $(x^2 - 4x + 4) \div (x - 2)$

29. $(x^2 - 49) \div (x - 7)$

30. $(x^2 - 4) \div (x + 2)$

31. $(2x^2 - 7x - 4) \div (x - 4)$

32. $(2x^2 - x - 3) \div (x + 1)$

33. $(2x^2 + 5x - 3) \div (2x - 1)$

34. $(2x^2 - 9x - 5) \div (2x + 1)$

35. $(4x^2 + 4x - 3) \div (2x - 1)$

36. $(4x^2 - 8x - 5) \div (2x + 1)$

Sample Problems

a. $(x^2 - 3x + 1) \div (x + 2)$

$$\begin{array}{r l}
 & x - 5 \\
x + 2 & \overline{)\, x^2 - 3x + 1} \\
 & \underline{x^2 + 2x} \\
 & -5x + 1 \\
 & \underline{-5x - 10} \\
 & 11
\end{array}$$

Ans. $x - 5 + \dfrac{11}{x + 2}$

b. $(2x^2 + 3x - 3) \div (2x - 1)$

$$\begin{array}{r l}
 & x + 2 \\
2x - 1 & \overline{)\, 2x^2 + 3x - 3} \\
 & \underline{2x^2 - x} \\
 & 4x - 3 \\
 & \underline{4x - 2} \\
 & -1
\end{array}$$

Ans. $x + 2 + \dfrac{-1}{2x - 1}$

37. $(x^2 + 3x + 1) \div (x + 2)$

38. $(x^2 - x + 3) \div (x + 1)$

39. $(x^2 + 3x - 9) \div (x + 5)$
40. $(x^2 - 2x - 2) \div (x - 3)$
41. $(x^2 - 7) \div (x + 6)$
42. $(x^2 - 10) \div (x - 7)$
43. $(2x^2 + x - 2) \div (x + 1)$
44. $(3x^2 - 8x - 1) \div (x - 3)$
45. $(4x^2 - 4x - 5) \div (2x + 1)$
46. $(6x^2 + x + 2) \div (3x + 2)$

5.4 LOWEST COMMON DENOMINATOR

The smallest natural number that is a multiple of each of the denominators of a set of fractions is called the **lowest common denominator** (L.C.D.) of the set of fractions. For example, the lowest common denominator for $5/12$, $3/10$, and $1/6$ contains among its factors the factors of 6, 10, and 12.

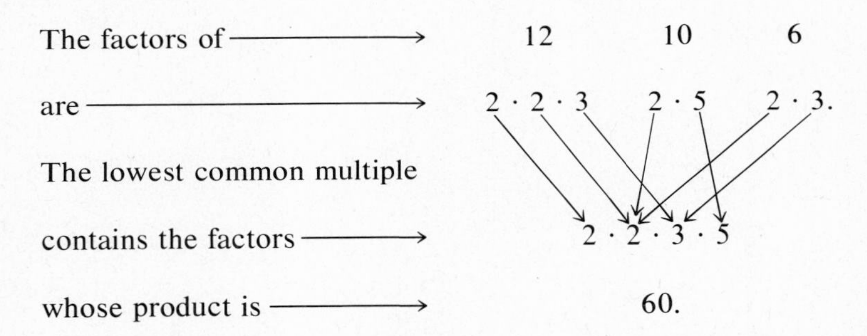

The L.C.D. of a set of algebraic fractions is the simplest algebraic expression that is a multiple of each of the denominators in the set. Thus, the L.C.D. of the fractions

$$\frac{3}{x}, \quad \frac{2}{x+1}, \quad \text{and} \quad \frac{1}{x^2(x-1)}$$

is

$$x^2(x+1)(x-1),$$

because this is the simplest expression that is a multiple of each of the denominators.

In many cases, the lowest common denominator may be found by inspection. Where the L.C.D. is not evident by inspection, it may be found by the following procedure:

1. Completely factor each denominator.
2. Include in the L.C.D. each of these factors the greatest number of times it occurs in any single denominator.

EXERCISES 5.4

■ *Find the least common denominator for each set of fractions.*

Sample Problem

$\frac{1}{6}, \quad \frac{1}{8}, \quad \frac{1}{36}$

Completely factor each denominator.

$2 \cdot 3 \qquad 2 \cdot 2 \cdot 2 \qquad 2 \cdot 2 \cdot 3 \cdot 3$

Write as a product each different factor occurring in the denominator. Include each factor the greatest number of times it occurs in any *single* denominator.

$2 \cdot 2 \cdot 2 \cdot 3 \cdot 3$

Ans. 72

1. $\frac{1}{4}, \frac{1}{6}, \frac{1}{12}$
2. $\frac{1}{3}, \frac{1}{15}, \frac{1}{5}$
3. $\frac{1}{6}, \frac{1}{9}, \frac{1}{18}$
4. $\frac{1}{5}, \frac{1}{10}, \frac{1}{15}$
5. $\frac{1}{3}, \frac{1}{14}, \frac{1}{28}$
6. $\frac{1}{2}, \frac{1}{16}, \frac{1}{20}$
7. $\frac{3}{4}, \frac{5}{8}, \frac{7}{18}$
8. $\frac{1}{6}, \frac{1}{5}, \frac{1}{14}$
9. $\frac{2}{3}, \frac{4}{8}, \frac{5}{6}$
10. $\frac{3}{4}, \frac{5}{8}, \frac{7}{18}$
11. $\frac{a}{15}, \frac{b}{20}, \frac{c}{6}$
12. $\frac{a}{27}, \frac{b}{12}, \frac{c}{18}$

Sample Problem

$\frac{1}{x}, \quad \frac{3}{x^2y}, \quad \frac{5}{x^2y^3}$

Completely factor each denominator.

$x \qquad x \cdot x \cdot y \qquad x \cdot x \cdot y \cdot y \cdot y$

Write as a product each different factor occurring in the denominators. Include each factor the greatest number of times it occurs in any *single* denominator.

$xxyyy$

Ans. x^2y^3

13. $\frac{2}{x}, \frac{3}{x^2}, \frac{1}{y}$

14. $\frac{a}{x^2}, \frac{a}{x^2y}, \frac{1}{z}$

15. $\frac{1}{xy}, \frac{2}{yz}, \frac{3}{xz}$

16. $\frac{a}{x^2y}, \frac{2}{xyz}, \frac{3}{yz^2}$

17. $\frac{3}{n}, \frac{2}{m^2n}, \frac{4}{mn^3}$

18. $\frac{3}{4x}, \frac{2}{6x^2}, \frac{1}{2y}$

19. $\frac{1}{8xy}, \frac{2}{3x^2}, \frac{2}{4xy}$

20. $\frac{2}{4xy}, \frac{3}{6yz^2}, \frac{1}{3xy^2z}$

Sample Problem

$$\frac{1}{x}, \quad \frac{1}{x^2-1}, \quad \frac{1}{x^2+2x+1}$$

Completely factor each denominator.

$$x \qquad (x-1)(x+1) \qquad (x+1)(x+1)$$

Write as a product each different factor occurring in the denominators. Include each factor the greatest number of times it occurs in any *single* denominator.

$$x(x-1)(x+1)(x+1)$$

Ans. $x(x-1)(x+1)^2$

21. $\frac{2}{x^2-y^2}, \frac{1}{x-y}$

22. $\frac{3}{x^2+2x}, \frac{4}{x+2}$

23. $\frac{3}{x^2}, \frac{4}{x^2+2x}$

24. $\frac{5}{x^2+2x+1}, \frac{3}{x^2+4x+3}$

25. $\frac{2}{x^2+3x-4}, \frac{3}{(x-1)^2}, \frac{2}{x+4}$

26. $\frac{3}{a^2-a-6}, \frac{a+2}{a^2+7a+10}, \frac{3}{(a-3)^2}$

5.5 BUILDING FRACTIONS

Just as it is often convenient to reduce fractions to lower terms, it is also often convenient to build fractions to higher terms. In algebra, as in arithmetic, we use the fundamental principle in the following form.

If both the numerator and denominator of a given fraction are multiplied by the same nonzero number, the resulting fraction is equivalent to the given fraction.

Using symbols, this principle appears as

$$\frac{\mathbf{a}}{\mathbf{b}} = \frac{\mathbf{a}}{\mathbf{b}}\left(\frac{\mathbf{c}}{\mathbf{c}}\right) = \frac{\mathbf{a} \cdot \mathbf{c}}{\mathbf{b} \cdot \mathbf{c}}, \quad \mathbf{c} \neq \mathbf{0}.$$

When applying this principle, we are in effect multiplying a quantity by 1 because

$$\frac{2}{2}, \frac{3}{3}, \frac{4}{4}, \ldots \text{ are equivalent to 1.}$$

If we wish to express $3/4$ as a fraction with a denominator of 8, we multiply the numerator and denominator by 2 to obtain

$$\frac{3(2)}{4(2)} = \frac{6}{8}.$$

In general, to change $\frac{a}{b}$ to a fraction with a denominator bc:

1. Divide b, the denominator of the given fraction, into bc, the denominator to be obtained, to find the *building factor* c.
2. Multiply the numerator and denominator of the given fraction by the building factor c.

For example, to change

$$\frac{3}{x^2y} \quad \text{to} \quad \frac{?}{x^3y^2},$$

we can divide x^2y into x^3y^2 to obtain the building factor xy. Then, we can multiply the numerator and denominator of the first fraction by this building factor to obtain

$$\frac{3(xy)}{x^2y(xy)} = \frac{3xy}{x^3y^2}.$$

If negative signs are attached to any part of the fraction, it is usually convenient to write the fraction in standard form before building it.

EXERCISES 5.5

■ *Express each fraction as a fraction with the indicated denominator.*

Sample Problem

$$\frac{3}{4} = \frac{?}{20}$$

Obtain building factor.

$(20 \div 4 = 5)$

Multiply numerator and denominator of given fraction by building factor 5.

$$\frac{3(5)}{4(5)}$$

Ans. $\frac{15}{20}$

1. $\frac{2}{5} = \frac{?}{10}$
2. $\frac{3}{7} = \frac{?}{21}$
3. $\frac{2}{3} = \frac{?}{18}$
4. $\frac{4}{9} = \frac{?}{36}$
5. $\frac{11}{6} = \frac{?}{36}$
6. $\frac{4}{3} = \frac{?}{33}$
7. $\frac{7}{3} = \frac{?}{42}$
8. $\frac{9}{8} = \frac{?}{64}$

Sample Problem

$$\frac{2}{-5a} = \frac{?}{10a^2}$$

Write in standard form.

$$\frac{-2}{5a}$$

Obtain building factor.

$(10a^2 \div 5a = 2a)$

Multiply numerator and denominator of given fraction by building factor $2a$.

$$\frac{-2(2a)}{5a(2a)}$$

Ans. $\frac{-4a}{10a^2}$

9. $\frac{5}{3x} = \frac{?}{6x}$
10. $\frac{6}{-7a} = \frac{?}{14a^2}$
11. $\frac{-a}{b} = \frac{?}{12b^3}$
12. $-\frac{3a}{5b} = \frac{?}{15ab}$
13. $\frac{-x^2}{y^2} = \frac{?}{3y^3}$
14. $\frac{-ax}{by} = \frac{?}{ab^2y}$

Sample Problem

$$ab = \frac{?}{ab^2}$$

Obtain building factor.

$$(ab^2 \div 1 = ab^2)$$

Multiply numerator and denominator of given fraction by building factor ab^2.

$$\frac{ab(ab^2)}{1(ab^2)}$$

Ans. $\frac{a^2b^3}{ab^2}$

15. $2 = \frac{?}{36}$ **16.** $-x = \frac{?}{y^2}$ **17.** $y = \frac{?}{xy}$

18. $3a = \frac{?}{9b^2}$ **19.** $x^2 = \frac{?}{3x^2y}$ **20.** $-2b^2 = \frac{?}{4a^2b}$

Sample Problem

$$\frac{1}{3} = \frac{?}{3(x - a)}$$

Obtain building factor.

$$[3(x - a) \div 3 = (x - a)]$$

Multiply numerator and denominator of given fraction by building factor $(x - a)$.

$$\frac{1(x - a)}{3(x - a)}$$

Ans. $\frac{x - a}{3(x - a)}$

21. $\frac{1}{2} = \frac{?}{2(x + y)}$ **22.** $\frac{2}{6} = \frac{?}{6(x - y)}$ **23.** $\frac{-2a}{5} = \frac{?}{5(a + 4)}$

24. $\frac{3b}{-4} = \frac{?}{4(a + b)^2}$ **25.** $2a = \frac{?}{a + 3}$ **26.** $3x = \frac{?}{6(x - 2)}$

Sample Problem

$$\frac{2x}{x - y} = \frac{?}{(x - y)(x + y)}$$

Obtain building factor.

$$[(x-y)(x+y) \div (x-y) = (x+y)]$$

Multiply numerator and denominator of given fraction by building factor $x+y$

Ans. $\dfrac{2x\ x+y}{(x-y)\ x+y}$

27. $\dfrac{3}{x-y} = \dfrac{?}{(x-y)(x+y)}$

28. $\dfrac{2x}{-(x-y)} = \dfrac{?}{(x-y)(x-y)}$

29. $-\dfrac{3}{2x-1} = \dfrac{?}{(x+1)(2x-1)}$

30. $\dfrac{-1}{a+b} = \dfrac{?}{(2a-b)(a+b)}$

31. $\dfrac{7a}{b+2} = \dfrac{?}{(b-3)(b+2)}$

32. $\dfrac{6x^2}{3x-4} = \dfrac{?}{(3x-4)(2x+5)}$

Sample Problem

$$\frac{3}{x-3} = \frac{?}{x^2-7x+12}$$

Factor denominator of each fraction.

$$\frac{3}{x-3} = \frac{?}{(x-3)(x-4)}$$

Obtain building factor.

$$[(x-3)(x-4) \div (x-3) = (x-4)]$$

Multiply numerator and denominator of given fraction by building factor $x-4$

Ans. $\dfrac{3\ x-4}{(x-3)\ x-4}$

33. $\dfrac{a}{a-3} = \dfrac{?}{a^2-3a}$

34. $\dfrac{2}{b} = \dfrac{?}{b+b^2}$

35. $\dfrac{-3}{x+y} = \dfrac{?}{x^2-y^2}$

36. $\dfrac{-2}{a-b} = \dfrac{?}{a^2-b^2}$

37. $\dfrac{y}{y-1} = \dfrac{?}{y^2+y-2}$

38. $\dfrac{-1}{x-1} = \dfrac{?}{2x^2-4x+2}$

39. $\dfrac{x+1}{x^2-x} = \dfrac{?}{x^3-2x^2+x}$

40. $\dfrac{y+1}{y^2-1} = \dfrac{?}{(y-1)(y^2+2y+1)}$

■ *In Exercises 41–60, change both fractions to equivalent fractions with lowest common denominator.*

Sample Problem

$$\frac{1}{2x}, \qquad \frac{1}{15x^2},$$

Determine L.C.D.

$$2 \cdot 3 \cdot 5 \cdot x \cdot x = 30x^2$$

Obtain building factors.

$$(30x^2 \div 2x = 15x),\ (30x^2 \div 15x^2 = 2)$$

Multiply numerator and denominator of each fraction by the respective building factors: $15x$ for the first fraction and 2 for the second.

$$\frac{1(15x)}{2x(15x)}, \qquad \frac{1(2)}{15x^2(2)}$$

Ans. $\frac{15x}{30x^2}, \frac{2}{30x^2}$

41. $\frac{1}{2}, \frac{1}{3}$ **42.** $\frac{2}{3}, \frac{3}{4}$ **43.** $\frac{1}{7}, \frac{3}{5}$ **44.** $\frac{-5}{6}, \frac{3}{4}$

45. $\frac{-5}{12}, \frac{3}{8}$ **46.** $\frac{6}{15}, \frac{2}{21}$ **47.** $\frac{-1}{a}, \frac{1}{3b}$ **48.** $\frac{a}{b}, \frac{3}{2a}$

49. $\frac{5a}{4b^2}, \frac{-2}{ab}$ **50.** $\frac{x}{y^2}, \frac{-3}{xy}$ **51.** $\frac{-a}{b^2}, \frac{2}{a^2b}$ **52.** $\frac{-b}{3}, \frac{-3}{b}$

Sample Problem

$$\frac{1}{a-1}, \qquad \frac{2a}{a^2-1}$$

Factor denominators where possible.

$$(a-1) \qquad (a+1)(a-1)$$

Determine L.C.D.

$$(a-1)(a+1)$$

Build first fraction to an equivalent fraction with the required denominator.

$$[(a-1)(a+1) \div (a-1) = (a+1)]$$

$$\frac{1(a+1)}{(a-1)(a+1)}$$

Ans. $\dfrac{a+1}{(a-1)(a+1)}, \dfrac{2a}{(a-1)(a+1)}$

53. $\dfrac{-3}{x-a}, \dfrac{2}{3}$

54. $\dfrac{2}{xy}, \dfrac{x}{x-y}$

55. $\dfrac{3}{ab}, \dfrac{a}{a+b}$

56. $\dfrac{-5}{a^2-b^2}, \dfrac{2}{(a+b)}$

57. $\dfrac{-x}{x^2+3x+2}, \dfrac{2x}{x^2+5x+4}$

58. $\dfrac{x+1}{x^2+x-2}, \dfrac{x}{x^2+5x+6}$

59. $\dfrac{3}{x^2-1}, \dfrac{x-2}{x^2+4x+3}$

60. $\dfrac{-5}{2x^2+3x-2}, \dfrac{2}{2x^2-3x+1}$

5.6 SUMS OF FRACTIONS WITH LIKE DENOMINATORS

We define the sum of two or more arithmetic or algebraic fractions as follows:

The sum of two or more fractions with common denominators is a fraction with the same denominator and a numerator equal to the sum of the numerators of the original fractions.

Thus,

$$\frac{4}{8}+\frac{9}{8}=\frac{13}{8},$$

$$\frac{2}{x}+\frac{5}{x}+\frac{3}{x}=\frac{10}{x},$$

$$\frac{3}{x+1}+\frac{x}{x+1}=\frac{3+x}{x+1}.$$

In general,

$$\frac{\mathbf{a}}{\mathbf{c}}+\frac{\mathbf{b}}{\mathbf{c}}=\frac{\mathbf{a}+\mathbf{b}}{\mathbf{c}}.$$

POSITIVE SIGN, FRACTION & DENOMINATOR

A fraction not in standard form should be changed to standard form before adding.

EXERCISES 5.6

■ *Write each sum as a single term.*

Sample Problems

a. $\frac{x}{5} + \frac{y}{5}$

$Ans.\ \frac{x + y}{5}$

b. $\frac{2x}{7} - \frac{3y}{7}$

$\frac{2x}{7} + \frac{-3y}{7}$ Write in standard form.

Add numerators.

$Ans.\ \frac{2x - 3y}{7}$

1. $\frac{2}{9} + \frac{5}{9}$ **2.** $\frac{4}{7} + \frac{2}{7}$ **3.** $\frac{5x}{11} - \frac{3}{11}$

4. $\frac{2}{13} - \frac{3y}{13}$ **5.** $\frac{2}{5} + \frac{1}{5} - \frac{x}{5}$ **6.** $\frac{4}{7} - \frac{x}{7} + \frac{y}{7}$

Sample Problem

$\frac{3}{2a} - \frac{1}{2a}$

$\frac{3}{2a} + \frac{-1}{2a}$ Write in standard form.

Add numerators and simplify.

$\frac{\not{2}}{\not{2}a}$

$Ans.\ \frac{1}{a}$

7. $\frac{5}{2a} + \frac{3}{2a}$ **8.** $\frac{6}{5a} + \frac{4}{5a}$ **9.** $\frac{7}{2b} - \frac{5}{2b}$

10. $\frac{5}{3a} - \frac{8}{3a}$ **11.** $\frac{7}{3x} + \frac{7}{3x} - \frac{2}{3x}$ **12.** $\frac{2}{7x} - \frac{10}{7x} - \frac{13}{7x}$

13. $\frac{4x}{5y} - \frac{x}{5y} + \frac{2x}{5y}$ **14.** $\frac{7x}{3y} - \frac{5x}{3y} - \frac{14x}{3y}$

Sample Problems

a. $\frac{x - y}{a} + \frac{y}{a}$ b. $\frac{x + y}{x} + \frac{x - y}{x}$

Add numerators.

$$\frac{x - y + y}{a} \qquad \frac{x + y + x - y}{x}$$

Simplify.

Ans. $\frac{x}{a}$ $\qquad \frac{2\not{x}}{\not{x}}$

Ans. 2

15. $\frac{x+1}{2} + \frac{3}{2}$

16. $\frac{x-2}{5} + \frac{3}{5}$

17. $\frac{x-2y}{3x} + \frac{x+3y}{3x}$

18. $\frac{3-x}{2y} + \frac{4-x}{2y}$

19. $\frac{x+1}{2a} + \frac{x-1}{2a}$

20. $\frac{2x-y}{3y} + \frac{2x+2y}{3y}$

21. $\frac{x^2-x}{2} + \frac{x^2}{2} + \frac{3x}{2}$

22. $\frac{2x-y}{3} + \frac{x-y}{3} + \frac{x+y}{3}$

23. $\frac{2x+y}{y} + \frac{x-2y}{y} + \frac{x+y}{y}$

24. $\frac{x-2y}{2x} + \frac{x+y}{2x} + \frac{2x+y}{2x}$

Sample Problem

$$\frac{x-2}{a+b} - \frac{2x+1}{a+b}$$

Insert parentheses and write in standard form.

$$\frac{(x-2)}{a+b} + \frac{-(2x+1)}{a+b}$$

Add numerators.

$$\frac{(x-2)-(2x+1)}{a+b}$$

Remove parentheses.

$$\frac{x-2-2x-1}{a+b}$$

Simplify.

Ans. $\frac{-x-3}{a+b}$

25. $\frac{2x+3}{2} - \frac{x-3}{2}$

26. $\frac{2x-y}{3} - \frac{3x-y}{3}$

27. $\frac{2a+b}{a-b} - \frac{a-2b}{a-b}$

28. $\frac{2a-b}{b} - \frac{a-2b}{b}$

29. $\frac{3}{a+b} - \frac{a+3}{a+b}$

30. $\frac{b-1}{a} - \frac{b+1}{a}$

31. $\frac{2x-y}{x+y} - \frac{x-3y}{x+y} + \frac{2x}{x+y}$

32. $\frac{2u-3v}{u+2} - \frac{u+2v}{u+2} + \frac{u}{u+2}$

33. $\frac{3}{x+2y} - \frac{x+3}{x+2y} + \frac{x+1}{x+2y}$

34. $\frac{2x-y}{x-y} + \frac{x-2y}{x-y} - \frac{3x-3y}{x-y}$

Sample Problem

$$\frac{2x-y}{2a+2b} + \frac{2x-3y}{2a+2b}$$

Add numerators.

$$\frac{(2x-y)+(2x-3y)}{2a+2b}$$

Remove parentheses and combine like terms.

$$\frac{4x-4y}{2a+2b}$$

Factor numerator and denominator and reduce.

$$\frac{\overset{2}{\cancel{4}}(x-y)}{\cancel{2}(a+b)}$$

Ans. $\frac{2(x-y)}{a+b}$

35. $\frac{2a+b}{3} + \frac{4a-2b}{3}$

36. $\frac{6x-6y}{5} + \frac{4x-4y}{5}$

37. $\frac{2x+y}{2} + \frac{4x+y}{2}$

38. $\frac{x-y}{4} + \frac{3x-7y}{4}$

39. $\frac{3u+2v}{4u-2v} - \frac{u+2v}{4u-2v}$

40. $\frac{u+7}{2u-4v} + \frac{u-5}{2u-4v}$

41. $\frac{x}{2x+4} - \frac{2-x}{2x+4}$

42. $\frac{x+y}{2(x-y)} + \frac{2x-2y}{2(x-y)} + \frac{x-3y}{2(x-y)}$

Sample Problem

$$\frac{3}{x^2+2x+1} - \frac{2-x}{x^2+2x+1}$$

Insert parentheses and write in standard form.

$$\frac{3}{x^2+2x+1} + \frac{-(2-x)}{x^2+2x+1}$$

$$\frac{3-(2-x)}{x^2+2x+1}$$

Add numerators and simplify.

$$\frac{x+1}{x^2+2x+1}$$

Factor denominator and reduce to lowest terms.

$$\frac{\cancel{(x+1)}}{\cancel{(x+1)}(x+1)}$$

Ans. $\dfrac{1}{x+1}$

43. $\dfrac{x+1}{x^2-2x+1}-\dfrac{5-3x}{x^2-2x+1}$

44. $\dfrac{2x+1}{x^2-x-6}+\dfrac{1-x}{x^2-x-6}$

45. $\dfrac{2x-3y}{x^2+3xy-4y^2}-\dfrac{x-7y}{x^2+3xy-4y^2}$

46. $\dfrac{u+3v}{u^2-v^2}+\dfrac{3u+v}{u^2-v^2}$

47. $\dfrac{x^2-2}{x^2-x}-\dfrac{2-4x}{x^2-x}-\dfrac{1}{x^2-x}$

48. $\dfrac{3x^2-4}{x^2-4}-\dfrac{x^2}{x^2-4}-\dfrac{4}{x^2-4}$

5.7 SUMS OF FRACTIONS WITH UNLIKE DENOMINATORS

We add fractions with unlike denominators by building the fractions to equivalent fractions with like denominators and adding as in the preceding section. The process is identical with that used in arithmetic. As an example, consider the sum $\frac{1}{2}+\frac{2}{5}$ and the sum $\frac{a}{2}+\frac{b}{5}$. In each case, we first determine the lowest common denominator, 10, and build each fraction to a fraction with that denominator. Thus,

$$\frac{(5)1}{(5)2}+\frac{(2)2}{(2)5} \quad \text{and} \quad \frac{(5)a}{(5)2}+\frac{(2)b}{(2)5}$$

are equivalent to

$$\frac{5}{10}+\frac{4}{10} \quad \text{and} \quad \frac{5a}{10}+\frac{2b}{10},$$

from which we obtain

$$\frac{9}{10} \quad \text{and} \quad \frac{5a+2b}{10}.$$

EXERCISES 5.7

■ *Rewrite each sum as a single term.*

Sample Problem

$$\frac{2}{5x} - \frac{1}{10x}$$

Write in standard form; find L.C.D. $10x$ and build each fraction to a fraction with denominator $10x$.

$$\frac{(2)2}{(2)5x} + \frac{-1}{10x}$$

$$\frac{4}{10x} + \frac{-1}{10x}$$

Add numerators.

Ans. $\frac{3}{10x}$

Odd 1–59

1. $\frac{3}{8} + \frac{1}{4}$
2. $\frac{1}{6} + \frac{1}{2}$
3. $\frac{1}{2} - \frac{3}{4}$
4. $\frac{3}{10} - \frac{4}{5}$
5. $\frac{x}{8} + \frac{5x}{2}$
6. $\frac{2y}{3} + \frac{5y}{12}$
7. $\frac{5}{ax} + \frac{3}{x}$
8. $\frac{2}{ax} + \frac{3}{a}$
9. $\frac{5x}{2y} - \frac{3x}{y}$
10. $\frac{7x}{3y} - \frac{x}{9y}$
11. $\frac{3}{x} + \frac{2}{x^2} - \frac{1}{x^3}$
12. $\frac{4}{x^3} - \frac{3}{x^2} - \frac{2}{x}$

Sample Problem

$$\frac{2}{3x} - \frac{3}{4x}$$

Write in standard form; find L.C.D. $12x$ and build each fraction to a fraction with denominator $12x$.

$$\frac{(4)2}{(4)3x} + \frac{-3(3)}{4x(3)}$$

$$\frac{8}{12x} + \frac{-9}{12x}$$

Add numerators.

Ans. $\frac{-1}{12x}$

13. $\frac{1}{3} + \frac{1}{4}$ **14.** $\frac{1}{2} + \frac{2}{5}$ **15.** $\frac{3}{4} - \frac{1}{5}$

16. $\frac{2}{3} - \frac{3}{2}$ **17.** $\frac{2x}{3} + \frac{x}{4}$ **18.** $\frac{2x}{5} + \frac{x}{3}$

19. $\frac{y}{7} - \frac{2y}{5}$ **20.** $\frac{3y}{2} - \frac{4y}{7}$ **21.** $\frac{2}{x} + \frac{4}{y}$

22. $\frac{3}{x} - \frac{2}{y}$ **23.** $\frac{1}{x} + \frac{1}{y} + \frac{1}{z}$ **24.** $\frac{1}{x} - \frac{2}{y} + \frac{3}{z}$

Sample Problem

$$\frac{x-1}{4} - \frac{2x+5}{2}$$

Insert parentheses and write in standard form; find L.C.D. 4 and build each fraction to a fraction with denominator 4.

$$\frac{(x-1)}{4} + \frac{-(2x+5)(2)}{2(2)}$$

Add numerators.

$$\frac{(x-1) - 2(2x+5)}{4}$$

Remove parentheses and simplify.

$$\frac{x - 1 - 4x - 10}{4}$$

Ans. $\frac{-3x - 11}{4}$

25. $\frac{x-2}{6} - \frac{x+1}{3}$ **26.** $\frac{2x+1}{3} - \frac{x-1}{9}$ **27.** $\frac{3y-2}{3} + \frac{2y-1}{6}$

28. $\frac{3x+4}{2} + \frac{4x-1}{4}$ **29.** $\frac{2-x}{6} + \frac{3+x}{2}$ **30.** $\frac{y+2}{3} - \frac{y-4}{6}$

31. $\frac{5x+1}{6x} + \frac{3x-2}{2x}$ **32.** $\frac{2b-c}{2c} + \frac{c+a}{c}$ **33.** $\frac{x-y}{2x} - \frac{x+y}{3x}$

34. $\frac{4y-9}{3y} - \frac{3y-8}{4y}$ **35.** $\frac{2a-b}{4b} - \frac{a-3b}{6a}$ **36.** $\frac{a-b}{ab} - \frac{b-c}{bc}$

Sample Problem

$$\frac{1}{x+1} - \frac{1}{2x+2}$$

Write in standard form.

$$\frac{1}{x+1} + \frac{-1}{2x+2}$$

(continued)

$$\frac{1}{x+1}+\frac{-1}{2(x+1)}$$

Factor denominators where possible.

Find L.C.D. $2(x+1)$ and build each fraction to a fraction with denominator $2(x+1)$.

$$\frac{(2)1}{(2)(x+1)}+\frac{-1}{2(x+1)}$$

Add numerators.

Ans. $\dfrac{1}{2(x+1)}$

37. $\dfrac{2}{x+y}-\dfrac{1}{2x+2y}$

38. $\dfrac{2}{x+1}-\dfrac{3}{2x+2}$

39. $\dfrac{5}{6x+6}-\dfrac{3}{2x+2}$

40. $\dfrac{7}{5y-10}-\dfrac{5}{3y-6}$

41. $\dfrac{3}{2a+b}-\dfrac{2}{4a+2b}+\dfrac{1}{8a+4b}$

42. $\dfrac{3}{2x+3y}-\dfrac{5}{4x+6y}+\dfrac{1}{8x+12y}$

Sample Problem

$$\frac{x}{x+2}-\frac{1}{x-1}$$

Write in standard form.

$$\frac{x}{x+2}+\frac{-1}{x-1}$$

Find L.C.D. $(x-1)(x+2)$ and build each fraction to a fraction with denominator $(x-1)(x+2)$.

$$\frac{(x-1)x}{(x-1)(x+2)}+\frac{-1(x+2)}{(x-1)(x+2)}$$

Add numerators.

$$\frac{x(x-1)-1(x+2)}{(x-1)(x+2)}$$

Remove parentheses and simplify.

$$\frac{x^2-x-x-2}{(x-1)(x+2)}$$

Ans. $\dfrac{x^2-2x-2}{(x-1)(x+2)}$

43. $\dfrac{x}{x+3}+\dfrac{x}{x-3}$

44. $\dfrac{2}{x+2}-\dfrac{3}{x+3}$

45. $\dfrac{3}{3x-4}-\dfrac{5}{5x+6}$

46. $\dfrac{1}{x+y}-\dfrac{1}{x-y}$

47. $\dfrac{1}{2a+1}-\dfrac{3}{a-2}+\dfrac{2}{2a+1}$

48. $\dfrac{x}{2x-y}+\dfrac{y}{x-2y}+\dfrac{y}{2x-y}$

Sample Problem

$$\frac{x+1}{x+2}-\frac{x-1}{x-2}$$

Enclose numerators and denominators in parentheses and write in standard form.

$$\frac{(x+1)}{(x+2)}+\frac{-(x-1)}{(x-2)}$$

Find L.C.D. $(x-2)(x+2)$ and build each fraction to a fraction with denominator $(x-2)(x+2)$.

$$\frac{(x-2)(x+1)}{(x-2)(x+2)}+\frac{-(x-1)(x+2)}{(x-2)(x+2)}$$

Add numerators.

$$\frac{(x-2)(x+1)-(x-1)(x+2)}{(x-2)(x+2)}$$

Perform indicated multiplication. Write products in parentheses.

$$\frac{(x^2-x-2)-(x^2+x-2)}{(x-2)(x+2)}$$

Remove parentheses and simplify.

$$\frac{x^2-x-2-x^2-x+2}{(x-2)(x+2)}$$

Ans. $\dfrac{-2x}{(x-2)(x+2)}$

49. $\dfrac{x-2}{x+2}-\dfrac{x+2}{x-2}$

50. $\dfrac{y-4}{y-2}-\dfrac{y-7}{y-5}$

51. $\dfrac{x+1}{x+2}-\dfrac{x+2}{x+3}$

52. $\dfrac{x+y}{x-y}-\dfrac{x-y}{x+y}$

53. $\dfrac{2x-3y}{x+y}+\dfrac{x+y}{x-y}$

54. $\dfrac{a+2b}{2a-b}-\dfrac{2a+b}{a-2b}$

Sample Problem

$$\frac{x}{x^2-9}-\frac{1}{x^2+4x-21}$$

Factor denominators and write in standard form.

$$\frac{x}{(x-3)(x+3)}+\frac{-1}{(x+7)(x-3)}$$

Find L.C.D. $(x+7)(x-3)(x+3)$ and build each fraction to a fraction with this denominator.

$$\frac{(x+7)x}{(x+7)(x-3)(x+3)}+\frac{-1(x+3)}{(x+7)(x-3)(x+3)}$$

Add numerators.

$$\frac{x(x+7)-(x+3)}{(x+7)(x-3)(x+3)}$$

Remove parentheses.

$$\frac{x^2+7x-x-3}{(x+7)(x-3)(x+3)}$$

Simplify.

Ans. $\dfrac{x^2+6x-3}{(x+7)(x-3)(x+3)}$

55. $\dfrac{1}{x^2-x-2}-\dfrac{1}{x^2+2x+1}$

56. $\dfrac{2}{x^2-5x+6}-\dfrac{5}{x^2+2x-15}$

57. $\dfrac{3x}{x^2+3x-10}-\dfrac{2x}{x^2+x-6}$

58. $\dfrac{2}{x^2-x-6}+\dfrac{3}{x^2-9}$

59. $\dfrac{5x}{x^2+3x+2}-\dfrac{3x-6}{x^2+4x+4}$

60. $\dfrac{8}{x^2-4y^2}+\dfrac{2}{x^2-5xy+6y^2}$

5.8 PRODUCTS OF FRACTIONS

The product of two fractions is defined as follows.

The product of two fractions is a fraction whose numerator is the product of the numerators and whose denominator is the product of the denominators of the given fractions.

Thus, in symbols we have

$$\frac{a}{b} \cdot \frac{c}{d} = \frac{ac}{bd}.$$

Any common factor occurring in both a numerator and a denominator of either fraction may be divided out either before or after multiplying. For example,

$$\frac{\overset{1}{\cancel{3}}}{\underset{2}{\cancel{4}}} \cdot \frac{\overset{1}{\cancel{2}}}{\underset{1}{\cancel{3}}} = \frac{1}{2} \cdot \frac{1}{1} = \frac{1}{2}$$

or

$$\frac{3}{4} \cdot \frac{2}{3} = \frac{\overset{1}{\cancel{6}}}{\underset{2}{\cancel{12}}} = \frac{1}{2}.$$

If any of the factors contain negative signs, it is advisable to proceed with the problem as if all the factors were positive and then attach the appropriate sign to the solution. A positive sign is attached to the solution if there are no negative signs or an even number of negative signs on the factors; a negative sign is attached to the solution if there are an odd number of negative signs on the factors.

In algebra, an expression such as $a\left(\frac{b}{c}\right)$ is often rewritten as an equivalent expression $\frac{ab}{c}$ as follows:

$$a\left(\frac{b}{c}\right) = \frac{a}{1} \cdot \frac{b}{c} = \frac{ab}{c}.$$

The form most convenient for use in a particular problem should be employed.

EXERCISES 5.8

■ *Change each of the following to the form* $\frac{ab}{c}$.

Sample Problems

a. $\frac{3}{4}a$ b. $\frac{5}{b}a$ c. $\frac{2}{3}(x-y)$ d. $-\frac{1}{3}x$

Ans. $\frac{3a}{4}$ *Ans.* $\frac{5a}{b}$ *Ans.* $\frac{2(x-y)}{3}$ *Ans.* $\frac{-x}{3}$

1. $\frac{2}{3}x$ **2.** $\frac{3}{4}y$ **3.** $-\frac{2}{5}a$ **4.** $-\frac{4}{7}b$

5. $\frac{3}{4}(a-b)$ **6.** $\frac{2}{3}(b-c)$ **7.** $-\frac{3}{5}(2x-y)$ **8.** $-\frac{4}{7}(x-2y)$

■ *Change each of the following to the form* $\frac{a}{b}c$.

Sample Problems

a. $\frac{2a}{5}$ b. $\frac{b}{3}$ c. $\frac{2(x+2y)}{5}$ d. $\frac{-3(2x-y)}{4}$

Ans. $\frac{2}{5}a$ *Ans.* $\frac{1}{3}b$ *Ans.* $\frac{2}{5}(x+2y)$ *Ans.* $\frac{-3}{4}(2x-y)$

9. $\frac{3x}{7}$ **10.** $\frac{4y}{3}$ **11.** $\frac{-5a}{7}$ **12.** $\frac{-b}{5}$

13. $\frac{5(a-b)}{2}$ **14.** $\frac{-3(a+2b)}{4}$ **15.** $\frac{x+y}{7}$ **16.** $\frac{-(x+y)}{5}$

■ *Write each product as a single term.*

Sample Problem

$$\frac{3}{8}\cdot\frac{12}{27}$$

Factor numerators and denominators and divide numerators and denominators by common factors.

$$\frac{\not{3}}{\not{2}\cdot\not{2}\cdot 2}\cdot\frac{\not{2}\cdot\not{2}\cdot\not{3}}{\not{3}\cdot\not{3}\cdot 3}$$

Multiply remaining factors of numerators and remaining factors of denominators.

Ans. $\frac{1}{6}$

PRIME FACTORS

17. $\frac{1}{2}\cdot\frac{3}{4}$ **18.** $\frac{16}{38}\cdot\frac{19}{12}$ **19.** $\frac{81}{121}\cdot\frac{99}{90}$

20. $\frac{3}{5}\cdot\frac{8}{12}$ **21.** $\frac{24}{30}\cdot\frac{20}{36}\cdot\frac{3}{4}$ **22.** $\frac{18}{30}\cdot\frac{6}{8}\cdot\frac{4}{20}$

Sample Problem

$$\frac{12x^2}{5y} \cdot \frac{-10y^3}{3x^2y}$$

Divide numerator and denominator by common factors.

$$\frac{\overset{4}{\cancel{12}\cancel{x^2}}}{\cancel{5}\cancel{y}} \cdot \frac{-\overset{2y}{\cancel{10}\cancel{y^3}}}{\cancel{3}\cancel{x^2}\cancel{y}}$$

Multiply remaining factors of numerators and remaining factors of denominators. Prefix appropriate sign to the answer.

Ans. $-8y$

23. $\frac{1}{3} \cdot \frac{3y}{1}$

24. $\frac{2}{3} \cdot \frac{9x^2}{4}$

25. $\frac{6x^3}{5} \cdot \frac{2}{3x}$

26. $\frac{7a}{3} \cdot \frac{1}{a^3}$

27. $6x^2y \cdot \frac{2}{3x^2}$

28. $5x^2y^2 \cdot \frac{1}{x^3y^3}$

29. $\frac{-6xy}{3} \cdot \frac{4x}{8xy^2}$

30. $\frac{-24ab^2}{8a} \cdot \frac{21a^2b}{14b}$

31. $\frac{-21r^2s}{8t} \cdot \frac{-14t^2}{3rs}$

32. $\frac{-12a^2b}{5c} \cdot \frac{10bc^2}{24a^3b}$

33. $\frac{-6xyz}{4a^2b} \cdot \frac{10ab^2}{15xyz^2}$

34. $\frac{-56x^3yz^2}{24xy^2} \cdot \frac{-48z}{28x^2z^3}$

Sample Problem

$$\frac{2x-4}{3x+6} \cdot \frac{2x+3}{x-2}$$

Factor numerators and denominators and divide numerators and denominators by common factors.

$$\frac{2\cancel{(x-2)}}{3(x+2)} \cdot \frac{(2x+3)}{\cancel{(x-2)}}$$

Multiply remaining factors of numerators and remaining factors of denominators.

Ans. $\frac{2(2x+3)}{3(x+2)}$

35. $\frac{3x-9}{5x-15} \cdot \frac{10x-5}{8x-4}$

36. $\frac{2x+4}{3x-9} \cdot \frac{x-3}{x+2}$

37. $\frac{5a+25}{2a} \cdot \frac{4a}{2a+10}$

38. $\frac{2a-4b}{8a+24b} \cdot \frac{2a+6b}{4a-8b}$

39. $\frac{2x+3y}{x-2y} \cdot \frac{3x-6y}{x-2y} \cdot \frac{x-2y}{6x+9y}$

40. $\frac{7x+14}{14x-28} \cdot \frac{2x-4}{x+2} \cdot \frac{x-3}{x+1}$

Sample Problem

$$\frac{x^2 - 2x - 3}{x^2 - 9} \cdot \frac{x^2 + 5x + 6}{x^2 - 1}$$

Factor numerators and denominators and divide numerators and denominators by common factors.

$$\frac{(x-3)(x+1)}{(x-3)(x+3)} \cdot \frac{(x+3)(x+2)}{(x-1)(x+1)}$$

Multiply remaining factors of numerators and remaining factors of denominators.

Ans. $\dfrac{x+2}{x-1}$

41. $\dfrac{x^2 - 3x - 10}{x^2 + 2x - 35} \cdot \dfrac{x^2 + 4x - 21}{x^2 + 9x + 14}$

42. $\dfrac{4y^2 - 1}{y^2 - 16} \cdot \dfrac{y^2 - 4y}{2y + 1}$

43. $\dfrac{6x^2 - x - 2}{12x^2 + 5x - 2} \cdot \dfrac{8x^2 - 6x + 1}{4x^2 - 1}$

44. $\dfrac{y^2 - y - 20}{y^2 + 7y + 12} \cdot \dfrac{y^2 + 9y + 18}{y^2 - 7y + 10}$

45. $\dfrac{x^2 + xy - 2y^2}{x^2 - 3xy + 2y^2} \cdot \dfrac{x^2 - xy - 2y^2}{x^2 + 5xy + 6y^2}$

46. $\dfrac{x^2 + x - 6}{2x^2 + 6x} \cdot \dfrac{8x^2}{x^2 - 5x + 6}$

47. $\dfrac{x^2 - 4}{x^2 - 1} \cdot \dfrac{x - 1}{2x^2 + 4x}$

48. $\dfrac{a^2 + a}{2a + 1} \cdot \dfrac{10a + 5}{3a + 3}$

49. $\dfrac{x^2 - 4}{x^2 - 5x + 6} \cdot \dfrac{x^2 - 2x - 3}{x^2 + 3x + 2}$

50. $\dfrac{x^2 + 3x}{x^2 - 3x - 4} \cdot \dfrac{x^2 - 5x + 4}{x^2 + 2x - 3}$

51. $\dfrac{y^2 - y - 20}{y^2 - 6y + 5} \cdot \dfrac{y^2 + 5y - 6}{y^2 + 7y + 12} \cdot \dfrac{y^2 - 9}{y^2 - 36}$

52. $\dfrac{x^2 - xy}{xy + y^2} \cdot \dfrac{x^2 - 4y^2}{x^2 - y^2} \cdot \dfrac{x^2 - 2xy - 3y^2}{x^2 - 5xy + 6y^2}$

5.9 QUOTIENTS OF FRACTIONS

In dividing one fraction by another, we seek a number which, when multiplied by the divisor, yields the dividend. This is precisely the same notion as that of dividing one integer by another; $b \div a$ means that we seek a number q, the quotient, such that $aq = b$.

To divide $\frac{1}{2}$ by $\frac{2}{3}$, we seek a number q such that $\frac{2}{3}q = \frac{1}{2}$. In order to solve this equation for q, we multiply each member of the equation by $\frac{3}{2}$, which is the reciprocal* of $\frac{2}{3}$. Thus,

*The reciprocal of any nonzero number a is the number $\frac{1}{a}$. The reciprocal of any fraction may be obtained by "inverting" the fraction.

$$\left(\frac{\cancel{3}}{\cancel{2}}\right)\frac{\cancel{2}}{\cancel{3}}\,q=\frac{1}{2}\left(\frac{3}{2}\right)$$
$$q=\frac{1}{2}\cdot\frac{3}{2},$$

and we have

$$q=\frac{1}{2}\div\frac{2}{3}=\frac{1}{2}\cdot\frac{3}{2}$$

which can be simplified to $\frac{3}{4}$.

This is a perfectly general procedure.

The quotient of two fractions equals the product of the dividend and the reciprocal of the divisor.

In symbols, we have

$$\frac{a}{b}\div\frac{c}{d}=\frac{a}{b}\cdot\frac{d}{c}.$$

As in multiplication, when fractions have signs attached, it is advisable to proceed with the problem as if all the factors were positive and then attach the appropriate sign to the solution.

EXERCISES 5.9

■ *Write each quotient as a single term.*

Sample Problems

a. $6\div\frac{3}{4}$

$$\frac{\overset{2}{\cancel{6}}}{1}\cdot\frac{4}{\cancel{3}}$$

Ans. 8

b. $\frac{1}{6}\div\frac{5}{8}$

$$\frac{1}{\underset{3}{\cancel{6}}}\cdot\frac{\overset{4}{\cancel{8}}}{5}$$

Ans. $\frac{4}{15}$

Invert divisor and multiply.

1. $\frac{3}{4}\div\frac{5}{6}$

2. $\frac{2}{7}\div\frac{4}{3}$

3. $4\div\frac{6}{5}$

4. $8 \div \frac{2}{3}$ **5.** $\frac{4}{5} \div 10$ **6.** $\frac{6}{7} \div 8$

Sample Problems

a. $\frac{3a^2x}{2by} \div \frac{6ax^2}{b^2y^2}$

b. $\frac{-a^2}{b} \div a^3$

Invert divisor and multiply.

$\frac{\overset{a}{\cancel{3}a^{\cancel{2}}\cancel{x}}}{2\cancel{b}\cancel{y}} \cdot \frac{\overset{b\ y}{b^{\cancel{2}}y^{\cancel{2}}}}{\underset{2\ x}{\cancel{6}\cancel{a}x^{\cancel{2}}}}$

$\frac{-\cancel{a^2}}{b} \cdot \frac{1}{\underset{a}{\cancel{a^3}}}$

Ans. $\frac{aby}{4x}$

Ans. $\frac{-1}{ab}$

7. $\frac{2c}{3d} \div \frac{4c}{6d}$ **8.** $\frac{c^2}{d} \div \frac{c^4}{d^2}$ **9.** $\frac{15}{27ab} \div \frac{16b}{9a}$

10. $\frac{a}{b^2} \div \frac{ab^2}{b^3}$ **11.** $\frac{-x^2y^2}{u^2v^2} \div \frac{xy^2}{u^2v}$ **12.** $\frac{14a^2b^3}{15x^2y} \div \frac{-21a^2b^2}{35xy}$

13. $16y^2 \div \frac{4y}{3}$ **14.** $ax^2 \div \frac{x^2}{b}$ **15.** $\frac{3}{4}xy \div (-12y^2)$

16. $\frac{36x^3}{7y} \div (-3x^2)$ **17.** $\frac{9x^2y}{ab} \div \frac{3xy^2}{b^2}$ **18.** $\frac{xy}{a^2b} \div \frac{x^3y}{ab}$

Sample Problem

$\frac{3xy + x}{y^2 - y} \div \frac{3y + 1}{y}$

Invert divisor.

$\frac{3xy + x}{y^2 - y} \cdot \frac{y}{3y + 1}$

Factor where possible and multiply.

$\frac{x\cancel{(3y + 1)}}{\cancel{y}(y - 1)} \cdot \frac{\cancel{y}}{\cancel{3y + 1}}$

Ans. $\frac{x}{y - 1}$

19. $\frac{a^2 - ab}{ab} \div \frac{2a - 2b}{ab}$ **20.** $\frac{2x - 2y}{xy} \div \frac{4x - 4y}{xy}$

21. $\frac{6a - 12}{3a + 9} \div \frac{4a - 8}{5a + 15}$ **22.** $\frac{x^2 + xy}{x^2 - xy} \div \frac{x + y}{4x - 4y}$

23. $\frac{10x^2 - 5x}{12x^3 + 24x^2} \div \frac{2x^2 - x}{2x^2 + 4x}$

24. $\frac{ax - ay}{bx + by} \div \frac{cx - cy}{dx + dy}$

Sample Problem

$$\frac{x^2 - 2x - 8}{x^2 + x - 2} \div \frac{2x^2 - 5x + 2}{x^2 - 3x + 2}$$

Invert divisor.

$$\frac{x^2 - 2x - 8}{x^2 + x - 2} \cdot \frac{x^2 - 3x + 2}{2x^2 - 5x + 2}$$

Factor where possible and multiply.

$$\frac{(x - 4)\cancel{(x + 2)}}{\cancel{(x + 2)}\cancel{(x - 1)}} \cdot \frac{\cancel{(x - 2)}\cancel{(x - 1)}}{(2x - 1)\cancel{(x - 2)}}$$

Ans. $\frac{x - 4}{2x - 1}$

25. $\frac{4x^2 - y^2}{x^2 - 4y^2} \div \frac{2x - y}{x - 2y}$

26. $\frac{x^2 - x - 6}{x^2 + 2x - 15} \div \frac{x^2 - 4}{x^2 - 25}$

27. $\frac{y^2 - 6y + 5}{y^2 + 8y + 7} \div \frac{y^2 - 3y - 10}{y^2 + 3y + 2}$

28. $\frac{x^2 - 8x + 15}{x^2 + 9x + 14} \div \frac{x^2 + 4x - 21}{x^2 - 6x - 16}$

29. $\frac{2x^2 - x - 28}{3x^2 - x - 2} \div \frac{4x^2 + 16x + 7}{3x^2 + 11x + 6}$

30. $\frac{y^2 + 7y + 10}{y^2 + 7y + 12} \div \frac{y^2 + 6y + 5}{y^2 + 8y + 16}$

31. $\frac{3y + 2}{5y^2 - y} \cdot \frac{2y^2 - y}{2y^2 - y - 1} \div \frac{6y^2 + y - 2}{10y^2 + 3y - 1}$

32. $\frac{x^2 + 4x + 3}{x^2 - 8x + 7} \cdot \frac{x^2 - 2x - 35}{x^2 - 7x - 8} \div \frac{x^2 + 8x + 15}{x^2 - 9x + 8}$

33. $\frac{3x^2 + 2x - 5}{2x^2 + x - 6} \cdot \frac{2x - 3}{3x + 5} \div \frac{x^2 + 2x - 3}{2x^2 + 3x - 2}$

34. $\frac{x^2 - x}{x^2 - 2x - 3} \cdot \frac{x^2 + 2x + 1}{x^2 + 4x} \div \frac{x^2 - 3x - 4}{x^2 - 16}$

35. $\frac{a^2 + 11a + 18}{a^2 + 4a - 5} \cdot \frac{a^2 - 6a - 7}{a^2 + 8a + 12} \div \frac{a^2 - 7a - 8}{a^2 + 2a - 15}$

36. $\frac{2y^2 - 5y - 3}{18y^2 + 3y - 1} \cdot \frac{6y^2 + 5y + 1}{2y + 1} \div \frac{2y^2 - 7y + 3}{36y^2 - 1}$

5.10 COMPLEX FRACTIONS

A fraction which contains a fraction or fractions in either its numerator or denominator or both is called a **complex fraction.** For example,

$$\frac{\frac{1}{2}}{\frac{3}{4}} \quad \text{and} \quad \frac{a + \frac{a}{3}}{2}$$

are complex fractions. Like simple fractions, complex fractions represent quotients. For example,

$$\frac{\frac{1}{2}}{\frac{3}{4}} = \frac{1}{2} \div \frac{3}{4} \quad \text{and} \quad \frac{a + \frac{a}{3}}{2} = \left(a + \frac{a}{3}\right) \div 2.$$

We may simplify a complex fraction by multiplying the numerator and denominator by the same number. If the number is the L.C.D. of all fractions in the numerator and denominator, the result of this multiplication is a simple fraction equivalent to the given complex fraction.

EXERCISES 5.10

■ *Simplify.*

Sample Problems

a. $\dfrac{\frac{2}{3}}{\frac{4}{9}}$ b. $\dfrac{\frac{3c^2}{4d^4}}{\frac{5c^3}{12d^4}}$

Find the L.C.D. for all fractions in numerator and denominator; 9 in problem *a* and $12d^4$ in problem *b*; multiply numerator and denominator by the L.C.D.

$$\frac{\overset{3}{(\cancel{9})}\,\frac{2}{\cancel{3}}}{(\cancel{9})\,\frac{4}{\cancel{9}}}, \qquad \frac{\overset{3}{(\cancel{12d^4})}\,\frac{3c^2}{\cancel{4d^4}}}{(\cancel{12d^4})\,\frac{5c^3}{\cancel{12d^4}}}$$

Divide numerator and denominator by common factors.

$$\frac{\overset{3}{\cancel{6}}}{\underset{2}{\cancel{4}}} \qquad \frac{9\cancel{c^2}}{\underset{c}{5\cancel{c^3}}}$$

Ans. $\dfrac{3}{2}$ *Ans.* $\dfrac{9}{5c}$

1. $\dfrac{\frac{3}{4}}{\frac{1}{2}}$ 2. $\dfrac{\frac{2}{3}}{\frac{4}{5}}$ 3. $\dfrac{\frac{5}{6}}{\frac{2}{3}}$ 4. $\dfrac{\frac{4}{9}}{\frac{2}{3}}$

5. $\dfrac{\frac{4}{5}}{\frac{7}{10}}$ 6. $\dfrac{\frac{8}{5}}{\frac{16}{7}}$ 7. $\dfrac{\frac{3x}{y}}{\frac{x}{2y}}$ 8. $\dfrac{\frac{2x}{y}}{\frac{x}{3y}}$

9. $\dfrac{\frac{3}{ax}}{\frac{9}{bx}}$ 10. $\dfrac{\frac{2a}{bx}}{\frac{3a}{cx}}$ 11. $\dfrac{\frac{2x}{y^2}}{\frac{6x^2}{5y}}$ 12. $\dfrac{\frac{4x^2}{3y}}{\frac{2x}{9y^2}}$

Sample Problem

$$\frac{1-\frac{1}{3}}{2+\frac{5}{6}}$$

Find the L.C.D. for all fractions in numerator and denominator, 6. Multiply each term in numerator and each term in denominator by 6.

$$\frac{(6)1-(\overset{2}{\not 6})\frac{1}{\not 3}}{(6)2+(\not 6)\frac{5}{\not 6}}$$

Simplify.

$$\frac{6-2}{12+5}$$

Ans. $\frac{4}{17}$

13. $\dfrac{\frac{2}{3}}{3-\frac{1}{3}}$ 14. $\dfrac{1+\frac{1}{5}}{\frac{2}{5}}$ 15. $\dfrac{1-\frac{1}{3}}{2+\frac{2}{3}}$ 16. $\dfrac{3+\frac{1}{10}}{2+\frac{3}{5}}$

17. $\dfrac{\frac{1}{2}-\frac{3}{8}}{\frac{5}{4}+\frac{1}{2}}$ 18. $\dfrac{\frac{2}{3}-\frac{1}{6}}{\frac{1}{3}+\frac{5}{6}}$ 19. $\dfrac{\frac{1}{2}+\frac{1}{3}}{\frac{1}{3}-\frac{1}{6}}$ 20. $\dfrac{\frac{3}{4}-\frac{1}{2}}{\frac{1}{6}+\frac{1}{3}}$

Sample Problem

$$\frac{x + \frac{y}{z}}{x - \frac{z}{y}}$$

Find the L.C.D. for all fractions in numerator and denominator, yz. Multiply each term in numerator and each term in denominator by yz.

$$\frac{(yz)x + (y\not{z})\frac{y}{\not{z}}}{(yz)x - (\not{y}z)\frac{z}{\not{y}}}$$

Simplify.

Ans. $\dfrac{xyz + y^2}{xyz - z^2}$

21. $\dfrac{\frac{2}{y} + \frac{1}{2y}}{y + \frac{y}{2}}$ 22. $\dfrac{4 - \frac{1}{x^2}}{2 - \frac{1}{x}}$ 23. $\dfrac{y - \frac{1}{y}}{y + \frac{1}{y}}$ 24. $\dfrac{a + \frac{a}{b}}{1 + \frac{1}{b}}$

25. $\dfrac{\frac{x}{3y} - \frac{1}{2}}{\frac{4}{3y} - \frac{2}{x}}$ 26. $\dfrac{2 - \frac{a}{b}}{2 - \frac{b}{a}}$ 27. $\dfrac{\frac{3}{2b} - \frac{1}{b}}{\frac{4}{a} + \frac{3}{2a}}$ 28. $\dfrac{\frac{1}{ab} - \frac{1}{b}}{\frac{1}{b} - \frac{1}{ab}}$

5.11 FRACTIONAL EQUATIONS

To solve an equation containing fractions, it is generally easiest to find first an equivalent equation which is free of fractions. We do this by multiplying each member of an equation by the lowest common denominator of the fractions. Thus, to solve the equation

$$\frac{x}{3} - 2 = \frac{4}{5},$$

each member can be multiplied by the L.C.D. 15 to obtain an equivalent equation which does not contain a fraction. Thus we have

$$\overset{5}{(\not{15})}\left(\frac{x}{\not{3}}\right) - 15(2) = \overset{3}{\not{15}}\left(\frac{4}{\not{5}}\right)$$

$$5x - 30 = 12,$$

$$5x = 42,$$

$$x = \frac{42}{5}.$$

The solution of any equation in which each member has been multiplied by an expression containing the variable should be checked. For example, multiplying each member of

$$6+\frac{4}{x-3}=\frac{x+1}{x-3}$$

by $x-3$, we obtain

$$(x-3)6+(\cancel{x-3})\frac{4}{\cancel{x-3}}=(\cancel{x-3})\frac{x+1}{\cancel{x-3}},$$
$$6x-18+4=x+1,$$
$$5x=15,$$
$$x=3.$$

Substituting 3 for x in the original equation yields

$$6+\frac{4}{0}=\frac{3+1}{0}.$$

Since division by zero is undefined, it follows that 3 is *not* a solution. The equation does not have a solution.

EXERCISES 5.11

■ *Solve.*

Sample Problem

$$5x-2=\frac{7}{2}-\frac{x}{2}$$

Multiply each member by L.C.D. 2.

$$(2)5x-(2)2=(\cancel{2})\frac{7}{\cancel{2}}-(\cancel{2})\frac{x}{\cancel{2}}$$

Complete the solution.

ODD PROB 1-41

$$10x-4=7-x$$
$$11x=11$$

Ans. $x=1$ *Check.* $5(1)-2=7/2-1/2$
$3=3$

1. $\frac{5x}{2}-1=x+\frac{1}{2}$
2. $y-\frac{3}{10}=\frac{1}{2}+\frac{3y}{5}$
3. $\frac{5y}{6}-\frac{1}{6}=\frac{2y}{3}+\frac{5}{6}$
4. $\frac{2t}{3}-\frac{1}{4}=\frac{25}{12}+\frac{t}{3}$
5. $\frac{x}{6}-\frac{7}{3}=\frac{2x}{9}-\frac{x}{4}$
6. $\frac{8x}{3}-3=\frac{2x}{3}-6$
7. $4+\frac{4}{y}=\frac{12}{y}$
8. $1+\frac{3}{x}=\frac{12}{x}$
9. $2+\frac{5}{z}=\frac{11}{z}$

10. $2+\frac{5}{2x}=\frac{3}{x}+\frac{3}{2}$ **11.** $3-\frac{1}{x}=\frac{7}{5x}-\frac{9}{5}$ **12.** $\frac{1}{3}-\frac{1}{4}=\frac{10}{3x}-\frac{1}{18}$

Sample Problem

$$\frac{x+11}{6}-\frac{11-x}{3}=1$$

Enclose numerators in parentheses and write in standard form.

$$\frac{(x+11)}{6}+\frac{-(11-x)}{3}=1$$

Multiply each member by L.C.D. 6

$$\frac{(\not{6})(x+11)}{\not{6}}+\frac{-(\overset{2}{\not{6}})(11-x)}{\not{3}}=(6)1$$

$$(x+11)-2(11-x)=6$$

Complete the solution.

$$x+11-22+2x=6$$
$$3x-11=6$$
$$3x=17$$

Ans. $x=\frac{17}{3}$

13. $\frac{y+12}{9}=\frac{y-9}{2}$

14. $\frac{y+1}{4}-\frac{3}{2}=\frac{2y-9}{10}$

15. $\frac{x-1}{10}+\frac{19}{15}=\frac{x}{3}$

16. $\frac{2x}{3}-\frac{2x+5}{6}=\frac{1}{2}$

17. $\frac{x+6}{2}-1=5$

18. $\frac{2x-2}{2}+2=\frac{1}{2}$

19. $\frac{x-2}{x}=\frac{14}{3x}-\frac{1}{3}$

20. $\frac{3}{2x}-\frac{x-3}{2x}=\frac{5}{2x}-1$

21. $\frac{2-y}{5y}=\frac{4}{15y}-\frac{1}{6}$

22. $\frac{2z-5}{z}-\frac{3}{z}=-\frac{2}{3}$

23. $\frac{x-3}{2x}+\frac{3x-7}{2x}=\frac{1}{3}$

24. $\frac{4}{x}+\frac{5}{2}=\frac{4x+5}{2x}-\frac{2x-3}{5x}$

Sample Problem

$$\frac{2}{x+10}=\frac{1}{x+3}$$

Multiply each term in the equation by L.C.D. $(x+10)(x+3)$.

$$\cancel{(x+10)}(x+3)\frac{2}{\cancel{(x+10)}} = (x+10)\cancel{(x+3)}\frac{1}{\cancel{(x+3)}}$$
$$2(x+3) = x+10$$

Complete the solution.

$$2x+6 = x+10$$

Ans. $x=4$ *Check.* $\frac{2}{4+10} = \frac{1}{4+3};\ \frac{1}{7} = \frac{1}{7}$

25. $\frac{3}{5} = \frac{x}{x+2}$ **26.** $\frac{2}{x+4} = \frac{2}{3x}$ **27.** $\frac{3}{2y-1} = \frac{7}{3y+1}$

28. $\frac{7}{4-x} = \frac{4}{7+x}$ **29.** $\frac{2}{x-9} = \frac{9}{x+12}$ **30.** $\frac{4}{y-5} = \frac{-5}{y+4}$

■ *Solve for x in terms of the other variables.*

Sample Problem

$$\frac{a+b}{x} = \frac{3}{c}$$

Multiply each member of the equation by L.C.D. xc.

$$(\cancel{x}c)\frac{(a+b)}{\cancel{x}} = (x\cancel{c})\frac{3}{\cancel{c}}$$

Simplify and complete solution.

$$ca+cb = 3x$$
$$3x = ca+cb$$

Ans. $x = \frac{ca+cb}{3}$ or $x = \frac{c(a+b)}{3}$

31. $\frac{a}{x} + \frac{b}{x} = 2$ **32.** $\frac{a+c}{x} = \frac{2}{b}$ **33.** $\frac{a}{x+1} = \frac{2a}{x-2}$

34. $\frac{2x+a}{x-b} = \frac{3}{2}$ **35.** $\frac{3}{4-x} = \frac{a}{b}$ **36.** $\frac{3}{x-a} - \frac{2}{x+b} = 0$

37. $\frac{x}{x-3} = \frac{3}{x-3} + 2$ **38.** $\frac{x}{x-2} - 7 = \frac{2}{x-2}$ **39.** $\frac{a}{x+6} - \frac{a}{x} = \frac{3}{x}$

40. $\frac{a+b}{x+4} - \frac{a}{x} = \frac{-b}{x}$ **41.** $\frac{x+2b}{x+b} = \frac{b}{x+b} + 2$ **42.** $\frac{a}{x-a} = 2 - \frac{x+2a}{x-a}$

5.12 APPLICATIONS

The word problems in the following exercises lead to equations which involve fractions.

EXERCISES 5.12

■ *Solve.*

Sample Problem

If two-thirds of a certain number is added to three-fourths of the number, the result is 17. Find the number.

Represent the unknown quantity symbolically.

Let $x =$ the number

Write an equation representing the word sentence.

$$\frac{2}{3}x + \frac{3}{4}x = 17$$

Solve the equation.

$$(\overset{4}{\cancel{12}})\frac{2}{\cancel{3}}x + (\overset{3}{\cancel{12}})\frac{3}{\cancel{4}}x = (12)(17)$$

$$8x + 9x = 204$$

$$17x = 204$$

$$x = 12$$

Ans. The number is 12.

1. If one-half of a certain number is added to three times the number, the result is $^{35}/_{2}$. Find the number.
2. If two-thirds of a certain number is subtracted from twice the number, the result is 20. Find the number.
3. Find two consecutive integers such that the sum of one-half the first and two-thirds of the next is 17.
4. Find two consecutive integers such that twice the second less one-half of the first is 14.
5. The denominator of a certain fraction is 6 more than the numerator and the fraction is equivalent to $^{3}/_{4}$. Find the numerator.
6. The denominator of a certain fraction is eight more than the numerator and the fraction is equivalent to $^{3}/_{5}$. Find the denominator.

7. A partner in a partnership receives 2/3 of the profits of the partnership. How much more than $200 per week must the business make in profits if this partner is to receive $160 per week from the partnership?

8. A man owns a three-eighths interest in a lot that was purchased for $4000. What should the lot sell for if the man is to obtain a $600 profit on his investment?

9. A man has 75 kg of ore of which 1/15 is copper. How many more kg of the same ore does he need to have 15 kg of copper?

10. A man saves one-sixth of his weekly wages. How much more than $100 per week must he make if he is to save $25 per week?

11. Fahrenheit (F) and Celsius (C) temperatures are related by the equation $F = 9/5\ C + 32$. What is the Celsius temperature in a room in which the Fahrenheit temperature is 68?

12. Using the equation in Exercise 11, find the Celsius temperature of boiling water at sea level (212°F).

13. A student received grades of 72, 78, 84, and 94 on the first four tests. What grade would he need on the next test in order to have an average of 80?

14. A boy got 17 hits on his first 60 times at bat. How many hits does he need in the next 20 times at bat to have an average of 0.300?

Sample Problem

An express train travels 180 miles in the same time that a freight train travels 120 miles. If the express goes 20 miles per hour faster than the freight, find the rate of each. *Given:* $\text{time} = \dfrac{\text{distance}}{\text{rate}}$.

Represent the unknown quantities in terms of r.

Let r = rate of freight train;
then $r + 20$ = rate of express train.

Since the fact that the times are equal is the significant equality in the problem, express the time of each train for the trip in terms of r.

Freight train's time: $t_1 = \dfrac{\text{distance}}{\text{rate}} = \dfrac{120}{r}$.

Express train's time: $t_2 = \dfrac{\text{distance}}{\text{rate}} = \dfrac{180}{r + 20}$.

Equate the expressions for time.

(continued)

$$t_1 = t_2$$

$$\frac{120}{r} = \frac{180}{r + 20}$$

Clear fractions by multiplying each term by the L.C.D. of all of the fractions, $r(r + 20)$.

$$\cancel{r}(r + 20)\,\frac{120}{\cancel{r}} = r(\cancel{r + 20})\,\frac{180}{\cancel{r + 20}}$$

Solve for r.

$$\begin{aligned} 120(r + 20) &= 180r \\ 120r + 2400 &= 180r \\ 2400 &= 60r \\ r &= 40 \end{aligned}$$

Ans. $r = 40$ mph, freight train's speed.
$r + 20 = 60$ mph, express train's speed.

15. A man drives 120 miles in the same time that another man drives 80 miles. If the speed of the first driver is 20 miles per hour greater than the speed of the second driver, find the speed of each.
16. An airplane travels 630 miles in the same time that an automobile covers 210 miles. If the speed of the airplane is 120 miles per hour greater than the speed of the automobile, find the speed of each.
17. A man rides 15 miles on his bicycle in the same time it takes him to walk 7 miles. If his rate riding is 2 miles per hour more than his rate walking, how fast does he walk?
18. A man rides 10 miles in a car and then walks 4 miles on foot. If his rate driving is 20 times his rate walking, and if the whole trip takes him $2\frac{1}{4}$ hours, how fast does he walk?
19. Two men drive from town A to town B, a distance of 300 miles. If one man drives twice as fast as the other, and arrives at town B 5 hours ahead of the other, how fast was each driving?
20. Two trains traveled from town A to town B, a distance of 400 miles. If one train traveled twice as fast as the other and arrived at town B 4 hours ahead of the other, how fast was each traveling?

5.13 RATIO AND PROPORTION

The quotient of two numbers, $a \div b$ or $\frac{a}{b}$, is sometimes referred to as a **ratio** and read "the ratio of a to b." This is a convenient way to compare two numbers. A statement that two ratios are equal, for example,

$$\frac{2}{3}=\frac{4}{6} \quad \text{or} \quad \frac{a}{b}=\frac{c}{d},$$

is called a **proportion** and read "2 is to 3 as 4 is to 6" and "a is to b as c is to d." The numbers a, b, c, and d respectively are called the first, second, third, and fourth **terms** of the proportion. The first and fourth terms are called the **extremes** of the proportion, and the second and third terms are called the **means** of the proportion.

If each ratio in the proportion $\frac{a}{b}=\frac{c}{d}$ is multiplied by bd, we have

$$(\cancel{b}d)\,\frac{a}{\cancel{b}}=(b\cancel{d})\,\frac{c}{\cancel{d}}.$$

Thus,

$$ad=bc.$$

In any proportion, the product of the extremes is equal to the product of the means.

A proportion is a special type of fractional equation. The above rule to obtain an equivalent equation without denominators is a special case of our general approach using the multiplication axiom.

Proportions can be used to make conversions from English units of measurement to metric units and vice versa. The following basic relationships (in addition to the table of metric measures on page 320 in the Appendix) will be helpful in setting up appropriate proportions to make other conversions.

1 meter (m) = 39.37 inches (in.)
1 kilogram (kg) = 2.2 pounds (lb)
1 kilometer (km) = 0.62 miles (mi)
1 liter (l) = 1.06 quarts (qt)
1 pound (lb) = 454 grams (g)
1 inch (in.) = 2.54 centimeters (cm)

For example, to change 8 inches to centimeters, we could set up the proportion

$$\frac{8}{1}=\frac{x}{2.54}$$

and solve for x, the number of centimeters. In this case

$$x = 8(2.54)$$
$$= 20.32,$$

so that 8 inches = 20.32 centimeters.

EXERCISES 5.13

■ *Express each of the following as a ratio.*

Sample Problems

a. 1 in. to 8 in.

Ans. $\frac{1}{8}$

b. 10 cm to 14 cm

$$\frac{\cancel{10}}{\cancel{40}} \quad 4$$

Ans. $\frac{1}{4}$

c. 12 to 30

$$\frac{\overset{2}{\cancel{12}}}{\underset{5}{\cancel{30}}}$$

Ans. $\frac{2}{5}$

1. 4 in. to 6 in. **2.** 8 cm to 12 cm **3.** 10 m to 18 m

4. 6 mm to 14 mm **5.** 12 g to 30 g **6.** 14 kg to 42 kg

7. 8 to 20 **8.** 10 to 30 **9.** 6 to 20

10. 12 to 30 **11.** 16 to 10 **12.** 20 to 6

■ *Express each of the following as a proportion.*

Sample Problems

a. 2 is to 5 as 4 is to 10.

Ans. $\frac{2}{5} = \frac{4}{10}$

b. 4 is to 9 as x is to 27.

Ans. $\frac{4}{9} = \frac{x}{27}$

13. 8 is to 3 as 24 is to 9.

14. 18 is to 4 as 9 is to 2.

15. 21 is to 24 as 7 is to 8.

16. 6 is to 12 as 4 is to 8.

17. 15 is to x as 10 is to 4.

18. 12 is to 6 as 4 is to x.

19. 6 is to 2 as x is to $x + 1$.

20. $x + 3$ is to x as 15 is to 3.

■ *Solve each proportion for x.*

Sample Problems

a. $\frac{4}{5} = \frac{x}{20}$

$(4)(20) = 5x$

$80 = 5x$

Ans. $x = 16$

b. $\frac{14}{12} = \frac{x}{x-1}$

Set the product of the extremes equal to the product of the means and solve equation.

$14(x-1) = 12x$

$14x - 14 = 12x$

$2x = 14$

Ans. $x = 7$

21. $\frac{3}{x} = \frac{1}{5}$

22. $\frac{2}{7} = \frac{x}{28}$

23. $\frac{x}{21} = \frac{5}{7}$

24. $\frac{6}{11} = \frac{x}{22}$

25. $\frac{3}{14} = \frac{x}{7}$

26. $\frac{12}{5} = \frac{6}{x}$

27. $\frac{x}{x+2} = \frac{2}{3}$

28. $\frac{x}{x-2} = \frac{14}{10}$

29. $\frac{1}{3} = \frac{x+3}{x+5}$

30. $\frac{1}{2} = \frac{x}{6-x}$

31. $\frac{3}{4} = \frac{x+2}{12-x}$

32. $\frac{x-7}{14+x} = \frac{-3}{4}$

Make each of the following conversions using a proportion. (Round off answers to the nearest hundredth when appropriate.)

Sample Problem

18.48 pounds to kilograms

Represent the unknown quantity symbolically.

Let $x =$ number of kilograms

Use data on page 151 and set up a proportion. The ratio of weights in pounds is equal to the ratio of weights in kilograms.

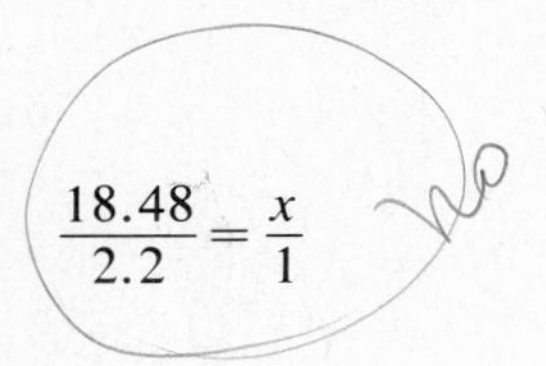

$\frac{18.48}{2.2} = \frac{x}{1}$

Solve equation.

$x = 8.4$

Ans. 18.48 pounds = 8.4 kilograms.

33. 22 inches to centimeters.

34. 3.6 pounds to grams.

35. 24.6 kilometers to miles.

36. 12.4 liters to quarts.

37. 36 pounds to kilograms.

38. 40 miles to kilometers.

39. 32 quarts to liters.

40. 100 inches to meters.

■ *Solve each problem using a proportion.*

Sample Problem

It takes 2 hours to address 70 envelopes. At the same rate, how many envelopes can be addressed in 5 hours?

Represent the unknown quantity symbolically.

Let $x =$ number of envelopes addressed in 5 hours.

Set up proportion. The ratio of the times is equal to the ratio of the number of envelopes addressed.

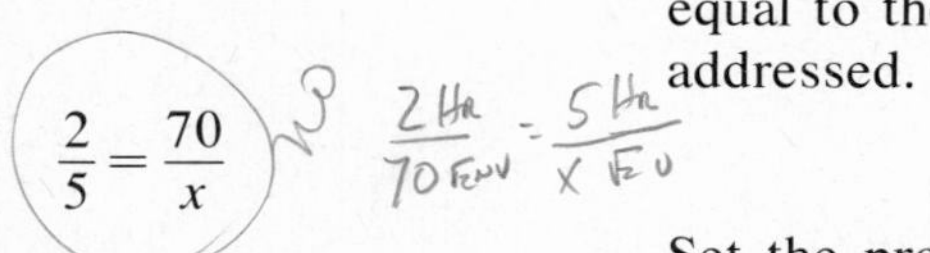

$$\frac{2}{5} = \frac{70}{x}$$

Set the product of the extremes equal to the product of the means and solve equation.

$2x = 350$

Ans. $x = 175$ envelopes

41. How many pounds of coffee will be required to make 3000 cups of coffee if 3 pounds will make 225 cups?

42. Five kilograms of sugar costs \$3.10. How many kilograms can be purchased for \$21.70?

43. A car uses 32 liters of gas to travel 184 kilometers. How many liters would be required to drive 460 kilometers?

44. A man earns \$4200 in 30 weeks. At the same rate of earnings, how much could he anticipate earning in one year (52 weeks)?

45. A family uses 3 liters of milk every 2 days. How many liters will be used in 3 months (90 days)?

46. If $3/4$ centimeters on a map represents 10 kilometers, how many kilometers does 6 centimeters represent?

47. If 660 bricks are required for 740 centimeters of a wall, how many bricks will be required for 925 centimeters?

48. The sum of two numbers is 48 and their ratio is $5/19$. What are the numbers?

49. If 20 pounds of apples cost $1.60, how much would 28 pounds of apples cost?

50. A typist takes 1 hour and 20 minutes to type 12 pages of manuscript. If she types at the same rate, how long would it take her (in hours) to type 50 pages?

CHAPTER REVIEW

1. Graph the following numbers.

$\frac{-27}{4}, \frac{-5}{2}, 2, \frac{11}{2}, \frac{37}{4}$

2. Change to equivalent fractions in standard form.

a. $-\frac{3}{x+y}$ b. $-\frac{-a}{x}$ c. $-\frac{b-2}{4}$

3. Rewrite in the form $\frac{ac}{b}$.

a. $\frac{2}{3}(x-3)$ b. $-\frac{1}{3}(x^2+1)$ c. $-\frac{3}{4}(2x+y)$

4. Reduce to lowest terms.

a. $\frac{x^2y^2z^2}{xy^3}$ b. $\frac{b-3}{b^2-2b-3}$ c. $\frac{a^2+a}{a^3-a}$

5. Build each fraction to an equivalent fraction with denominator shown.

a. $\frac{3}{x-y}, \frac{?}{2(x-y)}$ b. $\frac{3}{a+3}, \frac{?}{a^2+5a+6}$ c $\frac{x}{x-2}, \frac{?}{x^2-3x+2}$

6. Change both fractions to equivalent fractions with identical denominators.

a. $\frac{2}{3}, \frac{3}{5}$ b. $\frac{3}{x^2y}, \frac{-2}{xy^2}$ c. $\frac{a}{a^2-1}, \frac{3}{a+1}$

■ *Simplify.*

7. a. $\frac{2}{5}-\frac{1}{5}+\frac{3}{5}$ b. $\frac{x+3}{y}-\frac{3}{y}$ c. $\frac{a-2}{3}-\frac{a+3}{3}$

8. a. $\frac{3}{x}-\frac{2}{3x}$ b. $\frac{3}{r}+\frac{4}{2s}$ c. $\frac{2}{ab^2}-\frac{3}{a^2b}$

9. a. $\frac{3}{a-b}+\frac{1}{a+b}$ b. $\frac{a}{a^2-1}-\frac{1}{a^2+a}$ c. $\frac{1}{x^2-25}+\frac{5}{x^2-4x-5}$

10. a. $\frac{2xy^2}{3} \cdot \frac{x}{4y^2}$ b. $\frac{x^2 - 2x}{5} \cdot \frac{25}{x^2}$ c. $\frac{x^2 - 7x + 6}{x^2 - 1} \cdot \frac{x + 1}{x - 6}$

11. a. $\frac{2r}{3s} \div \frac{2r^2}{21s^2}$ b. $\frac{a^2 - b^2}{4} \div \frac{a^2 + ab}{4a - 4}$ c. $\frac{2x^2 - 5x - 3}{2x^2 + x} \div \frac{x - 3}{x^4}$

12. a. $\frac{\frac{3}{6}}{\frac{2}{9}}$ b. $\frac{\frac{2}{3} + \frac{1}{6}}{\frac{1}{3} + \frac{5}{6}}$ c. $\frac{1 + \frac{1}{2}}{3 - \frac{1}{4}}$

13. a. $\frac{1 - \frac{a}{b}}{1 + \frac{2}{b}}$ b. $\frac{x - \frac{x}{y}}{y - \frac{y}{x}}$ c. $\frac{\frac{1}{y} + 3}{2 - \frac{3}{y}}$

■ *Solve.*

14. a. $\frac{x}{2} = -1 + \frac{2x}{3}$ b. $\frac{x}{3} + \frac{7}{9} = \frac{1}{3}$ c. $\frac{x + 1}{2} = \frac{2x - 9}{5} + 3$

15. a. $\frac{6}{x} = \frac{16}{x + 5}$ b. $\frac{2 + y}{y} = \frac{3}{2}$ c. $\frac{y - 2}{2y} = \frac{5}{2}$

16. a. $\frac{10}{x + 4} - \frac{6}{x} = \frac{-4}{x}$ b. $\frac{14}{x - 1} + \frac{1}{x} = \frac{8}{x}$ c. $1 - \frac{3 + y}{2y} = \frac{3 - y}{y}$

■ *Solve for x.*

17. a. $\frac{b}{3} = \frac{2ax}{4}$ b. $\frac{b - x}{4} - \frac{b}{3} = \frac{x}{2}$ c. $\frac{a}{x - 1} = \frac{2a}{x}$

18. If three times a certain number is divided by 10 more than that number, the result is ½. What is the number?

19. A sample of 92 parts in a manufacturing plant proved to contain 3 defective parts. If the sample was a valid sample, how many defective parts would you expect to find in a run of 276 parts?

20. One car travels 90 miles in the same time that another car travels 60 miles. If the slower car is traveling 10 mph slower than the other car, find the rate of each.

Cumulative Review

1. The relation between centrigrade and Fahrenheit temperatures is given by $F = \frac{9}{5}C + 32$. Solve for C in terms of F.

2. Show by direct substitution that 4 is a solution of the equation

$$\frac{2x}{5} + \frac{2(x-3)}{5} = 2.$$

3. If a and b represent numbers, then $a + b = b + a$. This statement is called the __?__ law of addition.
4. If the number represented by $(24 \cdot 5) + (24 \cdot 8)$ is divided by 24, the quotient is __?__.
5. The sum of the integers between -5 and $+5$ is __?__.
6. Write $(300 \cdot 3) + (70 \cdot 3) + (5 \cdot 3)$ as $(? \cdot 3)$.
7. The reciprocal of $13/42$ is __?__.
8. If $x = -2$ and $y = 4$, find the numerical value of $x^3 - 5y$.
9. Simplify: $(3 - x) + (2 + x) - (1 - x)$.
10. Factor completely: $x^3 - x$.
11. Factor completely: $5b^3 + 10b^2 + 5b$.
12. Divide $(x^2 - 6x + 8)$ by $(x - 2)$.

■ *In Exercises 13–16, solve for x.*

13. $1 - \frac{x}{3} = \frac{2}{5}$.
14. $\frac{x+1}{2x+4} = \frac{3}{7}$.
15. $\frac{x}{3} + \frac{x}{2} = x - 1$.
16. $\frac{x}{b} = \frac{a}{c}$.
17. If $1/4$ centimeter on a map represents 8 kilometers, how many kilometers does 5 centimeters represent?
18. Fifty coins in nickels and dimes amount to \$3.50. How many of each are there?
19. The sum of three consecutive even integers is 16 more than the next even integer. What are the integers?
20. A man gave $1/3$ of his money to one son and $1/4$ to another son. He has \$10 left. How much did he have to start with?

6 FIRST-DEGREE EQUATIONS IN TWO VARIABLES

The language of mathematics is particularly effective in representing relationships between two or more variables. We have used many examples of such relationships in previous chapters. As another example, let us consider the distance traveled in a certain length of time by a car moving at a constant speed of 40 miles per hour. We can represent this relationship by:

1. A word sentence: The distance traveled in miles is equal to forty times the number of hours traveled.
2. An equation: $d = 40t$.
3. A tabulation of values.
4. A graph showing the relationship between time and distance.

We have already used word sentences and equations to describe such relationships; in this chapter we shall give our attention to tabular and graphical representations.

6.1 SOLUTIONS OF EQUATIONS IN TWO VARIABLES

The equation $d = 40t$ serves to pair with each time t a distance d. Thus,

$$\begin{aligned}
&\text{if} \quad t = 1, \quad \text{then} \quad d = 40,\\
&\text{if} \quad t = 2, \quad \text{then} \quad d = 80,\\
&\text{if} \quad t = 3, \quad \text{then} \quad d = 120, \text{ etc.}
\end{aligned}$$

The pair of numbers 1 and 40, considered together, is called a **solution** of the equation in two variables $d = 40t$. If we agree to refer to paired numbers in a specified order, we may abbreviate the above solutions as (1, 40), (2, 80), (3, 120), etc., where it is understood that the first number refers to time, whereas the second refers to distance. We call such pairs of numbers **ordered pairs,** and refer to the first and second numbers in the pairs as **components.** With this agreement, solutions of the equation $d = 40t$ are ordered pairs (t, d) whose components satisfy the equation. Some such ordered pairs are

$$(0, 0), (1, 40), (2, 80), (3, 120), (4, 160), \text{ and } (5, 200).$$

Such pairings are sometimes shown in tabular form, horizontally or vertically:

t	0	1	2	3	4	5
d	0	40	80	120	160	200

t	d
0	0
1	40
2	80
3	120
4	160
5	200

In a relationship of this type, we note that the variables d and t may be replaced by *any one* of a given set of numbers. On the other hand, symbols such as 40 are limited to one specific value, and are called **constants.** In any particular discussion involving two variables, when we assign a value to one of the variables, the value for the other variable is determined and therefore dependent upon the first. It is convenient to speak of the variable associated with the first component of an ordered pair as the **independent variable** and the variable associated with the second component of an ordered pair as the **dependent variable.**

Equations in two variables can be rewritten equivalently in a variety of forms just as we obtained equivalent equations in Section 3.7.

1. If the same quantity is added to or subtracted from equal quantities, the resulting quantities are equal.
2. If equal quantities are multiplied by equal quantities, their products are equal.
3. If equal quantities are divided by the same nonzero quantities the quotients are equal.

For example, we can solve

$$2y - 3x = 4$$

for y in terms of x by first adding $3x$ to each member to obtain

$$2y - 3x + 3x = 4 + 3x,$$
$$2y = 4 + 3x,$$

and then divide each member by 2 to obtain

$$\frac{2y}{2} = \frac{4 + 3x}{2},$$
$$y = \frac{4 + 3x}{2}.$$

EXERCISE 6.1

1. The equation $d = 4t$ relates the distance traveled by a person walking 4 miles per hour to the length of time he walks.

a. Which symbols are constants?
b. Which symbols are variables?
c. What is the effect on d of increasing t?

2. In Exercise 1,

a. Which symbol is the independent variable if a solution is given by (t, d)?
b. Which symbol is the dependent variable.
c. If t is assigned the value 3, what is the value of d?

■ *In Exercises 3–16, find the value for the dependent variable y that is associated with each given value x. Express your answer as an ordered pair (x, y).*

3. $y = 2x + 3$

Sample Problem

$(-4, \ )$ Replace x with -4.
$y = 2(-4) + 3$
$y = -5$

Ans. $(-4, -5)$

a. $(2, \ )$ b. $(3, \ )$ c. $(0, \ )$ d. $(-2, \ )$ e. $(-3, \ )$ f. $(-6, \ )$

4. $y = 3x - 2$
a. $(2, \ )$ b. $(4, \ )$ c. $(0, \ )$ d. $(-1, \ )$ e. $(-4, \ )$ f. $(-5, \ )$

5. $y = 4x - 1$
a. $(3, \ )$ b. $(1, \ )$ c. $(0, \ )$ d. $(-3, \ )$ e. $(-4, \ )$ f. $(-5, \ )$

6. $y = 4 - 2x$
a. (2,) b. (1,) c. (0,) d. (−1,) e. (−2,) f. (−3,)

7. $y = 3x$
a. (2,) b. (1,) c. (0,) d. (−1,) e. (−2,) f. (−3,)

8. $y = -x$
a. (2,) b. (1,) c. (0,) d. (−1,) e. (−2,) f. (−3,)

9. $y - 2x = 5$

Sample Problem

(2,)

Solve $y - 2x = 5$ for y in terms of x.

$y = 2x + 5$

Replace x with 2

$y = 2(2) + 5$
$y = 9$

Ans. (2, 9)

a. (4,) b. (1,) c. (0,) d. (−1,) e. (−2,) f. (−3,)

10. $y + 3x - 5 = 0$
a. (2,) b. (1,) c. (0,) d. (−1,) e. (−2,) f. (−3,)

11. $y + 2x + 4 = 0$
a. (5,) b. (3,) c. (0,) d. (−1,) e. (−3,) f. (−5,)

12. $y - 5x + 6 = 0$
a. (6,) b. (4,) c. (0,) d. (−2,) e. (−4,) f. (−6,)

13. $2x - 3y = 6$

Sample Problem

(2,)

Solve $2x - 3y = 6$ for y in terms of x.

$$2x - 6 = 3y$$
$$3y = 2x - 6$$
$$y = \frac{2x - 6}{3}$$

Replace x with 2.

$$y = \frac{2(2) - 6}{3}$$
$$y = -\frac{2}{3}$$

Ans. $\left(2, -\frac{2}{3}\right)$

a. (10,) b. (5,) c. (0,) d. (−5,) e. (−10,) f. (−15,)

14. $3x + 2y = 4$
a. (2,) b. (1,) c. (0,) d. (−1,) e. (−2,) f. (−3,)

15. $4x + 2y = 7$
a. (7,) b. (5,) c. (0,) d. (−3,) e. (−5,) f. (−7,)

16. $x - y = 6$
a. (6,) b. (4,) c. (0,) d. (−2,) e. (−4,) f. (−6,)

6.2 GRAPHS OF ORDERED PAIRS

Just as a correspondence exists between numbers and points in a line, a correspondence exists between ordered pairs of numbers and points in a plane. By constructing a pair of perpendicular line graphs, called **axes,** at some point in a plane, we may assign an ordered pair of numbers to each point in the plane by referring to the perpendicular distance of the point from each of the intersecting line graphs. If the first component is positive, the point lies to the right of the vertical axis; if negative, it lies to the left. If the second component is positive, the point lies above the horizontal axis; if negative, it lies below. The point of intersection of the axes is called the **origin,** the distance y that the point is located from the x-axis is called the **ordinate** of the point, and the distance x that the point is located from the y-axis is called the **abscissa** of the point. The abscissa and ordinate together are called the **rectangular** or **Cartesian coordinates** of the point. Each of the four regions into which the axes divide the plane is called a **quadrant,** generally referred to by the number indicated in Figure 6.1.

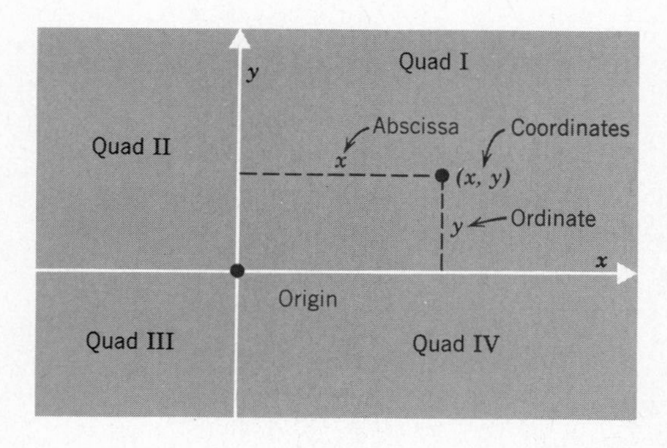

Figure 6.1

EXERCISES 6.2

■ *In Exercises 1–8, graph each set of ordered pairs on a rectangular coordinate system.*

Sample Problem

a. (2, 3)
b. (−2, 1)
c. (0, −2)
d. (−4, 4)
e. (3, −1)
f. (−3, −3)

Ans.

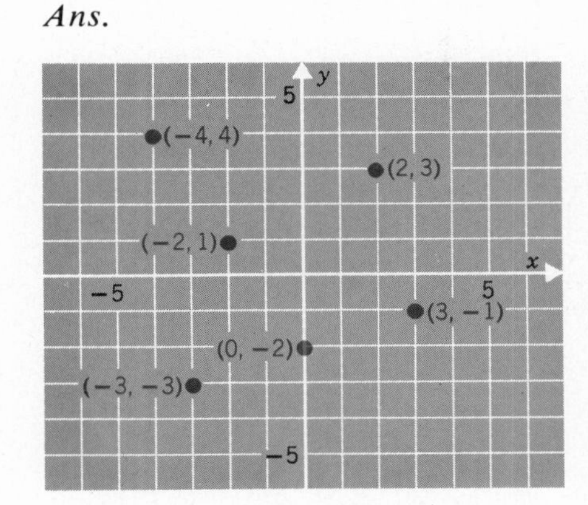

1. a. (1, 2) b. (−2, 3) c. (3, −1)
d. (−4, 5) e. (4, 4) f. (0, 5)

2. a. (3, 4) b. (−2, 0) c. (2, −1)
d. (0, 5) e. (4, −1) f. (−5, 1)

3. a. (0, 0) b. (0, 2) c. (0, 5)
d. (0, −2) e. (−5, 0) f. (5, 0)

■ *In Exercises 4 and 5, let the distance between successive marks on your axes represent 5 units.*

4. a. (10, 5) b. (−10, 5) c. (25, −5)
d. (0, 20) e. (−20, −20) f. (25, 15)

5. a. (0, 30) b. (−25, 0) c. (0, −20)
d. (30, −25) 3. (−5, 0) f. (5, −30)

■ *In Exercises 6, 7, and 8, let the distance between successive marks on the x-axis represent 1 unit and on the y-axis 5 units.*

6. a. (3, 25) b. (−1, 20) c. (5, 0)
d. (−2, −20) e. (0, −10) f. (−5, −30)

7. a. $(2, -10)$ b. $(1, -5)$ c. $(0, 0)$
d. $(-1, 5)$ e. $(-2, 10)$ f. $(-3, 15)$

8. a. $(3, -24)$ b. $(2, -16)$ c. $(1, -8)$
d. $(0, 0)$ e. $(-1, 8)$ f. $(-2, 16)$

9. Connect the points plotted in Exercise 7. What do you observe?

10. Connect the points plotted in Exercise 8. What do you observe?

11. In the rectangular coordinate system, name the (perpendicular) distance from a given point to: a. The x-axis. b. The y-axis.

12. Which component of the ordered pair (x, y) represents the (perpendicular) distance of a point from: The x-axis? The y-axis?

13. Name the point corresponding to $(0, 0)$.

14. Describe the location of the graphs of all ordered pairs of the form $(0, y)$ and $(x, 0)$.

15. Describe the location of the graphs of all ordered pairs (a, a); i.e., all points whose first and second components are equal.

16. Graph $(1, 2)$ and $(3, 6)$ and draw a line connecting the points.
a. Does the graph of $(2, 4)$ lie on the line?
b. If you extended the graph in both directions would the graphs of $(4, 8)$ and $(-1, -2)$ lie on the line?
c. Is the graph of $(0, 0)$ on this line?

17. Locate the graphs of $(2, 3)$ and $(2, 5)$ and draw a line connecting them.
a. Does the graph of $(2, 4)$ lie on this line?
b. Does the graph of $(1, 3)$ lie on this line?
c. If you extended the graph in both directions, would the graphs of $(2, 6)$ and $(2, -1)$ lie on the line?
d. Is the graph of $(0, 0)$ on the line?

18. How many points lie on any line in the plane?

19. What would be the least number of points necessary to determine a line?

20. Which axis, the horizontal or the vertical, is usually employed to represent:
a. The independent variable? b. The dependent variable?

6.3 GRAPHING FIRST-DEGREE EQUATIONS

In Section 6.1, we observed that a solution of an equation in two variables is an ordered pair. In Section 6.2, we observed that the components of an ordered pair are the coordinates of a point in a plane.

Thus, to graph an equation in two variables, we need only graph the set of ordered pairs that are solutions to the equation. For example, we may find some representative solutions to the first degree equation $y = x + 2$, say

$$(0, 2), \quad (-3, -1), \quad (-2, 0), \quad \text{and} \quad (3, 5),$$

which can be displayed in a tabular form as shown below.

If we graph the points determined by these ordered pairs and pass a straight line through them, we obtain the graph of all solutions of $y = x + 2$ as shown in Figure 6.2. The graphs of first-degree equations

x	y
0	2
−3	−1
−2	0
3	5

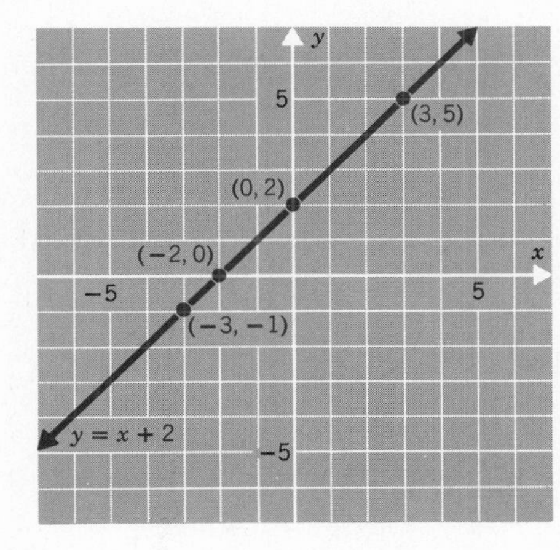

Figure 6.2

are *always* straight lines; therefore such equations are also referred to as **linear equations.**

Only two points are necessary to determine the graph of any first-degree equation. A third point is usually obtained as a check.

The following procedure is suggested in graphing a first-degree equation.

1. Construct a set of rectangular axes showing the scale and the variable represented by each axis.
2. Find two ordered pairs that are solutions of the equation to be graphed. Assign any convenient value to one variable and determine the corresponding value of the other variable.
3. Graph these ordered pairs.
4. Pass a straight line through the points.
5. Check by graphing a third ordered pair that is a solution of the equation and verify that it lies on the line.

EXERCISES 6.3

1. Given $d = 4t$ (see Section 6.1, Exercise 1), find the value of d corresponding to each value of t and express your answer in the form of

an ordered pair (t, d). Then graph each of the ordered pairs and connect them with a straight line.

a. (0, ?) b. (2, ?) c. (4, ?)
d. Where are all points located whose coordinates satisfy $d = 4t$?
e. Check by obtaining additional solutions (1, ?) and (3, ?) and graphing these.

2. What is the minimum number of points necessary to determine the line representing $d = 4t$?

■ *In Exercises 3–14:*

(1) *Find any two ordered pairs that are solutions of the given equation.*
(2) *Graph these points and draw a straight line through them.*
(3) *Check your result by finding a third solution of the equation and verify that its graph is a point on the line. Graph each equation on a separate set of axes.*

Sample Problem

$$2x + 3y = 6$$

Solve for y in terms of x.

$$y = \frac{6 - 2x}{3}$$

Take any two numbers for the first components, say 0 and 6 Find the second components of (0, ?) and (6, ?) so that the ordered pairs are solutions of the equation:

$$y = \frac{6 - 2(0)}{3} = 2$$

and

$$y = \frac{6 - 2(6)}{3} = -2.$$

Graph the ordered pairs (0, 2) and (6, −2) which are obtained and draw a straight line through them. Check by noting that the graph of a third pair, say with first component 3 and second component

$$y = \frac{6 - 2(3)}{3} = 0,$$

that satisfies the equation, also lies on the line.

x	y
0	2
6	−2
3	0

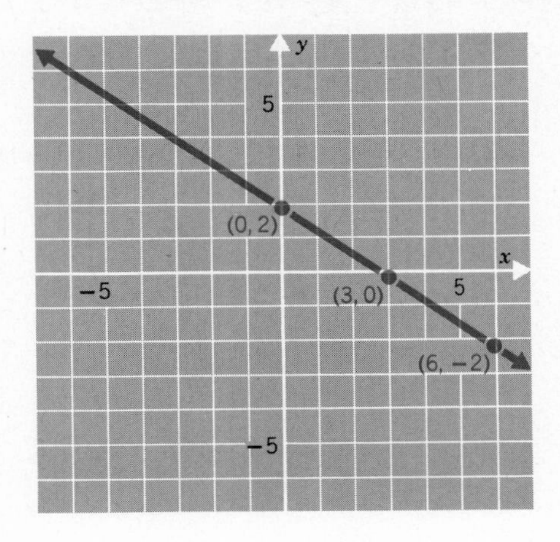

3. $y = x + 2$

4. $y = x - 2$

5. $y = 2x + 1$

6. $y = 3x - 1$

7. $y = 2x - 1$

8. $y + x = 4$

9. $y - 3x = 0$

10. $2y = 3x + 4$

11. $3y = 4 - x$

12. $3y + 2x = 12$

13. $2y + x - 6 = 0$

14. $y - 2x - 6 = 0$

15. Consider the two ordered pairs (5, 2) and (2, 2).

a. Graph these ordered pairs and draw a straight line through their graphs.
b. Are the graphs of (1, 2), (3, 2), (4, 2), (−2, 2) on the line?
c. Would the graph of $(x, 2)$ lie on the line for any (all) x?
d. Does the value of x have anything to do with the fact that a point lies on this line?
e. Is $y = 0x + 2$ an equation for the line?
f. Does $y = 2$ give a complete description of the line?

16. Consider the two ordered pairs (3, 5) and (3, 2).

a. Graph these ordered pairs and draw a straight line through their graphs.
b. Are the graphs of (3, −1), (3, 4), (3, 6) on the line?
c. Would the graphs of $(3, y)$ lie on the line for any (all) y?
d. Does the value of y have anything to do with the fact that a point lies on this line?
e. Is $x = 0y + 3$ an equation for the line?
f. Does $x = 3$ gives a complete description of the line?

■ *Graph each of the following equations.*

Sample Problem

$y = 3$

The equation $y = 3$ is equivalent to $y = 0x + 3$. For example,

if $x = 1$, $\quad y = 0(1) + 3 = 3$;

if $x = 2$, $\quad y = 0(2) + 3 = 3$;

if $x = 5$, $\quad y = 0(5) + 3 = 3$, etc.

Thus, the graph is a line parallel to the x-axis at a distance of 3 units above this axis.

x	y
1	3
2	3
5	3

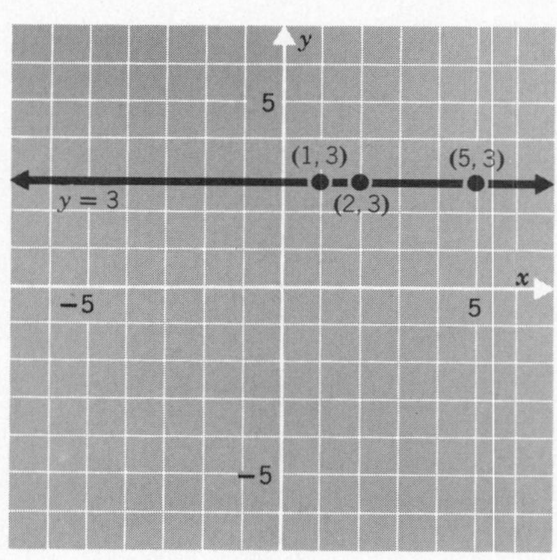

17. $x + 0y = 4$ **18.** $0x + y = -1$ **19.** $x = -3$ **20.** $y = 5$

21. $x = -2$ **22.** $y = -5$ **23.** $x = 0$ **24.** $y = 0$

6.4 INTERCEPT METHOD OF GRAPHING

In Section 6.3 we assigned values to x in equations in two variables to find the corresponding values of y. The solutions of an equation in two variables that are generally easiest to find are those in which either the first component or the second component is 0. For example, if we substitute 0 for x in the equation

$$3x + 4y = 12 \qquad (1)$$

we have

$$3(0) + 4y = 12,$$
$$4y = 12,$$
$$y = 3.$$

Hence a solution of Equation (1) is (0, 3). We can also find ordered pairs that are solutions of equations in two variables by assigning values to y and determining the corresponding values of x. In particular, if we substitute 0 for y in Equation (1) we obtain

$$3x + 4(0) = 12,$$
$$3x = 12,$$
$$x = 4,$$

and a second solution of the equation is (4, 0). The ordered pairs (0, 3) and (4, 0) can now be used to graph Equation (1). The graph is shown in Figure 6.3. The numbers 4 and 3 are called the **x-intercept,** and the **y-intercept,** of the graph, respectively.

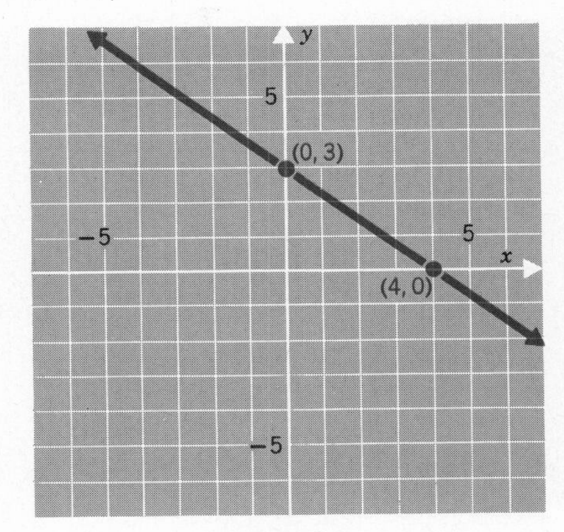

Figure 6.3

The foregoing method of drawing the graph of a linear equation is called the **intercept method of graphing.**

If the graph intersects the axes at the origin or near the origin, the intercept method is not satisfactory. It is then necessary to graph an ordered pair that is a solution of the equation and whose graph is not the origin or is not too close to the origin.

EXERCISES 6.4

■ *Graph each equation by the intercept method.*

Sample Problem

$$2x - y = 6 \qquad (1)$$

Substitute 0 for x and solve for y.

$$2(0) - y = 6$$
$$-y = 6$$
$$y = -6$$

(continued)

The ordered pair $(0, -6)$ is a solution of Equation (1). Substitute 0 for y in Equation (1) and solve for x.

$$2x - (0) = 6$$
$$2x = 6$$
$$x = 3$$

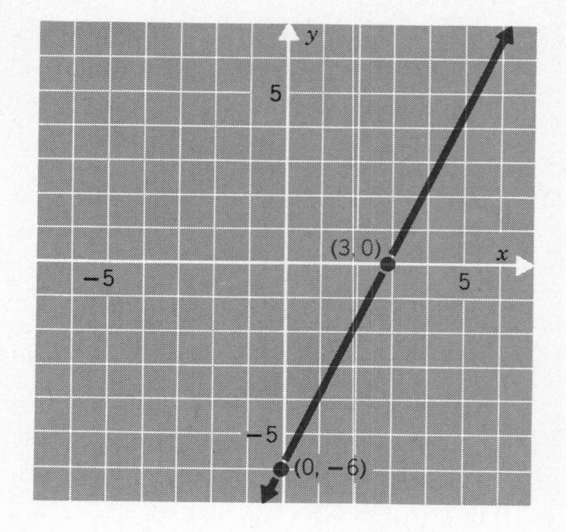

The ordered pair $(3, 0)$ is a solution of Equation (1).
Graph $(0, -6)$ and $(3, 0)$ and complete the graph of Equation (1).

1. $x + y = 5$ **2.** $x - y = 4$ **3.** $2x + y = 8$
4. $x + 2y = 6$ **5.** $3x - y = 6$ **6.** $x - 2y = 4$
7. $2x + 3y = 12$ **8.** $3x + 5y = 15$ **9.** $3x - 4y = 12$
10. $4x - 5y = 20$ **11.** $y = x + 6$ **12.** $y = x + 4$
13. $y = 2x - 4$ **14.** $y = 3x + 9$ **15.** $y = 2x + 5$
16. $y = 3x - 7$ **17.** $x = 4 + y$ **18.** $x = 6 - 2y$
19. $x = 3y - 10$ **20.** $x = 5y + 5$ **21.** $2x - y = 0$
22. $x - 3y = 0$ **23.** $2x + 3y = 1$ **24.** $4x - 3y = 1$

6.5 DIRECT VARIATION

The relationship

$$\mathbf{y = kx} \quad (k \text{ is a constant}) \tag{1}$$

that we studied in previous sections of this chapter, is given a special name. It is called a **direct variation.** The variable y is said to *vary directly as* x. For example, in the formula

$$d = 40t,$$

which relates the distance traveled at a constant rate of 40 miles per hour to the time traveled, d varies directly as t. As the time increases, the distance increases.

The constant k involved in Equation 1 is called the **constant of vari-**

ation. It is also called the **slope** of the graph of the equation $y = kx$. For example, the graphs of several direct relationships are shown in Figure 6.4 along with their respective slopes.

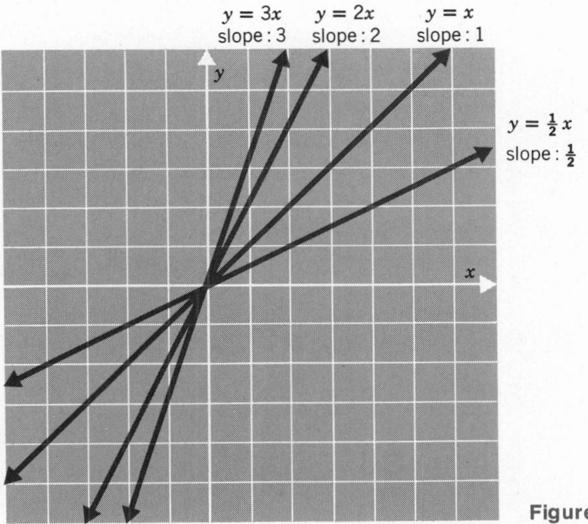

Figure 6.4

Note that the slope of the line is related to its "steepness."

If we have one set of associated values for the variables in a direct variation, we can find a value for each of the variables for any given value of the other. For example, if we know that the pressure P in a special liquid *varies directly* as the depth d below the surface of the liquid, the statement of the relationship is given in symbols by

$$P = kd.$$

Now solving for k in terms of P and d, by dividing both members of the equation by d, yields

$$\frac{P}{d} = k,$$

which can be written as

$$k = \frac{P}{d}.$$

Since k is a *constant,* the ratio $\frac{P}{d}$ is constant. Thus, if we know that $P = 40$ when $d = 10$, we can, for example, find P when $d = 15$. Substituting the appropriate values for P and d, we obtain the proportion (equal ratios),

$$\frac{40}{10} = \frac{P}{15},$$

from which

$$40 \cdot 15 = 10 \cdot P,$$

$$\frac{40 \cdot 15}{10} = P,$$

$$60 = P.$$

Hence $P = 60$ when $d = 15$.

EXERCISES 6.5

■ *Graph and specify the slope of each of the following relationships.*

Sample Problem

$$y = \frac{2}{3}x$$

Select any two values of x, such as 0 and 3.

If $x = 0$, then $y = 0$, and $(0, 0)$ is a solution of the equation.

If $x = 3$, then $y = 2$, and $(3, 2)$ is a solution.

Graph the ordered pairs and draw the line through the points.

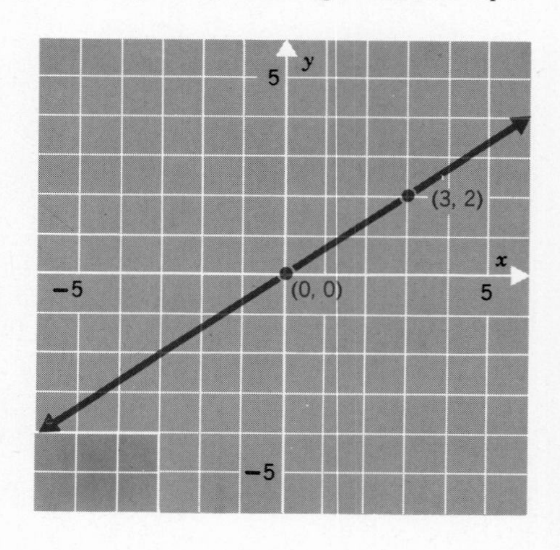

Ans. The slope of the line is equal to the constant of variation, $2/3$.

1. $y = 4x$
2. $y = 5x$
3. $y = 6x$
4. $y = 7x$
5. $y = \frac{3}{5}x$
6. $y = \frac{3}{4}x$
7. $y = \frac{5}{4}x$
8. $y = \frac{7}{3}x$
9. $y - 2x = 0$ (Hint: first solve for y in terms of x)
10. $y - 4x = 0$
11. $2y - x = 0$
12. $3y - 5x = 0$

Sample Problem

If y varies directly as x, and $y = 22$ when $x = 33$, find y if $x = 40$.

The relationship is given by

$$y = kx,$$

from which by dividing each member by x, the constant k is given by the ratio

$$\frac{y}{x} = k.$$

Since the ratio y/x is a constant, the ratios are equal for different sets of conditions. Hence, substituting appropriate values in the ratio yields the proportion (equal ratios)

$$\frac{22}{33} = \frac{y}{40},$$

from which

$$(22)(40) = 33y,$$

$$\frac{(22)(40)}{33} = y,$$

Ans. $y = \dfrac{80}{3}$

13. If y varies directly with x, and $y = 6$ when $x = 4$, find y when $x = 14$.

14. If z varies directly with x, and $z = 9$ when $x = 12$, find z when $x = 3$.

15. If R varies directly with L, and $R = 42$ when $L = 30$, find R when $L = 45$.

16. If I varies directly with d, and $I = 640$ when $d = 120$, find I when $d = 600$.

17. If the distance (d) in miles that a car travels varies directly with the time (t) in hours, and $d = 96$ when $t = 2$, find the distance traveled in 3½ hours.

18. If the cost (C) in dollars of a certain vitamin varies directly with the weight (w) in pounds, and $C = 60$ when $w = 24$, find the cost of 54 pounds.

19. If the resistance (R) in an electric circuit varies directly with the voltage E, and $R = 4.2$ when $E = 60$, find E when $R = 6.6$.

20. If the tension (T) on a spring varies directly as the distance (s) it is stretched, and $T = 54$ when $s = 12$, find s if $T = 20$.

6.6 GRAPHICAL SOLUTION OF SYSTEMS OF LINEAR EQUATIONS

It is frequently useful to be able to find a single ordered pair that is a solution to each of two different equations. One means of obtaining such an ordered pair is by graphing the two equations on the same set of axes and determining the coordinates of the point where they intersect. For example, consider the equations $x + y = 5$ and $x - y = 1$. Using the intercept method of graphing, we find that two ordered pairs that are solutions of $x + y = 5$ or $y = 5 - x$ are

$$(0, 5) \quad \text{and} \quad (5, 0).$$

Two ordered pairs that are solutions of $x - y = 1$ or $y = x - 1$ are

$$(0, -1) \quad \text{and} \quad (1, 0).$$

The graphs of the equations are shown in Figure 6.5. The coordinates of the point of intersection are the components of $(3, 2)$.* Hence, $(3, 2)$ should satisfy each equation.

Check. $y = 5 - x$. $\quad (2) = 5 - (3)$ $\quad 2 = 2$; $\qquad y = x - 1$, $\quad (2) = (3) = 1$, $\quad 2 = 2$.

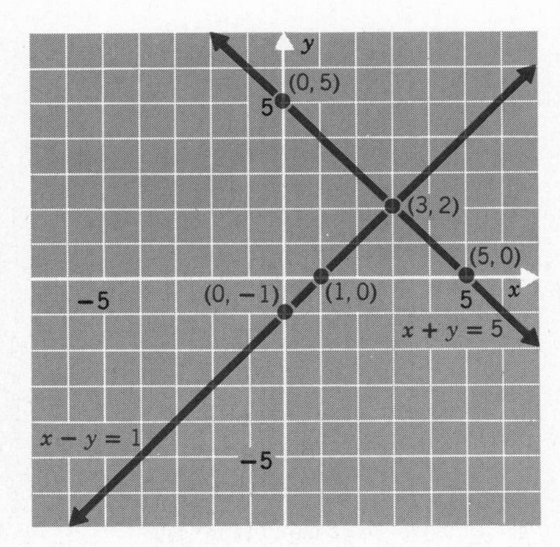

Figure 6.5

Equations considered together in this fashion are said to form a **system of equations** and, in general, the solution of such a system is a single ordered pair. The components of this ordered pair satisfy each of the two equations. If the graphs of the equations do not intersect, that

* It should be noted that, in general, graphical solutions are only approximate. We shall develop methods for exact solutions in a later section.

is, the lines are parallel (see Figure 6.6*a*) the equations are said to be **inconsistent,** and there is no ordered pair that will satisfy both equations. If the graphs of the equations are the same line (see Figure 6.6*b*) the equations are said to be **dependent,** and each ordered pair which satisfies one equation will satisfy the other.

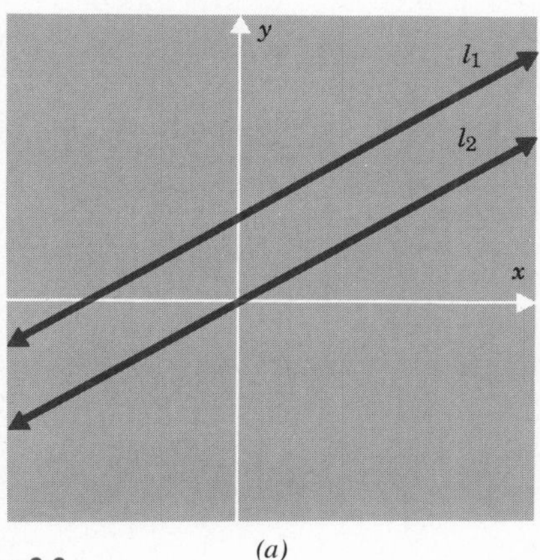

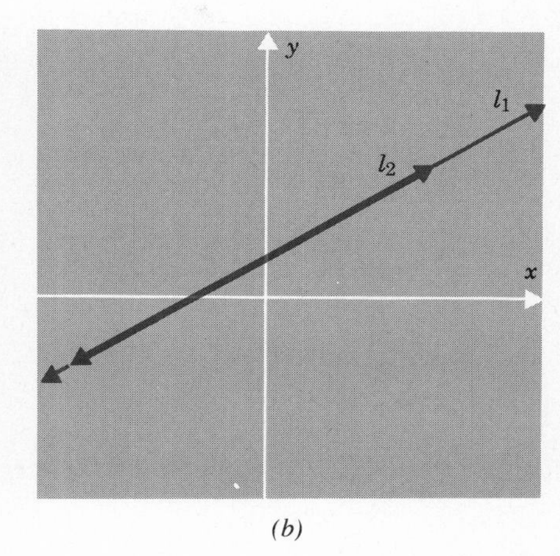

Figure 6.6 (*a*) (*b*)

EXERCISES 6.6

■ *Find the solution of each system by graphical methods. If no solution exists, so state.*

Sample Problem

$$x + y = 8$$
$$5x - 2y = 5$$

Using the intercept method of graphing, we find that two solutions of $x + y = 8$ are

$$(0, 8) \quad \text{and} \quad (8, 0).$$

Two solutions of $5x - 2y = 5$ are

$$\left(0, -\frac{5}{2}\right) \quad \text{and} \quad (1, 0).$$

The graphs of the two equations intersect at the point corresponding to (3, 5).

Ans. (3, 5)

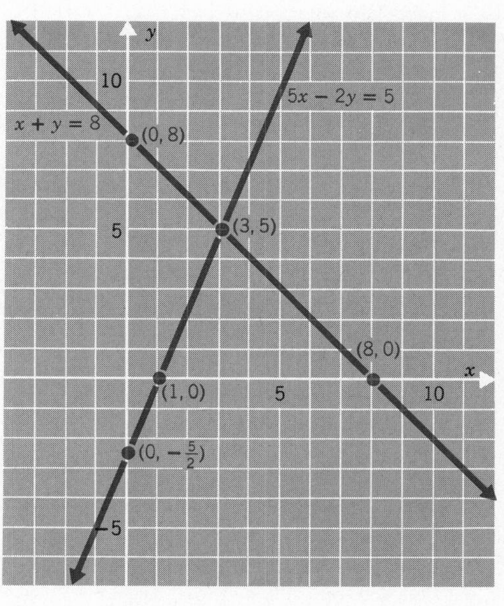

1. $y = 3x$, $x + y = 8$

2. $x - y = -2$, $3x - y = 8$

3. $2x - y = 0$, $2x - 3y = -4$

4. $y + 3x = 0$, $4x - 3y = 13$

5. $x - 3y = 0$, $2x - y = -5$

6. $2y + x = 0$, $y - x = 6$

7. $y = x$, $y = 4 - x$

8. $5y - x = 0$, $-y + x = 4$

9. $x + y = 6$, $x - y = 2$

10. $y = -x$, $2x + y = 3$

11. $y - x = 1$, $y + x = -5$

12. $x + y = 4$, $2x + 2y = 8$

13. $x - 2y = 3$, $3x - 6y = 9$

14. $3x = 4y$, $x = 2 - y$

15. $y = 2x + 5$, $x - 3y = -20$

16. $2x + 7y - 12 = 0$, $x + 5y = 0$

17. $x + 3y = 5$, $2x + 6y = 5$

18. $x = 3y - 4$, $6y = 2x + 5$

19. $5x - 2y + 8 = 0$, $3x + y + 7 = 0$

20. $7y - 2x + 6 = 0$, $8y - 5x - 4 = 0$

6.7 ALGEBRAIC SOLUTION OF SYSTEMS I

Systems of equations can also be solved algebraically. The solutions obtained by such a method will not be approximations.

Let us illustrate the algebraic solution of a system by using the example of the preceding section.

$$x + y = 5 \quad (1)$$
$$x - y = 1 \quad (2)$$

Obtain an equation in one variable by adding Equations 1 and 2.

$$2x = 6$$
$$x = 3$$

Substitute 3 for x in either Equation 1 or 2 to obtain the corresponding value of y. In this case we have selected Equation 1.

$$(3) + y = 5$$
$$y = 2$$

Ans. $x = 3$, $y = 2$; or $(3, 2)$

Check the solution by direct substitution in Equations 1 and 2.

$$x + y = 5 \qquad x - y = 1$$
$$(3) + (2) = 5; \qquad (3) - (2) = 1$$

Notice that this method of solution is simply an application of the addition axiom in order to obtain an equation containing a single variable. The equation in one variable in conjunction with either of the original equations then forms an equivalent system whose solution is readily obtained.

EXERCISES 6.7

■ *Solve each system.*

Sample Problem

$$\begin{aligned} 2x - y &= 0 \quad (1) \\ 2x + y &= 4 \quad (2) \\ \hline 4x \quad &= 4 \end{aligned}$$

Obtain an equation in one variable by adding Equations 1 and 2.

Solve the resulting equation for x.

$$x = 1$$

Substitute 1 for x in either Equation 1 or Equation 2. Here we select Equation 1, and solve for y.

$$\begin{aligned} 2(1) - y &= 0 \\ 2 &= y \end{aligned}$$

Ans. $x = 1$, $y = 2$; or $(1, 2)$

1. $x + y = 5$, $x - y = 1$
2. $x + y = 10$, $x - y = 4$
3. $x + y = 3$, $-x + y = 5$
4. $-x + y = 7$, $x + y = 5$
5. $x + 2y = 10$, $x - 2y = 2$
6. $2x + y = 3$, $-2x + y = -1$
7. $3x + 2y = 5$, $-2x - 2y = -4$
8. $5x - 3y = -1$, $3x + 3y = 9$
9. $3x + 3y = 15$, $3x - 3y = 27$
10. $6x - y = 4$, $2x + y = 4$
11. $7x - 3y = -10$, $x + 3y = 2$
12. $3x - 2y = 11$, $3x + 2y = 19$

Sample Problem

$$\begin{aligned} 2a + b &= 4 \quad (1) \\ a + b &= 3 \quad (2) \end{aligned}$$

Multiply each member of Equation 2 by -1 to obtain $2'$.

Add Equations 1 and 2′.* Alternatively, subtract Equation 2 from Equation 1.

$$\begin{array}{rll} -a - b = & -3 & (2') \\ 2a + b = & 4 & (1) \\ \hline a \quad = & 1 & \end{array}$$

Substitute 1 for a in either Equation 1 or Equation 2. We select Equation 2.

$$(1) + b = 3$$
$$b = 2$$

Ans. $a = 1$, $b = 2$; or $(1, 2)$

[Where the variables are a and b, the ordered pair is given in the form (a, b).]

13. $2a + b = 3$
$a + b = 2$

14. $a + b = 6$
$3a + b = 10$

15. $a + b = 7$
$a + 3b = 11$

16. $a + 2b = -1$
$3a + 2b = 1$

17. $3a + 2b = 1$
$2a + 2b = 0$

18. $6a + 5b = 6$
$-4a + 5b = -4$

19. $4a + 2b = 4$
$3a + 2b = 8$

20. $4a + 2b = 4$
$6a + 2b = 8$

21. $a + 3b = 2$
$2a + 3b = 7$

22. $3a - 4b = 19$
$-2a - 4b = -6$

23. $3a - 4b = 0$
$3a + 2b = 18$

24. $3a - 2b = 5$
$-a - 2b = -15$

25. $x + y = -1$
$2x + y = -5$

26. $2x + 3y = -5$
$2x - 4y = 16$

27. $3x + 4y = -10$
$x + 4y = -6$

28. $4x - 4y = -4$
$4x - 3y = -3$

29. $7x + 6y = 12$
$9x + 6y = 12$

30. $3a + 2b = -12$
$a + 2b = -4$

Sample Problem

$$\frac{x}{6} + \frac{y}{4} = \frac{3}{2} \qquad (1)$$

$$\frac{2x}{3} - \frac{y}{2} = 0 \qquad (2)$$

Remove fractions by multiplying each equation by L.C.D. of fractions in that equation.

$$\overset{2}{(\cancel{12})}\,\frac{x}{\cancel{6}} + \overset{3}{(\cancel{12})}\,\frac{y}{\cancel{4}} = \overset{6}{(\cancel{12})}\,\frac{3}{\cancel{2}}$$

$$\overset{2}{(\cancel{6})}\,\frac{2x}{\cancel{3}} - \overset{3}{(\cancel{6})}\,\frac{y}{\cancel{2}} = (6)0$$

* The symbol ′, called "prime," indicates an equivalent equation; that is, Equation 2′ is equivalent to Equation 2.

$$2x + 3y = 18 \qquad (1')$$
$$4x - 3y = 0 \qquad (2')$$

Solve resulting system.

$$6x = 18$$
$$x = 3$$

Substitute 3 for x in Equation 2′.

$$4(3) - 3y = 0$$
$$-3y = -12$$
$$y = 4$$

Ans. $x = 3$, $y = 4$; or $(3, 4)$

31. $y + \frac{1}{2}x = 0$
$\frac{1}{3}y - \frac{1}{3}x = 2$

32. $\frac{a}{5} + \frac{2b}{5} = 2$
$\frac{a}{2} - b = 1$

33. $\frac{2x}{3} + y = 3$
$\frac{x}{2} - \frac{y}{4} = \frac{5}{4}$

34. $x - \frac{2}{3}y = -4$
$\frac{x}{4} + \frac{y}{2} = -1$

35. $\frac{a}{4} - \frac{b}{3} = 0$
$\frac{a}{2} + \frac{b}{3} = 3$

36. $a + \frac{b+1}{3} = 0$
$\frac{5a}{2} - \frac{b+8}{4} = -5$

6.8 ALGEBRAIC SOLUTION OF SYSTEMS II

The solution of a system of equations by addition depends upon one of the variables having identical coefficients in both of the equations. If such is not the case, we may find equations that are equivalent to the equations in the system and do have identical coefficients on one of the variables. Consider the system:

$$-5x + 3y = 1 \qquad (1)$$
$$-7x - 2y = 3. \qquad (2)$$

If we multiply each member of Equation 1 by 2 and each member of Equation 2 by 3, we obtain an equivalent system

$$-10x + 6y = 2 \qquad (1')$$
$$-21x - 6y = 9. \qquad (2')$$

which may be solved as in the preceding section.

In all our preceding exercises and examples, the terms involving the variables have been in the left-hand member and the constant term in the right-hand member. We shall refer to this arrangement as the **standard form** for systems. It is convenient to arrange equations in such form before proceeding with the solution of a system of equations.

EXERCISES 6.8

■ *Solve.*

Sample Problem

$3x + 2y = 11$ (1)
$5x - 4y = 11$ (2)

Multiply each member of Equation 1 by 2. Add Equations 1′ and 2.

$6x + 4y = 22$ (1′)
$5x - 4y = 11$ (2)
$11x = 33$

Solve for x.

$x = 3$

Substitute 3 for x in Equation 1 and solve for y.

$3(3) + 2y = 11$
$2y = 2$
$y = 1$

Ans. $x = 3$, $y = 1$; or (3, 1)

1. $3x + 2y = 7$
$x + y = 3$

2. $2x - 3y = 8$
$x + y = -1$

3. $2x - y = 2$
$3x + 2y = 10$

4. $a - 4b = 9$
$3a + 2b = 13$

5. $3a - b = -5$
$2a + 3b = -7$

6. $-x + 3y = -1$
$-6x + y = -6$

7. $3a - 3b = -3$
$-6a + 2b = 14$

8. $3x - 6y = 6$
$x - 2y = 3$

9. $5x + 3y = 19$
$2x - y = 12$

10. $5x - 3y = 32$
$2x + 6y = -16$

11. $3x - 5y = -1$
$x + 2y = 18$

12. $3x + 3y = 0$
$6x + 9y = -6$

Sample Problem

$2y = 11 - 3x$
$5x = 11 + 4y$

Arrange equations in standard form.

$3x + 2y = 11$
$5x - 4y = 11$

Solve as described in the previous sample problem.

13. $3x + 2y = 7$
$x = y - 1$

14. $3x + 9 = -2y$
$x = -3y + 25$

15. $a + 5b = 0$
$-3a + 10 = 10b$

16. $x = 8 - 2y$
$2x = 6 + y$

17. $8b - 3a = 5$
$a = b$

18. $6x = 22 + 2y$
$8x = 33 - y$

19. $8a = 4b + 4$
$3a = 2b + 3$

20. $x + 3y = 35$
$0 = 2x - y$

21. $5b = 3a + 8$
$2b = -a + 1$

22. $x - y = 15$
$2x = -3y$

23. $3x - 2y = 3$
$2x = y + 2$

24. $2x = 2y + 2$
$4x = 5 + 4y$

25. $2x + 3y = -1$
$3x + 5y = -2$

26. $3x - 2y = 13$
$7x + 3y = 15$

27. $2x = 3y - 1$
$3x + 4y = 24$

28. $5x - 2y = 0$
$2x - 3y = -11$

29. $3b = 9$
$2b + a = 10$

30. $b = 5$
$2b = 4a - 2$

31. $2a + 3b = 0$
$5a - 2b = -19$

32. $8x - 7y = 0$
$7x = 8y + 15$

Sample Problem

$$\frac{x+1}{2} - \frac{2y-1}{6} = \frac{7}{6} \qquad (1)$$

$$\frac{2x-1}{4} - \frac{y-3}{4} = 1 \qquad (2)$$

Express each equation without fractions by multiplying by L.C.D. of fractions in that equation.

$$(6)\,\frac{x+1}{2} - (6)\,\frac{2y-1}{6} = (6)\,\frac{7}{6}$$

$$(4)\,\frac{2x-1}{4} - (4)\,\frac{y-3}{4} = (4)\,1$$

$$3(x+1) - (2y-1) = 7$$

$$(2x-1) - (y-3) = 4$$

Write equations in standard form.

$$3x - 2y = 3 \qquad (1')$$

$$2x - y = 2 \qquad (2')$$

Solve the system 1′ and 2′. Multiply Equation 2′ by −2, and solve for x.

$$3x - 2y = 3 \qquad (1')$$

$$-4x + 2y = -4 \qquad (2')$$

$$-x = -1$$

$$x = 1$$

(continued)

Substitute 1 for x in Equation 1′.

$$3(1) - 2y = 3$$
$$y = 0$$

Ans. $x = 1$, $y = 0$; or $(1, 0)$

33. $\frac{5a}{4} + b = \frac{11}{2}$
$a + \frac{b}{3} = 3$

34. $2a - \frac{5b}{2} = 13$
$\frac{a}{3} + \frac{b}{5} = \frac{14}{15}$

35. $\frac{x}{4} + \frac{y}{5} = 1$
$\frac{2x}{9} - \frac{y}{9} = -2$

36. $\frac{5x}{8} + y = \frac{1}{4}$
$\frac{5x}{4} - \frac{3y}{2} = 4$

37. $\frac{x}{3} - \frac{y}{2} = 1$
$\frac{2x + 3}{5} - \frac{5y + 1}{11} = 2$

38. $\frac{2x + 1}{7} + \frac{3y + 2}{5} = \frac{1}{5}$
$\frac{3x - 2}{2} + \frac{y + 4}{4} = 4$

39. $\frac{y + 4}{3} = \frac{x + 3}{7}$
$\frac{x}{2} - \frac{7y}{6} = \frac{19}{6}$

40. $\frac{2x}{3} - \frac{y - 4}{2} = 5$
$2x - \frac{3y}{2} = 9$

6.9 SOLVING WORD PROBLEMS USING TWO VARIABLES

If two variables are related by a single first-degree equation, there are infinitely many ordered pairs that are solutions of the equation. If the two variables are also related by another first-degree equation independent of the first, we have observed that there can be only one ordered pair that is a solution of both equations. Therefore, in order to solve problems using two variables, it is necessary to represent two independent relationships using two equations. Many problems can be solved more easily by using a system of equations than by using a single equation involving one variable.

EXERCISES 6.9

■ *In each of the following exercises:*

a. *Represent two independent conditions of the problem by a system of equations using two variables.*
b. *Solve the system.*

Sample Problem

A 12-foot board is cut into two parts so that one part is 2 feet longer than the other. How long is each part?

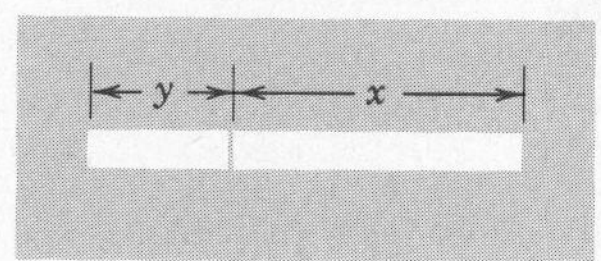

Draw sketch.

Identify the variables both in symbols and in words.

Let $x =$ the longer part;
$y =$ the shorter part.

Obtain two equations and solve the system.

$$\begin{aligned} x + y &= 12 \quad (1) \\ x - y &= 2 \quad (2) \\ \hline 2x &= 14 \\ x &= 7 \end{aligned}$$

Substitute 7 for x in Equation 1 or 2, say 2.

$$\begin{aligned} (7) - y &= 2 \\ y &= 5 \end{aligned}$$

Ans. Longer piece is 7 feet; shorter piece is 5 feet.

1. The sum of two numbers is 25 and their difference is 9. What are the numbers?
2. The sum of two numbers is 21 and their difference is 13. What are the numbers?
3. A 20-meter board is cut into two pieces, one of which is 2 meters longer than the other. How long is each piece?
4. A 30-meter board is cut into two pieces, one of which is 6 meters shorter than the other. How long is each piece?
5. Two packages weighed together total 28 kilograms. One of the packages weighs 8 kilograms less than twice the other. How much does each weigh?
6. Two packages weighed together total 45 kilograms. One of the packages weighs 11 kilograms more than the other. How much does each weigh?
7. A house and lot together sold for \$12,000. The house was valued at \$5000 more than the lot. What was the value of each separately?
8. A washer and a dryer together cost \$356. The dryer cost \$20 more than two times the washer. What was the price of each?
9. It took 24 working hours to paint the outside walls and trim of a house. If it took 6 more hours to paint the trim than it did the walls, how long did it take to paint each?

10. A loaf of bread and a can of corn together cost 72 cents. The bread cost a dime more than the corn. What was the cost of each?

11. A certain fishing spot is located 45 kilometers from town. Part of the distance can be driven in a car, but part of it must be traveled on foot. If it is possible to drive 19 more kilometers than must be walked, how far must be walked?

12. Two trains left towns A and B, which are 240 kilometers apart, at the same time and proceeded toward each other on parallel tracks. At the time they met, the train from A had traveled 10 kilometers farther than the train from B. How many kilometers from A were the trains when they met?

13. A freight train is made up of 92 cars, not counting the engine and its caboose. These cars are partly flat cars and partly box cars. There are 28 more flat cars than box cars. How many flat cars are there in the train?

14. An investor bought some stock in a water company and some stock in a uranium mine. After holding them a month, he sold them. He found he had made $45. "If I don't count the brokerage," he told his wife, "I made four times as much on the water stock as I lost on the uranium stock." How much did he lose on the uranium stock?

15. A man bought a toy train and a doll as presents for his children. He paid a total of $55. If the train cost $1 less than three times the doll, what was the price of each?

16. A mixture of coarse and fine powder weighs 450 grams. If the fine powder weighs 120 grams less than twice the weight of the coarse powder, how many grams of each is in the mixture?

17. The sum of two numbers is 24. One half of one number is 3 more than the other number. What are the numbers?

18. The difference of two numbers is 13. If the smaller number is 2 more than one fourth of the larger, what are the numbers?

19. A collection of 34 coins consists of dimes and quarters. How many coins of each kind are in the collection if the total value is $5.50?

20. The total income from two investments is $380. One investment yields 4% and the second investment yields 5%. How much is invested at each rate if the total investment was $8000?

21. A sum of $3600 is invested, part at 4% and the remainder at 6%. Find the amount of each investment if the interest on each investment is the same.

22. One solution contains 60% alcohol and a second solution contains

30% alcohol. How much of each solution is needed to make 33 liters that is 50% alcohol?

CHAPTER REVIEW

1. Given $I = 0.05P$. The yearly interest (I) on an investment equals the rate (0.05) times the principal (P).

a. Which of the symbols are constants?
b. Which symbols are variables?
c. If the principal increases, what happens to the interest?

2. Solve $2x - y = 4$ explictly for y.

3. Solve $2y - 3x = 6$ explicitly for y.

4. If $y = 2x + 1$, find the solutions with specified first components.

a. (3, ?) b. (−2, ?) c. (0, ?) d. (−½, ?)

5. If $x - y = 5$, find the solutions with specified first components.

a. (4, ?) b. (−2, ?) c. (0, ?) d. (−6, ?)

6. Graph the following ordered pairs on a set of rectangular axes.

a. (3, 4) b. (−2, 3) c. (3, −2) d. (0, 4)

■ *In Exercises 7–10, graph each equation.*

7. $x + y = 3$ **8.** $2y - x = 4$ **9.** $x = 3$ **10.** $3x + 2y = 6$

11. Where does the graph of $x - y = 8$ cross the x-axis?

12. Where does the graph of $x - y = 8$ cross the y-axis?

13. What is the slope of the graph of the equation $2y - 5x = 0$?

14. If y varies directly with x, and $y = 20$ when $x = 6$, find x when $y = 44$.

■ *In Exercises 15 and 16, find the solution of each system by algebraic methods and check by graphing.*

15. $x + y = 3$
$-x + y = 5$

16. $5x - 2y = -9$
$5x + 3y = 26$

17. Two packages together weigh 84 pounds, and one of the packages weighs 20 pounds more than the other. How much does each weigh?

18. A collection of 25 coins consists of dimes and quarters. How many coins of each kind are in the collection if the total value is $3.55?

19. The sum of two numbers is −40, and their difference is −8. What are the numbers?

20. One number is eight more than four times the other and their sum is −2. What are the numbers?

Cumulative Review

1. Graph all integers between $-3/2$ and $15/4$ on a line graph.
2. If $a = 0$, $b = 1$, $c = -1$, $d = 2$, find the value of $\frac{bc + d}{ab + d}$.
3. If $x = 1$, $y = -1$, and $z = -2$, find the value of $-xy^2 + (xz)^2 - yz^2$.
4. Given $s = \frac{1}{2} gt^2 + c$. Find s, if $c = 2000$, $g = 32$, and $t = 2$.

■ *Simplify.*

5. $\frac{3x + 6}{2x + 4}$
6. $\frac{a^2 - b^2}{a^2 - 2ab + b^2}$
7. $\frac{6a - 5}{8} + \frac{3a + 5}{12}$
8. $\frac{2}{x - 2} - \frac{3}{x + 1}$
9. $\frac{7a - 14}{5a - 10} \cdot \frac{3a - 3}{7a - 7}$
10. $\frac{3a + 3}{2a - 6} \div \frac{6a + 6}{3a - 9}$
11. $\dfrac{x - \frac{x}{y}}{1 - \frac{1}{y}}$
12. $(x^2 - 6x + 8) \div (x - 2)$

■ *Solve each equation for x.*

13. $\frac{3x}{4} - 9 = 0$
14. $x - \frac{3x - 2}{2} = \frac{1}{2}$
15. $\frac{3}{5} = \frac{x + 1}{x + 3}$
16. $\frac{x + a}{b} = a$
17. Graph: $y = 2x - 6$.
18. By algebraic methods solve the system: $2x - 2y = 10$
    $x + 2y = 2$.
19. The sum of two numbers is 28. One of them is 6 more than ten times the other. Find the numbers, first by using one equation with one variable and then by using a system of equations with two variables.
20. Find two consecutive integers such that twice the first less half the second is 10. Solve first by using one equation with one variable and then by using a system of equations with two variables.

Review of Factoring

The following exercises review processes we studied earlier. The ability to factor will be helpful to you in Chapter 7. If you have difficulty in factoring these expressions you should review the appropriate sections in Chapter 4.

■ *Factor each polynomial completely.*

[4.2]

1. $5x + 10y$ **2.** $3x^2 + 6x - 3$ **3.** $-2x^2 - 4$

4. $ab - ac$ **5.** $abc + ab - ac$ **6.** $x^2y - xy^2 + xy$

[4.4]

7. $x^2 + 5x + 6$ **8.** $y^2 - 6y + 9$ **9.** $y^2 - 7y - 8$

10. $y^2 + 2y - 35$ **11.** $2x^2 + 6x - 20$ **12.** $3y^2 + 9y - 12$

[4.6]

13. $2y^2 - y - 3$ **14.** $6y^2 + y - 1$ **15.** $6x^2 - 13x + 6$

16. $3x^2 - 8x - 35$ **17.** $6x^2 + 5x - 4$ **18.** $8y^2 - 2y - 1$

19. $4y^2 + 6y + 2$ **20.** $6x^2 + 21x + 9$ **21.** $18x^2 - 42x - 16$

22. $80x^2 - 10x - 25$ **23.** $4x^2 - 10x - 6$ **24.** $18y^2 - 3y - 6$

25. $x^2 + 2ax + a^2$ **26.** $x^2 - 2ax + a^2$ **27.** $y^2 + 6by + 9b^2$

28. $y^2 - 4by + 4b^2$ **29.** $x^2 + 8ax + 16a^2$ **30.** $y^2 - 10by + 25b^2$

[4.7]

31. $x^2 - 16$ **32.** $y^2 - 36$ **33.** $y^2 - b^2$ **34.** $x^2 - c^2$

35. $4x^2 - 25$ **36.** $9y^2 - 4c^2$ **37.** $12y^2 - 48$ **38.** $50x^2 - 32$

A **quadratic equation** is an equation which, in simplest form, contains the second but no higher power of the variable. To facilitate working with quadratic equations, we shall designate as **standard form** for such equations

$$ax^2 + bx + c = 0.$$

where a, b, and c are constants and $a \neq 0$. Observe that in standard form, the right-hand member is 0 and the terms in the left-hand member are in order of descending powers of the variable. If all three terms are present, for example,

$$2x^2 + 3x - 1 = 0 \quad \text{and} \quad 3x^2 - 2x + 1 = 0,$$

the equation is called a **complete quadratic equation.** If either $b=0$ or $c=0$, that is, if one or the other of the last two terms is missing, for example,

$$x^2 - 3 = 0 \quad \text{and} \quad x^2 + 4x = 0,$$

the equation is called an **incomplete quadratic equation.** In this chapter, we shall study one method of finding solutions of quadratic equations.

7.1 SOLUTION OF EQUATIONS IN FACTORED FORM

You know that a solution or root of an equation is a number which, when substituted for the variable, results in a true statement. Now, suppose we have an equation of the form

$$(x - 3)(x - 2) = 0.$$

This equation asserts that the product of two numbers, $(x-3)$ and $(x-2)$, is 0. As solutions we seek numbers which, when substituted for x, result in a true statement. To find such numbers, we utilize the following principle:

If the product of two or more factors is 0, at least one of the factors is 0.

In symbols we have

$$\text{If} \quad ab = 0, \quad \text{then} \quad a = 0 \quad \text{or} \quad b = 0 \quad \text{or both} \quad a \text{ and } b = 0.$$

Therefore, $(x-3)(x-2) = 0$ will be true only if $(x-3)$ is 0 or if $(x-2)$ is 0. For what values of x will $x-3=0$? Clearly, if $x=3$, then $x-3=0$. For what values of x will $x-2=0$? For $x=2$. Similarly, to solve any equation in *factored form for which one member is* 0, we can proceed as follows:

1. Set each factor equal to 0.
2. Solve each of the resulting equations.

Sometimes the values of the variable for which one or both factors equal zero can be determined by inspection. This should be done whenever possible.

EXERCISES 7.1

■ *For what values of x will each of the following expressions equal* 0?

Sample Problem

$x + 3$

Determine the value by inspection or set $(x+3)$ equal to 0 and solve the resulting equation.

$x + 3 = 0$

Ans. $x = -3$

1. $x - 7$ **2.** $x - 2$ **3.** $3x - 6$ **4.** $2x + 8$

5. $3x - 1$ **6.** $4x + 1$ **7.** $5x + 15$ **8.** $5x - 5$

■ *For what values of x will each of the following products equal* 0?

Sample Problem

$(x-4)(x+2)$

Determine values by inspection or set each factor equal to 0 and solve the resulting equations.

$x-4=0 \qquad x+2=0$

Ans. $x=4, \quad x=-2$

9. $(x-3)(x-5)$

10. $(x+2)(x+5)$

11. $x(x+4)$

12. $x(x-6)$

13. $(2x-6)(3x+9)$

14. $(5x+5)(4x-8)$

15. $(3x+1)(2x-3)$

16. $(4x+3)(3x-2)$

17. $2x(5x+4)(2x-7)$

18. $3x(x-5)(6x+9)$

■ *Solve each of the following equations.*

Sample Problem

$(x-3)(x+4)=0$

Determine values by inspection or set each factor equal to 0 and solve the resulting equations.

$x-3=0 \qquad x+4=0$

Ans. $x=3, \quad x=-4$

Check. $(3-3)(3+4)=0 \qquad [(-4)-3][(-4+4)]=0$

$(0)(7)=0 \qquad (-7)(0)=0$

19. $(x-2)(x-3)=0$

20. $(x+2)(x-4)=0$

21. $y(y-4)=0$

22. $p(p-7)=0$

23. $(r+3)(r)=0$

24. $(x+6)(2x)=0$

25. $(x-2)(x+3)=0$

26. $(x+5)(x-5)=0$

27. $(x-1)(x+8)=0$

28. $(x+3)(x+7)=0$

29. $(t+2)(t-3)=0$

30. $(x+1)(x-3)=0$

31. $u(u-4)=0$

32. $x(x+3)=0$

33. $(b+4)(b-3)=0$

34. $(b-7)(b-1)=0$

35. $(2x-3)(4x+3)=0$

36. $(3u-1)(u+1)=0$

37. $(3y - 2)(3y + 2) = 0$

38. $(x + 8)(3x - 8) = 0$

39. $2z(2z + 3) = 0$

40. $r(3r + 7) = 0$

41. $(x - 3)(x - 2)(x - 1) = 0$

42. $(x + 6)(x + 5)(x + 4) = 0$

43. $x(x + 2)(x - 1) = 0$

44. $x(x - 3)(x + 2) = 0$

45. $2x(x + 4)(x - 3) = 0$

46. $x(x + 1)(x - 1) = 0$

47. $x(2x + 1)(2x - 1) = 0$

48. $(2x - 3)(3x + 2)(2x + 2) = 0$

■ *Solve for x, y, or z.*

Sample Problem

$$x(3x + a)(x - a) = 0$$

Determine solutions by inspection or set each factor equal to 0 and solve the resulting equations.

$x = 0 \qquad 3x + a = 0 \qquad x - a = 0$

Ans. $x = 0, \quad x = -\frac{a}{3}, \quad x = a$

49. $(x - a)(x + a) = 0$

50. $(y - b)(y - 3b) = 0$

51. $(2x + a)(x - a) = 0$

52. $(2x + a)(2x - a) = 0$

53. $x(4x - b) = 0$

54. $ax(bx + c) = 0$

55. $y(2y + 3b)(3y - 2b) = 0$

56. $z(z - a)(2z + 3a) = 0$

57. $bz(z - b)(8z + b) = 0$

58. $bz(2z + 3b)(3z + 2b) = 0$

7.2 SOLUTION OF INCOMPLETE QUADRATIC EQUATIONS BY FACTORING

We observe from the preceding section that if we can write an equation so that one member is in the form of a product of linear factors and the other member is zero, we can find solutions for the equation. In general, there is one solution for each linear factor.

We begin our study of solving quadratic equations by first solving incomplete quadratic equations, that is, quadratic equations such as $y^2 - 2y = 0$, $x^2 = 4$, etc. To solve such a quadratic equation by factoring:

1. Write the equation in standard form.
2. Factor the left-hand member.
3. Set each factor equal to 0.
4. Solve each of the resulting equations.

If the equation is not factorable, other methods of solution must be used. These methods are discussed in Chapter 9.

We can check solutions for quadratic equations in the same manner in which we check solutions for linear equations—by direct substitution in the original equation.

EXERCISES 7.2

■ *Solve by factoring.*

Sample Problem

ODD PROB 1–45

$x^2 = 5x$	
	Write in standard form.
$x^2 - 5x = 0$	
	Factor left-hand member.
$x(x - 5) = 0$	
	Determine solutions by inspection or set each factor equal to 0 and solve the resulting equations.
$x = 0 \qquad x - 5 = 0$	

Ans. $x = 0, \quad x = 5$

Check. $(0)^2 = 5\,(0) \qquad (5)^2 = 5\,(5)$
$\qquad 0 = 0 \qquad\qquad 25 = 25$

1. $x^2 + 3x = 0$ **2.** $x^2 - 2x = 0$ **3.** $2y^2 - 5y = 0$

4. $3y^2 - 7y = 0$ **5.** $2y^2 = 9y$ **6.** $4y^2 = 3y$

7. $4x^2 = 16x$ **8.** $5x^2 = 10x$

Sample Problem

$x^2 = 25$	
	Write in standard form.
$x^2 - 25 = 0$	
	Factor left-hand member.
$(x - 5)(x + 5) = 0$	
	Determine solutions by inspection or set each factor equal to 0 and solve.
$x - 5 = 0 \qquad x + 5 = 0$	

Ans. $x = 5, \quad x = -5$

9. $x^2 - 1 = 0$ **10.** $x^2 = 36$ **11.** $x^2 - 4 = 0$

12. $x^2 = 9$ **13.** $x^2 - 16 = 0$ **14.** $x^2 - 100 = 0$

15. $x^2 = 64$ **16.** $x^2 = 49$ **17.** $3x^2 - 27 = 0$

18. $2x^2 = 32$ **19.** $5x^2 = 45$ **20.** $7x^2 = 63$

Sample Problem

$$\frac{15}{2}x^2 - \frac{10}{3} = 0$$

Multiply each term by the L.C.D. 6

$$(\overset{3}{\cancel{6}})\frac{15}{\cancel{2}}x^2 - (\overset{2}{\cancel{6}})\frac{10}{\cancel{3}} = (6)0$$

$$45x^2 - 20 = 0$$

Factor left-hand member completely.

$$5(9x^2 - 4) = 0$$

$$5(3x - 2)(3x + 2) = 0$$

Determine solutions by inspection or set each factor containing a variable equal to 0 and solve the resulting equations. The constant 5 has no effect on the solution.

$$3x - 2 = 0 \qquad 3x + 2 = 0$$

$$3x = 2 \qquad 3x = -2$$

Ans. $x = \frac{2}{3}, \quad x = -\frac{2}{3}$

21. $x^2 - \frac{1}{9} = 0$ **22.** $\frac{1}{3}x^2 - \frac{4}{3} = 0$ **23.** $\frac{2}{3}x^2 - \frac{3}{2} = 0$

24. $\frac{5}{2}y^2 - 10 = 0$ **25.** $\frac{x^2}{2} = 8$ **26.** $3x^2 = \frac{75}{4}$

27. $\frac{x^2}{2} + x = 0$ **28.** $\frac{x^2}{3} - 2x = 0$ **29.** $\frac{x^2}{4} + \frac{x}{2} = 0$

30. $\frac{x^2}{18} + \frac{x}{3} = 0$ **31.** $\frac{x^2}{5} = x$ **32.** $\frac{x^2}{6} = \frac{x}{2}$

■ *Solve for x, y,* or *z.*

Sample Problems

a. $x^2 = -ax$ b. $a^2x^2 = b^2$

Write in standard form.

(Continued)

$x^2 + ax = 0$ $\qquad$ $a^2x^2 - b^2 = 0$

Factor left-hand member.

$x(x + a) = 0$ $\qquad$ $(ax - b)(ax + b) = 0$

Determine solutions by inspection or set each factor equal to 0 and solve the resulting equations.

$x = 0 \qquad x + a = 0$ $\qquad$ $ax - b = 0 \qquad ax + b = 0$

$\qquad\qquad\qquad\qquad\qquad ax = b \qquad\qquad ax = -b$

Ans. $x = 0,\ x = -a$ $\qquad$ *Ans.* $x = \frac{b}{a},\ x = \frac{-b}{a}$

33. $x^2 - a^2 = 0$ **34.** $z^2 = bz$ **35.** $ax^2 - ac^2 = 0$

36. $d^2y^2 - a^2 = 0$ **37.** $ab^2x^2 = ab^2x$ **38.** $4x^2 - c^2 = 0$

39. $cy - 3y^2 = 0$ **40.** $ay = y^2$ **41.** $bx - b^2x^2 = 0$

42. $ax + a^3x^2 = 0$ **43.** $\frac{x^2}{a^2} - 4 = 0$ **44.** $\frac{y^2}{2} + cy = 0$

45. $\frac{a^2x^2}{b^2} - 9 = 0$ **46.** $\frac{y^2}{c} + by = 0$

7.3 SOLUTION OF COMPLETE QUADRATIC EQUATIONS BY FACTORING

The methods developed in Section 7.2 are also applicable to the solution of complete quadratic equations. As before, if the equation is not factorable, other methods must be used.

EXERCISES 7.3

■ *Solve by factoring.*

Sample Problem

$x^2 - 4x - 5 = 0$

Factor left-hand member.

$(x - 5)(x + 1) = 0$

Determine the solutions by inspection or set each factor equal to 0 and solve the resulting equations.

$$x - 5 = 0 \qquad x + 1 = 0$$

Ans. $x = 5, \quad x = -1$

Check.
$$(5)^2 - 4(5) - 5 = 0 \qquad (-1)^2 - 4(-1) - 5 = 0$$
$$25 - 20 - 5 = 0 \qquad 1 + 4 - 5 = 0$$
$$0 = 0 \qquad 0 = 0$$

ODD PROB 1-73

1. $x^2 - 3x + 2 = 0$ **2.** $x^2 + 3x + 2 = 0$ **3.** $y^2 + 4y + 4 = 0$
4. $y^2 - 8y + 12 = 0$ **5.** $y^2 - 3y - 4 = 0$ **6.** $y^2 - 3y - 10 = 0$
7. $x^2 + 4x - 21 = 0$ **8.** $x^2 + x - 42 = 0$ **9.** $x^2 + 5x - 14 = 0$
10. $x^2 + 8x + 15 = 0$ **11.** $x^2 + 12x + 36 = 0$ **12.** $x^2 + 14x + 49 = 0$
13. $x^2 - 16x + 15 = 0$ **14.** $x^2 + 14x - 15 = 0$

Sample Problem

$$2x^2 - 6x = 8$$

Write equation in standard form.

$$2x^2 - 6x - 8 = 0$$

Factor left-hand member completely.

$$2(x^2 - 3x - 4) = 0$$
$$2(x - 4)(x + 1) = 0$$

Determine solutions by inspection or set each factor containing a variable equal to 0 and solve the resulting equations.

$$x - 4 = 0 \qquad x + 1 = 0$$

Ans. $x = 4, \quad x = -1$

15. $2x^2 - 10x = 12$ **16.** $3t^2 - 6t = -3$ **17.** $4s^2 - 12s = 16$
18. $4x^2 - 24x = 28$ **19.** $2x^2 + 2x = 60$ **20.** $3x^2 + 6x = 45$
21. $3x^2 - x = 4$ **22.** $4x^2 + 4x = 3$ **23.** $6x^2 = 11x - 3$
24. $4x^2 = 4x + 3$ **25.** $4x^2 = 4x - 1$ **26.** $12x^2 = 8x + 15$
27. $12x - 9 = 4x^2$ **28.** $x + 15 = 2x^2$ **29.** $-1 = 9x^2 - 6x$
30. $-1 = 16x^2 + 8x$ **31.** $2x^2 + 6x - 7 = x$ **32.** $2x^2 + 4x - 12 = 2x$
33. $3y^2 = 2y + 60 + y$ **34.** $8x^2 + x = 6 - x$

Sample Problem

$$y^2 = \frac{13}{6}y - 1$$

(Continued)

Multiply each member by L.C.D. 6

$$(6)y^2 = (6)\frac{13}{6}y - (6)1$$

$$6y^2 = 13y - 6$$

Write in standard form; factor left-hand member.

$$6y^2 - 13y + 6 = 0$$

$$(2y - 3)(3y - 2) = 0$$

Determine solutions by inspection or set each factor equal to 0 and solve the resulting equations.

$$2y - 3 = 0 \qquad 3y - 2 = 0$$

$$2y = 3 \qquad 3y = 2$$

Ans. $y = \frac{3}{2}, \quad y = \frac{2}{3}$

35. $\frac{2}{3}x^2 + \frac{1}{3}x - 2 = 0$ **36.** $\frac{3}{4}x^2 + \frac{5}{2}x - 2 = 0$ **37.** $x^2 + 3x + \frac{9}{4} = 0$

38. $\frac{3}{2}x^2 - \frac{1}{4}x - \frac{1}{2} = 0$ **39.** $4x^2 + 13x + \frac{15}{2} = 0$ **40.** $\frac{1}{3}x^2 - \frac{5}{2}x + 3 = 0$

41. $\frac{x^2}{2} + x = \frac{15}{2}$ **42.** $x - 1 = \frac{x^2}{4}$ **43.** $\frac{x^2}{3} + x = \frac{-2}{3}$

44. $\frac{21}{2} + 2y = \frac{y^2}{2}$ **45.** $\frac{x^2}{6} + \frac{x}{3} = \frac{1}{2}$ **46.** $\frac{x^2}{15} = \frac{x}{5} + \frac{2}{3}$

Sample Problem

$$x(x + 2) = 8$$

Remove parentheses.

$$x^2 + 2x = 8$$

Write in standard form.

$$x^2 + 2x - 8 = 0$$

Factor left-hand member.

$$(x + 4)(x - 2) = 0$$

Determine solutions by inspection or set each factor equal to 0 and solve the resulting equations.

$$x + 4 = 0 \qquad x - 2 = 0$$

Ans. $x = -4, \quad x = 2$

47. $y(2y - 3) = -1$ **48.** $x(x + 2) = 3$ **49.** $2(x^2 - 1) = 3x$

50. $r(r - 2) = 6 - r$ **51.** $x(x + 2) - 3x - 2 = 0$ **52.** $2p(p - 2) = p + 3$

Sample Problem

$$(x - 4)(x + 3) = -10$$

Multiply the factors in the left-hand member.

$$x^2 - x - 12 = -10$$

Write in standard form.

$$x^2 - x - 2 = 0$$

Factor left-hand member.

$$(x - 2)(x + 1) = 0$$

Determine solutions by inspection or set each factor equal to 0 and solve the resulting equations.

$$x - 2 = 0 \qquad x + 1 = 0$$

Ans. $x = 2, \quad x = -1$

53. $(x - 2)(x + 1) = 4$

54. $(x - 5)(x + 1) = -8$

55. $(x - 2)(x - 1) = 1 - x$

56. $(x + 3)^2 = 2x + 14$

57. $(2x + 5)(x - 4) = -18$

58. $(2x - 1)(x - 2) = -1$

59. $(6x + 1)(x + 1) = 4$

60. $(x - 2)(x + 1) = x(2 - x)$

Sample Problem

$$\frac{1}{8x^2} - \frac{13}{24x} = -\frac{1}{2}$$

Multiply each term by L.C.D. $24x^2$.

$$\overset{3}{(\cancel{24x^2})}\frac{1}{\cancel{8x^2}} - \overset{x}{(\cancel{24x^2})}\frac{13}{\cancel{24x}} = -\overset{12}{(\cancel{24}x^2)}\frac{1}{\cancel{2}}$$

$$3 - 13x = -12x^2$$

Write in standard form.

$$12x^2 - 13x + 3 = 0$$

Factor left-hand member.

$$(4x - 3)(3x - 1) = 0$$

Determine solutions by inspection or set each factor equal to 0 and solve the resulting equations.

$$4x - 3 = 0 \qquad 3x - 1 = 0$$

$$4x = 3 \qquad 3x = 1$$

Ans. $x = \frac{3}{4}, \quad x = \frac{1}{3}$

Note. Answer should be checked to insure that no denominator in original equation is zero.

61. $x + \frac{1}{x} = 2$

62. $\frac{x}{4} - \frac{3}{4} = \frac{1}{x}$

63. $1 - \frac{2}{x} = \frac{15}{x^2}$

64. $\frac{1}{2} + \frac{1}{2x} = \frac{1}{x^2}$

65. $1 + \frac{1}{x(x-1)} = \frac{3}{x}$

66. $\frac{4}{x} - 3 = \frac{5}{2x+3}$

67. $x - \frac{10}{x-3} = 0$

68. $\frac{4}{3x} + \frac{3}{3x+1} + 2 = 0$

Sample Problem

$$\frac{7}{x-3} - \frac{3}{x-4} = \frac{1}{2}$$

Multiply each term by L.C.D. $2(x-3)(x-4)$.

$$2(x-3)(x-4)\frac{7}{(x-3)} - 2(x-3)(x-4)\frac{3}{(x-4)} = 2(x-3)(x-4)\frac{1}{2}$$

$$14(x-4) - 6(x-3) = (x-3)(x-4)$$

Remove parentheses.

$$14x - 56 - 6x + 18 = x^2 - 7x + 12$$

$$8x - 38 = x^2 - 7x + 12$$

Write in standard form.

$$x^2 - 15x + 50 = 0$$

Factor left-hand member.

$$(x-10)(x-5) = 0$$

Determine solutions by inspection or set each factor equal to 0 and solve the equations.

$$x - 10 = 0 \qquad x - 5 = 0$$

Ans. $x = 10, \quad x = 5$

Note. Answer should be checked to insure that no denominator in original equation is zero.

69. $\frac{14}{x-6} - \frac{6}{x-8} = \frac{1}{2}$

70. $\frac{12}{x-3} + \frac{12}{x+4} = 1$

71. $\frac{2}{x-3} - \frac{6}{x-8} = -1$

72. $\frac{4}{x-2} - \frac{7}{x-3} = \frac{2}{15}$

73. $\frac{4}{x-1} - \frac{4}{x+2} = \frac{3}{7}$

74. $\frac{3}{x+6} - \frac{2}{x-5} = \frac{5}{4}$

7.4 APPLICATIONS

A variety of word problems lead to quadratic equations. Upon solution, both results should be checked against the original word problem to insure that they fulfill the physical conditions set forth in the problem. For example, if the height h above the ground of an object thrown at a time $t = 0$ is given by

$$h = 48 + 32t - 16t^2,$$

then the number of seconds t it would take the object to strike the ground could be determined by setting $h = 0$ in the given equation and solving for t. We have

$$0 = 48 + 32t - 16t^2$$

or

$$16t^2 - 32t - 48 = 0.$$

Dividing each member by 16 produces

$$t^2 - 2t - 3 = 0.$$

Factoring, we have

$$(t - 3)(t + 1) = 0.$$

Then

$$\begin{aligned} t - 3 &= 0 \qquad \text{or} \qquad & t + 1 &= 0 \\ t &= 3 & t &= -1. \end{aligned}$$

In this case, -1 does not meet the physical requirements of the problem. Hence the object would strike the ground 3 seconds after it was thrown.

EXERCISES 7.4

■ *Solve each word problem.*

Sample Problem

The square of an integer is 7 less than eight times the integer. Find the integer.

Let $x =$ the integer

Write an equation expressing the conditions of the problem.

$$x^2 = 8x - 7$$

(Continued)

Solve equation.

$$x^2 - 8x + 7 = 0$$
$$(x - 7)(x - 1) = 0$$
$$x - 7 = 0 \qquad x - 1 = 0$$
$$x = 7 \qquad x = 1$$

Since 1 and 7 both meet the conditions of the problem, both are valid solutions.

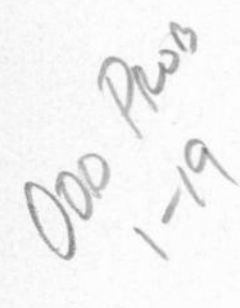

Ans. The integer is either 1 or 7.

1. The square of an integer is equal to five times the integer. Find the integer.
2. If three times the square of a certain integer is increased by the integer itself, the sum is 10. What is the integer?
3. Find two consecutive positive integers whose product is 72.
4. Find two consecutive positive integers whose product is 132.
5. The square of a positive integer increased by twice the square of the next consecutive integer, gives 66. Find the integer.
6. The square of a positive integer is 79 less than twice the square of the next consecutive integer. Find the integers.
7. The distance h above the ground of a certain projectile launched upward from the top of a 160 foot tall building at time $t = 0$, is given by $h = 160 + 48t - 16t^2$, where t is in seconds. Find the time at which the projectile will strike the ground.
8. In Problem 7, find the time at which the projectile is again at 160 ft. above the ground.
9. The Cost C of producing a certain television set is related to the number of hours t it takes to manufacture the set by

$$C = 8t^2 - 32t - 16.$$

How many hours would be devoted to producing a set at a cost of \$80?

10. In Problem 9, how many hours would be devoted to producing a set at a cost of \$24?

Sample Problem

The sum of a certain number and twice its reciprocal is $19/3$. Find the number.

Let $x =$ the number;

then $\frac{1}{x} =$ the reciprocal of the number.

Write an equation expressing the conditions of the problem.

$$x + 2\left(\frac{1}{x}\right) = \frac{19}{3}$$

Multiply by L.C.D. $3x$ to remove fractions.

$$(3x)x + (3\not{x})2\left(\frac{1}{\not{x}}\right) = (\not{3}x)\frac{19}{\not{3}}$$

$$3x^2 + 6 = 19x$$

Write in standard form.

$$3x^2 - 19x + 6 = 0$$

Solve equation.

$$(3x - 1)(x - 6) = 0$$

$$3x - 1 = 0 \qquad x - 6 = 0$$

$$3x = 1$$

$$x = \frac{1}{3} \qquad x = 6$$

Since both numbers meet the conditions of the problem, both are valid solutions.

Ans. The number is either $1/3$ or 6.

11. The sum of a certain number and its reciprocal is $25/12$. What is the number?
12. The sum of a certain number and its reciprocal is $53/14$. What is the number?
13. The sum of the reciprocals of two consecutive odd integers is $12/35$. What are the integers?
14. The sum of the reciprocals of two consecutive even integers is $9/40$. Find the integers.
15. The reciprocal of a positive integer is added to twice the reciprocal of the next consecutive integer and the sum is $25/72$. What are the integers?
16. Twice the reciprocal of a positive integer is subtracted from three times the reciprocal of the next successive integer and the difference is $2/21$. What are the integers?

Sample Problem

A boat travels 18 miles downstream and back in $4\frac{1}{2}$ hours. If the speed of the current is 3 miles per hour, what is the speed of the boat in still water?

(Continued)

Let $x =$ rate of boat in still water;
then $x + 3 =$ rate downstream
and $x - 3 =$ rate upstream.

Since $\text{time} = \dfrac{\text{distance}}{\text{rate}}$,

time downstream $= \dfrac{18}{x+3}$, and time upstream $= \dfrac{18}{x-3}$.

Write an equation expressing the fact that the sum of the times upstream and downstream is $4\frac{1}{2}$ or $\frac{9}{2}$ hours.

$$\frac{18}{x+3} + \frac{18}{x-3} = \frac{9}{2}$$

Multiply by L.C.D. $2(x+3)(x-3)$ to remove fractions.

$$2\cancel{(x+3)}(x-3)\frac{18}{\cancel{x+3}} + 2(x+3)\cancel{(x-3)}\frac{18}{\cancel{x-3}} = \cancel{2}(x+3)(x-3)\frac{9}{\cancel{2}}$$

$$36(x-3) + 36(x+3) = 9(x+3)(x-3)$$

$$36x - 108 + 36x + 108 = 9x^2 - 81$$

Write in standard form and solve.

$$9x^2 - 72x - 81 = 0$$

$$9(x^2 - 8x - 9) = 0$$

$$9(x-9)(x+1) = 0$$

$$x = 9 \qquad x = -1$$

Since -1 does not meet the conditions of the problem, only the positive solution is used.

Ans. The boat travels 9 miles per hour in still water.

17. A motor boat travels 24 miles downstream on a river and returns to its starting place. If the speed of the current is 2 miles per hour, and the round trip takes 5 hours, what is the speed of the boat in still water?

18. A crew rows a boat 6 miles downstream and then rows back to their starting place. If the speed of the current is 2 miles per hour, and the total trip takes 4 hours, how fast would the crew row in still water?

19. A man drove 180 miles from town A to town B and returned to town A. If he drove 15 miles per hour faster on the return trip than he did on the initial trip, and the initial trip took one hour longer, what was the man's speed on each trip?

20. A plane flew 480 miles at a certain speed, then increased its speed by 20 miles per hour and continued on the same course. After having flown a total distance of 840 miles in a total of 5 hours, it landed. What was its original speed?

CHAPTER REVIEW

■ *In Exercises 1–11 solve for x or y.*

1. a. $(x-2)(x+5)=0$ b. $y(y+3)=0$
2. a. $x^2-2x=0$ b. $3x^2=6x$
3. a. $x^2-49=0$ b. $3y^2=27$
4. a. $\frac{x^2}{4}-9=0$ b. $3x^2=\frac{27}{25}$
5. a. $a^2y^2-b^4=0$ b. $\frac{x^2}{c}=a^2x$
6. a. $y^2-4y-5=0$ b. $y^2-7y-18=0$
7. a. $x^2-2x=3$ b. $x^2=4x-4$
8. a. $\frac{y^2}{6}+\frac{y}{6}=2$ b. $\frac{y}{2}-1=\frac{y^2}{16}$
9. a. $y(y-6)=16$ b. $x(x+2)=8$
10. a. $(x+5)(x-8)=-36$ b. $(x+1)(x-2)=4$
11. a. $\frac{15}{y^2}+\frac{2}{y}=1$ b. $\frac{2}{x}-\frac{1}{6}=\frac{2}{x+2}$
12. The difference of two positive numbers is 7 and their product is 60. Find the numbers.
13. The sum of two positive numbers is 11 and their product is 30. Find the numbers.
14. The sum of the reciprocals of two consecutive integers is $^{11}/_{30}$. Find the integers.

Cumulative Review

1. Simplify: $\frac{4^2-2^2}{3}-\frac{4^2+2^2}{4}$.
2. If $a=0$, $b=-1$, and $c=2$, find the value of $a(b^2+c^2)$.
3. Arrange $-3, \frac{3}{8}, \frac{3}{7}, 3, 0, \frac{-5}{2}$, and $\frac{-5}{3}$ in order from smallest to largest.

4. Factor: $x^3 - 3x^2 + 2x$.

5. Multiply: $(2x - 3)(x + 4)$.

6. Write $-\dfrac{x+2}{-3}$ in standard form.

7. Simplify: $(x - 3)(x + 2) - (x^2 - 3x)$.

8. Represent $\dfrac{3}{x^2 + x} - \dfrac{2}{x + 1}$ as a single fraction.

9. Express $\dfrac{3}{x - 1}$ as a fraction with a denominator of $x^2 + 2x - 3$.

10. Simplify: $\dfrac{3x^2 - x - 2}{2x^2 + x - 3}$.

11. Solve the system $\begin{aligned} 2x - 3 &= y \\ 3x + 2y &= 15 \end{aligned}$ by algebraic methods.

12. Graph the equation: $x + 2y = 4$.

13. Solve: $\dfrac{7}{8} = \dfrac{42}{y + 4}$.

14. Express 120 pounds in kilograms.

15. The numerator of a certain fraction is five more than the denominator, and the fraction is equivalent to $4/3$. Find the fraction.

16. Two packages together weigh 140 kilograms. If one package weighs 30 kilograms more than the other, what is the weight of each?

■ *In Exercises 17–19, solve each equation.*

17. $3y^2 - 2y = y$

18. $25x^2 - 4 = 5$

19. $x(x + 3) = 5x + 3$

20. If S varies directly with T, and $S = 4.2$ when $T = 12.3$, find T when $S = 10.6$.

8.1 RADICALS

Many quadratic equations, such as

$$x^2 - 3 = 0 \quad \text{and} \quad x^2 + 4x - 1 = 0,$$

are not factorable by using any of the numbers we have studied. In this chapter we propose to investigate a new kind of number—one which will enable us to find solutions to some nonfactorable quadratic equations.

First, let us define a new representation for numbers with which you are already familiar. For all nonnegative numbers, a, we define $\sqrt{a}$ (read "**square root** of a") to be the nonnegative number such that

$$\sqrt{a} \cdot \sqrt{a} = a.$$

For example

$$\sqrt{4} = 2 \quad \text{because} \quad 2 \cdot 2 = 4,$$
$$\sqrt{9} = 3 \quad \text{because} \quad 3 \cdot 3 = 9,$$

and

$$\sqrt{16} = 4 \quad \text{because} \quad 4 \cdot 4 = 16.$$

The symbol $\sqrt{\ }$ is called a **radical.** The numbers 4, 9, and 16 as used above are called **radicands.**

Each positive number has two square roots, since we have $(a)(a) = a^2$ and $(-a)(-a) = a^2$. The symbol $\sqrt{a}$ is always used to represent the positive square root; to represent the negative square root of a, the symbol $-\sqrt{a}$ is used. For example, 3 has two square roots; $\sqrt{3}$ represents the positive root, sometimes called the **principal square root,** and $-\sqrt{3}$ represents the negative root. The symbol $\pm\sqrt{a}$ is sometimes used to denote both the positive and negative square roots of a.

Variables in all radicands in this book are assumed to represent positive numbers.

EXERCISES 8.1

■ *Find each square root.*

Sample Problems

a. $\sqrt{49}$ *Ans.* 7

b. $-\sqrt{\frac{4}{81}}$ *Ans.* $-\frac{2}{9}$

c. $\pm\sqrt{\frac{4}{25}}$ *Ans.* $\pm\frac{2}{5}$

d. $\sqrt{0}$ *Ans.* 0

1. $\sqrt{16}$ **2.** $\sqrt{36}$ **3.** $-\sqrt{81}$ **4.** $-\sqrt{121}$

5. $\pm\sqrt{144}$ **6.** $\pm\sqrt{225}$ **7.** $\sqrt{9}$ **8.** $\sqrt{1}$

9. $\sqrt{\frac{1}{36}}$ **10.** $-\sqrt{\frac{1}{4}}$ **11.** $\pm\sqrt{\frac{4}{9}}$ **12.** $\pm\sqrt{\frac{9}{25}}$

■ *Write each of the following as an equivalent radical expression.*

Sample Problems

a. 5 *Ans.* $\sqrt{25}$

b. -10 *Ans.* $-\sqrt{100}$

c. $\frac{2}{3}$ *Ans.* $\sqrt{\frac{4}{9}}$

13. 3 **14.** 8 **15.** -7 **16.** -6

17. 13 **18.** 12 **19.** -5 **20.** -4

21. $\frac{1}{2}$ **22.** $\frac{8}{9}$ **23.** $-\frac{7}{8}$ **24.** $-\frac{2}{7}$

■ *Find each square root.*

Sample Problems

a. $\sqrt{y^6}$ *Ans.* y^3

b. $\sqrt{49x^2y^6}$ *Ans.* $7xy^3$

c. $\pm\sqrt{(c+d)^2}$ *Ans.* $\pm(c+d)$

25. $\sqrt{x^2}$ **26.** $\sqrt{y^4}$ **27.** $\sqrt{4x^2}$ **28.** $\sqrt{a^2b^2}$

29. $-\sqrt{a^2c^4}$ **30.** $-\sqrt{9x^6y^6}$ **31.** $\pm\sqrt{36a^6}$ **32.** $\pm\sqrt{100x^{10}}$

33. $\sqrt{121a^2b^2}$ **34.** $\sqrt{(x+y)^2}$ **35.** $-\sqrt{(a+b)^2}$ **36.** $-\sqrt{4(x+y)^2}$

37. $\sqrt{\frac{a^2}{b^2}}$ **38.** $-\sqrt{\frac{b^4}{100}}$ **39.** $\pm\sqrt{\frac{9}{x^2y^2}}$ **40.** $\pm\sqrt{\frac{4x^2}{y^2}}$

■ *Write each of the following as equivalent radical expressions.*

Sample Problems

a. $3x^3y$

Ans. $\sqrt{9x^6y^2}$

b. $-(x+y)$

Ans. $-\sqrt{(x+y)^2}$

c. $\pm\frac{3x}{4}$

Ans. $\pm\sqrt{\frac{9x^2}{16}}$

41. x **42.** y^2 **43.** xy^2 **44.** $3x$

45. $4y^2$ **46.** $7xy$ **47.** $-8x^2y^2$ **48.** $-7x^7$

49. $-10x^{10}y^{10}$ **50.** $-(x+y)$ **51.** $\pm(2x+y)$ **52.** $\pm(3x+2y)$

53. $\frac{1}{2}$ **54.** $\frac{1}{3}$ **55.** $\frac{3}{4}$ **56.** $\frac{2}{3}a$

57. $\pm\frac{4}{5}b^2$ **58.** $\pm\frac{3}{2}ab$ **59.** $\frac{1}{a}$ **60.** $-\frac{1}{ab}$

61. $-\frac{3x}{y}$ **62.** $-\frac{y}{4x}$ **63.** $\frac{a+b}{a}$ **64.** $\frac{b+1}{b}$

8.2 IRRATIONAL NUMBERS

In Chapter 5, we discussed fractions. If the numerator of an arithmetic fraction is an integer and the denominator is a nonzero integer, the fraction represents a **rational number.** A number that cannot be expressed as the quotient of two integers is an **irrational number.** The study of irrational numbers in any detail is beyond the scope of this book. We shall content ourselves with observing that any radical with a radicand which is not, itself, the square of a rational number represents an irrational number. Thus,

$$\sqrt{2},\ \sqrt{3},\ \sqrt{5},\ \text{and}\ \sqrt{\frac{5}{7}},$$

are irrational numbers, whereas

$$\sqrt{4},\ \sqrt{\frac{9}{25}},\ \text{and}\ \sqrt{16},$$

are rational numbers that can also be written as 2, 3/5, and 4, respectively.

Irrational numbers cannot be exactly represented by common fractions or decimal fractions. However, we can approximate irrational numbers to any desired degree of accuracy. For example, correct to two decimal places, $\sqrt{2} = 1.41$. We can obtain approximations to the irrational square roots of positive integers by various means. In this book, we use a prepared table of square roots which can be found on page 321.

Note that the square roots of 1, 4, 9, 16, 25, 36, 49, 64, 81, and 100 are rational numbers equal to 1, 2, 3, 4, 5, 6, 7, 8, 9, and 10, respectively. The square roots of all other integers between 1 and 100 are irrational numbers and the entries shown for these numbers are only approximations to their true value. For example, $\sqrt{2}$ is approximately equal to 1.414 and $\sqrt{3}$ is approximately equal to 1.732. The symbol "$\approx$" is often used for the phrase "is approximately equal to." Thus,

$$\sqrt{2} \approx 1.414 \qquad \text{and } \sqrt{3} \approx 1.732.$$

In studying operations with radicals in this and the following sections, we assume that all laws valid for the fundamental operations with rational numbers hold for irrational numbers and that the symbols for the fundamental operations are unchanged. For example, the expression $2\sqrt{3}$ represents the product of 2 and $\sqrt{3}$. The expression $4 + \sqrt{7}$ represents the sum of 4 and $\sqrt{7}$.

The set of numbers made up of all rational and all irrational numbers is called the set of **real numbers.** This set of numbers fills the number line completely.

EXERCISES 8.2

■ *Which of the following numbers are rational and which are irrational? If the number is rational, express it as an integer or as a quotient of two integers.*

Sample Problems

a. $\sqrt{\frac{4}{9}}$ b. $\sqrt{3}$ c. $5 + \sqrt{3}$

Ans. rational, since $\sqrt{\frac{4}{9}} = \frac{2}{3}$ *Ans.* irrational *Ans.* irrational

1. 6 **2.** 8.61 **3.** $\sqrt{2}$ **4.** $\sqrt{4}$ **5.** $\sqrt{6}$

6. $\sqrt{9}$ **7.** $\sqrt{25}$ **8.** $\sqrt{100}$ **9.** $3\sqrt{16}$ **10.** $\sqrt{7}$

11. $-\sqrt{13}$ **12.** $-2\sqrt{100}$ **13.** $\sqrt{\frac{4}{9}}$ **14.** $-\sqrt{\frac{16}{25}}$ **15.** $-\sqrt{\frac{2}{3}}$

16. $\sqrt{\frac{4}{5}}$ **17.** $1 + \sqrt{4}$ **18.** $3.1 + \sqrt{4}$ **19.** $2 + \sqrt{5}$ **20.** $1 + \sqrt{3}$

■ *Using the table of square roots on page 321, find a decimal approximation for each of the following. Round off answers to two decimal places. For the sake of uniformity, if the digit in the third decimal place is 5, round to the next higher digit in the second decimal place.*

Sample Problems

a. $\sqrt{23}$

4.796

Ans. 4.80

b. $-2\sqrt{46}$

$-2(6.782)$

-13.564

Ans. -13.56

c. $\frac{1}{2}\sqrt{62}$

$\frac{7.874}{2}$

3.937

Ans. 3.94

21. $\sqrt{57}$ **22.** $\sqrt{83}$ **23.** $\sqrt{3}$ **24.** $\sqrt{17}$
25. $\sqrt{5}$ **26.** $\sqrt{92}$ **27.** $-\sqrt{26}$ **28.** $-\sqrt{54}$
29. $2\sqrt{3}$ **30.** $3\sqrt{2}$ **31.** $-5\sqrt{3}$ **32.** $-6\sqrt{5}$
33. $\frac{1}{3}\sqrt{18}$ **34.** $\frac{1}{4}\sqrt{48}$ **35.** $-\frac{1}{5}\sqrt{75}$ **36.** $-\frac{2}{3}\sqrt{21}$

Sample Problems

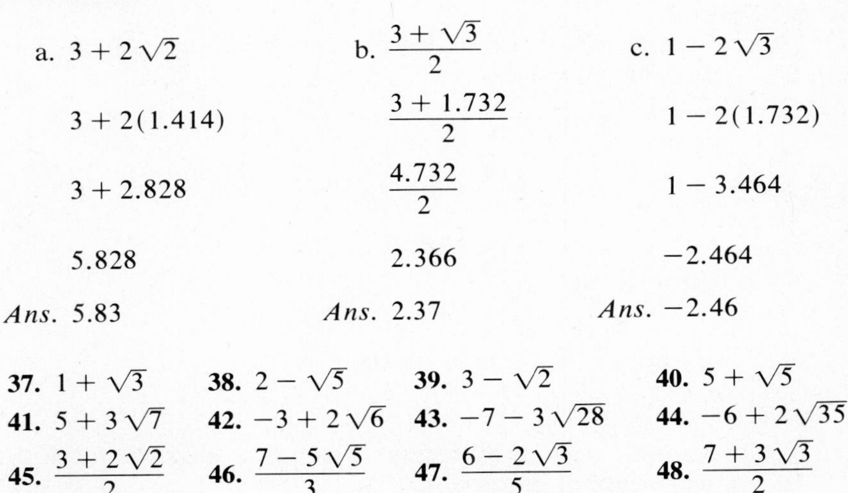

a. $3 + 2\sqrt{2}$

$3 + 2(1.414)$

$3 + 2.828$

5.828

Ans. 5.83

b. $\frac{3 + \sqrt{3}}{2}$

$\frac{3 + 1.732}{2}$

$\frac{4.732}{2}$

2.366

Ans. 2.37

c. $1 - 2\sqrt{3}$

$1 - 2(1.732)$

$1 - 3.464$

-2.464

Ans. -2.46

37. $1 + \sqrt{3}$ **38.** $2 - \sqrt{5}$ **39.** $3 - \sqrt{2}$ **40.** $5 + \sqrt{5}$
41. $5 + 3\sqrt{7}$ **42.** $-3 + 2\sqrt{6}$ **43.** $-7 - 3\sqrt{28}$ **44.** $-6 + 2\sqrt{35}$
45. $\frac{3 + 2\sqrt{2}}{2}$ **46.** $\frac{7 - 5\sqrt{5}}{3}$ **47.** $\frac{6 - 2\sqrt{3}}{5}$ **48.** $\frac{7 + 3\sqrt{3}}{2}$
49. $\sqrt{3} - \sqrt{2}$ **50.** $\sqrt{5} - \sqrt{7}$ **51.** $3\sqrt{3} - 2\sqrt{5}$ **52.** $2\sqrt{2} - 5\sqrt{5}$

■ *Graph each set of numbers on a separate line graph. Estimate the location of graphs of numbers between integers.*

Sample Problem *Ans.*

$\sqrt{13}$, 4, $\sqrt{7}$

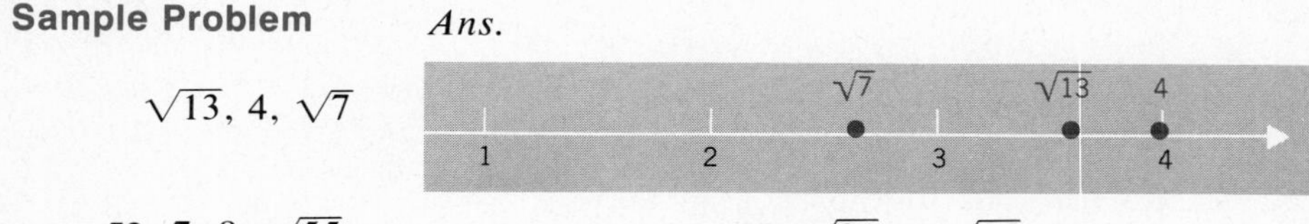

53. 7, 8, $\sqrt{55}$

54. $\sqrt{21}, -\sqrt{25}$, 6

55. $\sqrt{3}, \sqrt{5}, -\sqrt{7}$

56. $\sqrt{9}, -\sqrt{16}, \sqrt{25}$

57. $\sqrt{1}, -\sqrt{2}, \sqrt{3}$

58. $-\sqrt{6}, \sqrt{8}, \sqrt{10}$

59. $-\sqrt{4}, \sqrt{3}, -\sqrt{2}$

60. $\sqrt{2}, 0, -\sqrt{2}$

61. $-\sqrt{40}, \sqrt{30}, -\sqrt{20}$

62. $-\sqrt{1}, 0, \sqrt{1}$

63. $\sqrt{35}, -\sqrt{16}, -\sqrt{37}$

64. $\sqrt{21}, \sqrt{27}, \sqrt{30}$

8.3 SIMPLIFICATION OF RADICAL EXPRESSIONS—MONOMIALS

A radical expression is considered to be in *simplest form* if no prime factor of the radicand occurs more than once. We may determine whether a radical is in simplest form by examining the prime factors of the radicand; e.g., $\sqrt{78}$ is in simplest form because none of the factors in its completely factored form, $\sqrt{13 \cdot 3 \cdot 2}$, are repeated. The radical expression $\sqrt{20}$ is not in simplest form because the factor 2 occurs more than once when the radicand is in completely factored form, $\sqrt{2 \cdot 2 \cdot 5}$.

To find a means of simplifying a radical which has a repeated factor in the radicand, note that

$$\sqrt{4 \cdot 9} = \sqrt{36} = 6,$$

and

$$\sqrt{4}\,\sqrt{9} = 2 \cdot 3 = 6.$$

Hence,

$$\sqrt{4 \cdot 9} = \sqrt{4} \cdot \sqrt{9}.$$

If we generalize, we have

$$\sqrt{\mathbf{ab}} = \sqrt{\mathbf{a}}\,\sqrt{\mathbf{b}}.$$

Stated in words, this equation asserts that:

The square root of a product is equal to the product of the square roots of its factors.

In simplifying radicals, we first express the radicand as two factors, one of which is a perfect square. We can then write the latter without using radical notation. For example, $\sqrt{216}$ may be expressed as $\sqrt{36}\,\sqrt{6}$ from which we obtain the simpler form $6\sqrt{6}$. If perfect square factors are difficult to determine, it is frequently helpful to write a radicand in completely factored form and examine this form for repeated factors. These factors determine perfect squares. Thus, by writing $\sqrt{216}$ in completely factored form, we observe that

$$\begin{aligned}\sqrt{216} &= \sqrt{2\cdot 2\cdot 2\cdot 3\cdot 3\cdot 3}\\ &= \sqrt{2^2}\,\sqrt{3^2}\,\sqrt{2\cdot 3}\\ &= 2\cdot 3\,\sqrt{6}\\ &= 6\sqrt{6}.\end{aligned}$$

By simplifying radical expressions, we can extend the scope of the table of square roots. For example, $\sqrt{216}$, which is not available in the tables, may be approximated by observing from the example above that

$$\begin{aligned}\sqrt{216} &= 6\sqrt{6},\\ &\approx 6(2.449) = 14.694.\end{aligned}$$

In simplifying radicals whose radicands contain variables we observe that any variable possessing an exponent that is an even natural number is a perfect square, and its square root is obtained by dividing the exponent by 2. For example,

$$\sqrt{x^6} = x^3,$$

and

$$\sqrt{x^{10}} = x^5.$$

A radical expression whose radicand has a variable as a factor with an odd natural number for an exponent can be simplified by factoring the variable factor into two factors, one factor having an exponent of 1 and the other having an exponent which is an even number. For example,

$$\sqrt{x^7} = \sqrt{x^6}\sqrt{x} = x^3\sqrt{x}.$$

EXERCISES 8.3

■ *Simplify.*

Sample Problems

a. $\sqrt{24}$ b. $\sqrt{1575}$

Factor radicand completely.

(Continued)

$\sqrt{2 \cdot 2 \cdot 2 \cdot 3}$

$\sqrt{2^2}\sqrt{6}$

Ans. $2\sqrt{6}$

$\sqrt{5 \cdot 5 \cdot 3 \cdot 3 \cdot 7}$

Simplify.

$\sqrt{5^2}\sqrt{3^2}\sqrt{7}$

$5 \cdot 3\sqrt{7}$

Ans. $15\sqrt{7}$

1. $\sqrt{8}$ **2.** $\sqrt{12}$ **3.** $\sqrt{18}$ **4.** $\sqrt{49}$

5. $-\sqrt{20}$ **6.** $-\sqrt{27}$ **7.** $-\sqrt{72}$ **8.** $-\sqrt{24}$

9. $\sqrt{64}$ **10.** $\sqrt{162}$ **11.** $\sqrt{288}$ **12.** $\sqrt{84}$

13. $\sqrt{125}$ **14.** $\sqrt{450}$ **15.** $-\sqrt{1080}$ **16.** $-\sqrt{882}$

17. $\pm\sqrt{720}$ **18.** $\pm\sqrt{588}$ **19.** $\pm\sqrt{1944}$ **20.** $\pm\sqrt{1125}$

Sample Problems

a. $\sqrt{x^2}$

Ans. x

b. $\sqrt{y^3}$

$\sqrt{y^2}\sqrt{y}$

Ans. $y\sqrt{y}$

c. $-\sqrt{x^9}$

$-\sqrt{x^8}\sqrt{x}$

Ans. $-x^4\sqrt{x}$

21. $\sqrt{x^3}$ **22.** $\sqrt{y^5}$ **23.** $\sqrt{y^7}$ **24.** $\sqrt{x^{10}}$

25. $-\sqrt{x^{11}}$ **26.** $-\sqrt{x^{13}}$ **27.** $\sqrt{x^6}$ **28.** $-\sqrt{x^4}$

29. $\pm\sqrt{x^8}$ **30.** $\pm\sqrt{x^{15}}$ **31.** $\sqrt{x^{12}}$ **32.** $\sqrt{x^{14}}$

Sample Problems

a. $\sqrt{12y^3}$

$\sqrt{4y^2}\sqrt{3y}$

Ans. $2y\sqrt{3y}$

b. $\sqrt{48x^2y^3}$

$\sqrt{16x^2y^2}\sqrt{3y}$

Ans. $4xy\sqrt{3y}$

c. $-\sqrt{20y^3}$

$-\sqrt{4y^2}\sqrt{5y}$

Ans. $-2y\sqrt{5y}$

33. $\sqrt{4x^2}$ **34.** $\sqrt{8x^2}$ **35.** $\sqrt{9x^3}$ **36.** $\sqrt{12x^3}$

37. $-\sqrt{24y^5}$ **38.** $-\sqrt{121x^5}$ **39.** $\sqrt{64y^4}$ **40.** $\sqrt{36x^5}$

41. $\sqrt{49x^7}$ **42.** $\sqrt{16x^2}$ **43.** $\pm\sqrt{32x^3}$ **44.** $\pm\sqrt{72y^3}$

45. $\sqrt{80x^2}$ **46.** $\sqrt{98y^3}$ **47.** $-\sqrt{64x}$ **48.** $-\sqrt{3x^2}$

49. $\pm\sqrt{5y^3}$ **50.** $\pm\sqrt{7x^2}$ **51.** $\sqrt{48x^2y}$ **52.** $\sqrt{20x^2y^2}$

53. $\sqrt{25x^3y^2}$ **54.** $\sqrt{50xy^2}$ **55.** $-\sqrt{45a^4b}$ **56.** $-\sqrt{40x^5y^2}$

57. $\sqrt{3^2x^2}$ **58.** $\sqrt{2^2y^3}$ **59.** $\sqrt{8^2x^2y^2}$ **60.** $\sqrt{4b^3c^4}$

61. $-\sqrt{9cd}$ **62.** $-\sqrt{10x^3y^2}$ **63.** $\sqrt{7y^2z^2}$ **64.** $\sqrt{6y^3z^2}$

65. $\sqrt{\frac{9}{16}x^2y^2}$ **66.** $\sqrt{\frac{4}{9}x^3y}$ **67.** $\pm\sqrt{\frac{25}{36}y^2}$ **68.** $\pm\sqrt{\frac{1}{4}ab^2c^3}$

Sample Problems

a. $3\sqrt{4x^3}$

$3\sqrt{4x^2}\sqrt{x}$

$3(2x)\sqrt{x}$

Ans. $6x\sqrt{x}$

b. $\pm\frac{3x}{y}\sqrt{18x^3y^3}$

$\pm\frac{3x}{y}\sqrt{9x^2y^2}\sqrt{2xy}$

$\pm\frac{3x}{y}(3xy)\sqrt{2xy}$

Ans. $\pm 9x^2\sqrt{2xy}$

69. $2\sqrt{x^2}$ **70.** $3\sqrt{x^2y}$ **71.** $3\sqrt{4x}$ **72.** $4\sqrt{5x^2}$

73. $7\sqrt{49y^3}$ **74.** $-4\sqrt{16x^2}$ **75.** $2x\sqrt{x^2y}$ **76.** $3x\sqrt{9x^3}$

77. $-\frac{1}{3}\sqrt{9a^2}$ **78.** $\frac{1}{5}\sqrt{25y^3}$ **79.** $\pm\frac{1}{2}x\sqrt{16x^4}$ **80.** $\pm\frac{2a}{3}\sqrt{36a^3b^3}$

81. $\frac{1}{xy}\sqrt{x^3y^2}$ **82.** $\frac{x^2}{y}\sqrt{xy^3}$ **83.** $\frac{bc}{a}\sqrt{a^3b^2c}$ **84.** $\frac{a^2c}{b}\sqrt{b^4c}$

■ *Use the table of square roots, page 321, to approximate each expression. Round off answers to two decimal places.*

Sample Problems

a. $\sqrt{243}$

$\sqrt{81}\sqrt{3}$

$9\sqrt{3}$

$9(1.732)$

15.588

Ans. 15.59

b. $3 + 2\sqrt{200}$

$3 + 2\sqrt{100}\sqrt{2}$

$3 + 2(10)\sqrt{2}$

$3 + 20(1.414)$

$3 + 28.28$

Ans. 31.28

85. $\sqrt{108}$ **86.** $\sqrt{162}$ **87.** $\sqrt{275}$

88. $\sqrt{207}$ **89.** $3 + \sqrt{125}$ **90.** $24 - \sqrt{176}$

91. $5 - \sqrt{300}$ **92.** $11 - \sqrt{242}$ **93.** $-2\sqrt{243}$

94. $-5\sqrt{120}$ **95.** $6 + 3\sqrt{104}$ **96.** $1 + 2\sqrt{128}$

8.4 SIMPLIFICATION OF RADICAL EXPRESSIONS–POLYNOMIALS

You should recall from Section 1.7 that like terms of algebraic expressions are added by adding the numerical coefficients of the terms, e.g.,

$$2r + 3r = 5r,$$

where r represents any number. In particular, if r represents an irrational number, say $\sqrt{2}$, we have

$$2\sqrt{2}+3\sqrt{2}=5\sqrt{2}.$$

Thus, we may add radical expressions, by adding their numerical coefficients, provided the radicands involved are identical. If the radicands differ, we can only indicate addition, e.g., $3\sqrt{2}+4\sqrt{3}$. As before, if no numerical coefficient is written before a radical, it is understood that the coefficient is 1. Thus, $\sqrt{3}$ means $1\sqrt{3}$. Radicals should be written in simplest form before attempting to combine like terms.

Products involving radicals can also be rewritten in equivalent forms. From the distributive law

$$a(\sqrt{3}+\sqrt{2})=a\sqrt{3}+a\sqrt{2},$$

where a may be either rational or irrational. By the symmetric property of equality, we may reverse the multiplication to assert that

$$a\sqrt{3}+a\sqrt{2}=a(\sqrt{3}+\sqrt{2}),$$

where the right-hand member is in factored form.

EXERCISES 8.4

■ *Simplify.*

Sample Problems

a. $\sqrt{5}+3\sqrt{5}$

Ans. $4\sqrt{5}$

b. $\sqrt{5}-3\sqrt{5}$

Ans. $-2\sqrt{5}$

1. $\sqrt{3}+2\sqrt{3}$

2. $\sqrt{7}-3\sqrt{7}$

3. $3\sqrt{5}-2\sqrt{5}$

4. $8\sqrt{5}-2\sqrt{5}+3\sqrt{5}$

5. $2\sqrt{3}-4\sqrt{3}+2\sqrt{3}$

6. $\sqrt{5}-3\sqrt{5}+7\sqrt{5}$

Sample Problems

a. $5\sqrt{2}-\sqrt{8}+\sqrt{12}$
$5\sqrt{2}-2\sqrt{2}+2\sqrt{3}$

Ans. $3\sqrt{2}+2\sqrt{3}$

b. $2\sqrt{3a}+\sqrt{27a}-2\sqrt{12a}$
$2\sqrt{3a}+3\sqrt{3a}-4\sqrt{3a}$

Ans. $\sqrt{3a}$

7. $2\sqrt{3}+\sqrt{27}$

8. $\sqrt{8}+\sqrt{18}$

9. $\sqrt{50}-2\sqrt{32}$

10. $2\sqrt{6}-2\sqrt{24}+\sqrt{54}$

11. $\sqrt{12} + 2\sqrt{27} - 3\sqrt{48}$

12. $\sqrt{20} + \sqrt{45} - 2\sqrt{80}$

13. $3\sqrt{2} - 4\sqrt{3} + \sqrt{2}$

14. $2\sqrt{3} - \sqrt{4} + 3\sqrt{3}$

15. $\sqrt{3} + 2\sqrt{12} + \sqrt{18}$

16. $\sqrt{36} - 2\sqrt{32} + \sqrt{49}$

17. $3\sqrt{144} - 4\sqrt{49} + 3\sqrt{24}$

18. $\sqrt{12} - \sqrt{27} + 2\sqrt{8}$

19. $\sqrt{4a} + \sqrt{9a}$

20. $\sqrt{12a} - \sqrt{3a}$

21. $2\sqrt{x} + 2\sqrt{25x}$

22. $3\sqrt{2x} - \sqrt{8x}$

23. $\sqrt{16b^3} - b\sqrt{25b} + 3b\sqrt{b}$

24. $\sqrt{xy^2} + 2\sqrt{xy^2} - \sqrt{4xy^2}$

■ *Express without parentheses.*

a. $2(3 + 4\sqrt{2})$

Ans. $6 + 8\sqrt{2}$

b. $a(4\sqrt{3} - 6\sqrt{a})$

Ans. $4a\sqrt{3} - 6a\sqrt{a}$

25. $4(\sqrt{3} + 1)$

26. $2(3 - \sqrt{2})$

27. $-5(6 + \sqrt{7})$

28. $-2(\sqrt{6} - 3)$

29. $4(\sqrt{2} - \sqrt{3})$

30. $3(\sqrt{3} + \sqrt{7})$

31. $3(1 + 3\sqrt{a})$

32. $2(4\sqrt{a} - 5)$

33. $2(\sqrt{2} - 3\sqrt{3} - 5)$

34. $-3(6 + \sqrt{5} - 2\sqrt{3})$

35. $-(\sqrt{a} + \sqrt{b} - \sqrt{c})$

36. $-(2\sqrt{a} - \sqrt{b} + 2\sqrt{c})$

37. $x(\sqrt{x} + 2\sqrt{y})$

38. $y(\sqrt{xy} - \sqrt{y})$

39. $xy(y\sqrt{x} + 2)$

40. $xy(4 - x\sqrt{x})$

■ *Simplify radical expressions where possible and factor.*

Sample Problems

a. $8\sqrt{3} - 10$

Ans. $2(4\sqrt{3} - 5)$

b. $\sqrt{12} + 4$

$2\sqrt{3} + 4$

Ans. $2(\sqrt{3} + 2)$

c. $y\sqrt{x} - y^2\sqrt{y}$

Ans. $y(\sqrt{x} - y\sqrt{y})$

41. $2 + 2\sqrt{3}$

42. $6 - 3\sqrt{2}$

43. $4\sqrt{2} - 12$

44. $3\sqrt{7} - 3$

45. $4\sqrt{5} + 8$

46. $8 + 32\sqrt{5}$

47. $8 - 32\sqrt{5}$

48. $4 + 2\sqrt{3}$

49. $6 + 24\sqrt{2}$

50. $5\sqrt{5} - 10$

51. $3 + \sqrt{18}$

52. $2 - \sqrt{32}$

53. $4 - 2\sqrt{8}$

54. $6 + 2\sqrt{27}$

55. $21 + \sqrt{18}$

56. $6 + \sqrt{72}$

57. $4 + \sqrt{16y}$

58. $3 - 2\sqrt{9y}$

■ *Simplify fractions.*

Sample Problems

a. $\frac{6 - 3\sqrt{7}}{3}$

b. $\frac{-2 - \sqrt{72}}{4}$

Simplify radicals.

$\frac{-2 - 6\sqrt{2}}{4}$

Factor numerator and simplify.

$\frac{\cancel{3}(2 - \sqrt{7})}{\cancel{3}}$

$\frac{\cancel{2}(-1 - 3\sqrt{2})}{\underset{2}{\cancel{4}}}$

Ans. $2 - \sqrt{7}$

Ans. $\frac{-1 - 3\sqrt{2}}{2}$

59. $\frac{4 + 6\sqrt{3}}{2}$

60. $\frac{3 - 3\sqrt{2}}{3}$

61. $\frac{6 - 2\sqrt{5}}{2}$

62. $\frac{9 - 3\sqrt{5}}{3}$

63. $\frac{-2 + \sqrt{8}}{2}$

64. $\frac{-6 + \sqrt{54}}{3}$

65. $\frac{4 + 3\sqrt{12}}{2}$

66. $\frac{2 - \sqrt{8}}{4}$

67. $\frac{3 + \sqrt{18}}{6}$

68. $\frac{8 + \sqrt{32}}{16}$

69. $\frac{5 - \sqrt{75}}{10}$

70. $\frac{16 - 2\sqrt{48}}{16}$

■ *Write each sum or difference as a single fraction.*

Sample Problems

a. $\frac{2}{3} - \frac{\sqrt{7}}{3}$

b. $\frac{2}{3} + \frac{5\sqrt{7}}{6}$

In problem b, build $2/3$ to a fraction with denominator 6.

$\frac{(2)2}{(2)3} + \frac{5\sqrt{7}}{6}$

Add numerators.

Ans. $\frac{2 - \sqrt{7}}{3}$

Ans. $\frac{4 + 5\sqrt{7}}{6}$

71. $\frac{2}{3} + \frac{\sqrt{2}}{3}$

72. $\frac{5}{2} - \frac{\sqrt{3}}{2}$

73. $\frac{\sqrt{3}}{5} - \frac{1}{5}$

74. $\frac{\sqrt{2}}{7} + \frac{1}{7}$

75. $\frac{2\sqrt{10}}{3} - \frac{\sqrt{3}}{3}$

76. $\frac{\sqrt{17}}{5} - \frac{3\sqrt{7}}{5}$

77. $\frac{\sqrt{11}}{a}+\frac{1}{a}$ **78.** $\frac{\sqrt{5}}{b}-\frac{3}{b}$ **79.** $\frac{5}{4}+\frac{3\sqrt{2}}{2}$

80. $\frac{1}{10}-\frac{2\sqrt{3}}{5}$ **81.** $\frac{1}{2}+\frac{\sqrt{3}}{6}$ **82.** $\frac{\sqrt{5}}{3}-\frac{5}{6}$

Sample Problems

a. $\frac{1}{2}-\frac{\sqrt{3}}{3}$

$\frac{(3)1}{(3)2}-\frac{\sqrt{3}(2)}{3(2)}$

Ans. $\frac{3-2\sqrt{3}}{6}$

b. $4-\frac{2\sqrt{3}}{5}$

$\frac{(5)4}{(5)1}-\frac{2\sqrt{3}}{5}$

Ans. $\frac{20-2\sqrt{3}}{5}$

83. $\frac{2}{5}+\frac{\sqrt{3}}{3}$ **84.** $\frac{3}{7}-\frac{\sqrt{2}}{2}$ **85.** $\frac{\sqrt{3}}{4}-\frac{1}{3}$

86. $\frac{\sqrt{5}}{2}-\frac{2}{3}$ **87.** $\frac{2\sqrt{3}}{3}-\frac{\sqrt{2}}{2}$ **88.** $\frac{3\sqrt{5}}{4}+\frac{\sqrt{3}}{5}$

89. $4+\frac{3\sqrt{2}}{2}$ **90.** $2-\frac{\sqrt{2}}{3}$ **91.** $\frac{\sqrt{2}}{5}+1$

92. $\frac{2\sqrt{2}}{3}-1$ **93.** $\frac{3\sqrt{3}}{2}+3$ **94.** $\frac{3\sqrt{7}}{2}-3$

8.5 PRODUCTS OF RADICAL EXPRESSIONS

In Section 8.3, we observed that

$$\sqrt{ab}=\sqrt{a}\,\sqrt{b}.$$

By the symmetric property of equality,

$$\sqrt{\mathbf{a}}\,\sqrt{\mathbf{b}}=\sqrt{\mathbf{ab}}.$$

Stated in words, we have:

The product of two square roots is equal to the square root of the product of the radicands.

EXERCISES 8.5

■ *Simplify.*

Sample Problems

a. $\sqrt{2}\sqrt{3}$

$\sqrt{2 \cdot 3}$

Ans. $\sqrt{6}$

b. $\sqrt{2x}\sqrt{6xy}$ or $\sqrt{2x}\sqrt{6xy}$

$\sqrt{2x \cdot 6xy}$ $\qquad$ $\sqrt{2x}\sqrt{2x}\sqrt{3y}$

$\sqrt{12x^2y}$

$\sqrt{4x^2}\sqrt{3y}$

Ans. $2x\sqrt{3y}$

1. $\sqrt{3}\sqrt{5}$ **2.** $\sqrt{2}\sqrt{7}$ **3.** $\sqrt{3}\sqrt{10}$ **4.** $\sqrt{5}\sqrt{13}$

5. $\sqrt{3}\sqrt{6}$ **6.** $\sqrt{2}\sqrt{10}$ **7.** $\sqrt{8}\sqrt{2}$ **8.** $\sqrt{27}\sqrt{3}$

9. $\sqrt{2x}\sqrt{3x}$ **10.** $\sqrt{5a}\sqrt{3a}$ **11.** $\sqrt{2xy}\sqrt{6xy^2}$ **12.** $\sqrt{6x^2}\sqrt{3x^2y}$

13. $\sqrt{8a}\sqrt{18a}$ **14.** $\sqrt{12b}\sqrt{32b}$ **15.** $\sqrt{10x^2}\sqrt{15y}$ **16.** $\sqrt{18a^2}\sqrt{6b}$

Sample Problem

$2\sqrt{6}\sqrt{8}$ or $2\sqrt{6}\sqrt{8}$

$2\sqrt{48}$ $\qquad$ $2\sqrt{3}\sqrt{2}\sqrt{2}\sqrt{2}\sqrt{2}$

$2\sqrt{16}\sqrt{3}$ $\qquad$ $2\sqrt{3}\,(2)(2)$

$2 \cdot 4\sqrt{3}$

Ans. $8\sqrt{3}$

17. $\sqrt{2}\sqrt{5}\sqrt{3}$

18. $\sqrt{5}\sqrt{3}\sqrt{7}$

19. $\sqrt{5}\sqrt{10}\sqrt{2}$

20. $\sqrt{6}\sqrt{3}\sqrt{2}$

21. $(2\sqrt{3})(\sqrt{2})(\sqrt{9})$

22. $(5\sqrt{5})(3\sqrt{10})(\sqrt{4})$

23. $(2\sqrt{x})(3\sqrt{x})(\sqrt{x})$

24. $(x\sqrt{2})(x\sqrt{3})(\sqrt{6})$

25. $(a\sqrt{b})(b\sqrt{c})(c\sqrt{a})$

26. $(b\sqrt{a})(a\sqrt{b})(a\sqrt{ab})$

27. $(x\sqrt{x})(\sqrt{x^2})(\sqrt{x^3})$

28. $(a^2\sqrt{a})(2a\sqrt{a})(a\sqrt{a^2})$

Sample Problems

a. $\sqrt{3}(2 + \sqrt{2})$

$(\sqrt{3})(2) + (\sqrt{3})(\sqrt{2})$

Ans. $2\sqrt{3} + \sqrt{6}$

b. $\sqrt{3}(\sqrt{6} - \sqrt{15})$

$(\sqrt{3})(\sqrt{6}) - (\sqrt{3})(\sqrt{15})$

$\sqrt{18} - \sqrt{45}$

Ans. $3\sqrt{2} - 3\sqrt{5}$

29. $\sqrt{2}(3 + \sqrt{3})$ **30.** $\sqrt{3}(5 + \sqrt{5})$ **31.** $\sqrt{3}(\sqrt{6} + 2)$

32. $\sqrt{2}(\sqrt{6} + 3)$ **33.** $\sqrt{5}(4 + \sqrt{10})$ **34.** $\sqrt{3}(2 - \sqrt{15})$

35. $\sqrt{3}(\sqrt{2} + \sqrt{6})$ **36.** $\sqrt{5}(\sqrt{3} - \sqrt{10})$ **37.** $\sqrt{3}(\sqrt{3} + \sqrt{2})$

38. $\sqrt{5}(\sqrt{5}+\sqrt{3})$ **39.** $\sqrt{2}(\sqrt{10}-\sqrt{2})$ **40.** $\sqrt{3}(\sqrt{3}+\sqrt{15})$

Sample Problem

$(2+\sqrt{3})(1-2\sqrt{3})$

$2-4\sqrt{3}+\sqrt{3}-2\sqrt{3}\,\sqrt{3}$ Apply distributive law.

$2-3\sqrt{3}-6$ Simplify.

Ans. $-4-3\sqrt{3}$

41. $(3+\sqrt{2})(1-\sqrt{2})$ **42.** $(2-\sqrt{2})(3+\sqrt{2})$
43. $(\sqrt{5}-1)(\sqrt{5}+3)$ **44.** $(\sqrt{7}+3)(\sqrt{7}-5)$
45. $(2+\sqrt{3})(2-\sqrt{3})$ **46.** $(3+\sqrt{2})(3-\sqrt{2})$
47. $(3-2\sqrt{5})(3+2\sqrt{5})$ **48.** $(4-3\sqrt{6})(4+3\sqrt{6})$
49. $(2\sqrt{3}+\sqrt{5})(\sqrt{3}-2\sqrt{5})$ **50.** $(2\sqrt{5}-3\sqrt{2})(\sqrt{5}+\sqrt{2})$
51. $(3\sqrt{7}-2\sqrt{5})(2\sqrt{7}+3\sqrt{5})$ **52.** $(5\sqrt{6}-2\sqrt{3})(\sqrt{6}-\sqrt{3})$

8.6 QUOTIENTS OF RADICAL EXPRESSIONS

In order to simplify a fraction that contains a radical in the denominator or in both the numerator and the denominator, we observe that

$$\sqrt{\frac{36}{9}}=\sqrt{4}=2,$$

and

$$\frac{\sqrt{36}}{\sqrt{9}}=\frac{6}{3}=2.$$

Generalizing, we have

$$\frac{\sqrt{a}}{\sqrt{b}}=\sqrt{\frac{a}{b}}, \quad b \neq 0.$$

Stated in words:

The quotient of two square roots is equal to the square root of the quotient of the radicands.

For example,

$$\frac{\sqrt{6}}{\sqrt{3}}=\sqrt{\frac{6}{3}}=\sqrt{2}.$$

It is often convenient, particularly in simple numerical computations, to express a fraction whose denominator contains a radical as an equivalent fraction with a denominator free of radicals. This process is called **rationalizing the denominator.** This can be done by building to a fraction with a perfect square in the denominator and then extracting the indicated root. For example, we can rationalize the denominator of the fraction $\frac{\sqrt{2}}{\sqrt{3}}$ by multiplying the numerator and the denominator by $\sqrt{3}$ to obtain

$$\frac{\sqrt{2}\,\sqrt{3}}{\sqrt{3}\,\sqrt{3}} = \frac{\sqrt{6}}{\sqrt{9}} = \frac{\sqrt{6}}{3}.$$

Alternatively, we may represent the fraction $\frac{\sqrt{2}}{\sqrt{3}}$ in the form $\sqrt{\frac{2}{3}}$ and then multiply the numerator and the denominator of the radicand by 3 to obtain

$$\sqrt{\frac{2 \cdot 3}{3 \cdot 3}} = \sqrt{\frac{6}{9}} = \frac{\sqrt{6}}{\sqrt{9}} = \frac{\sqrt{6}}{3}.$$

The result is the same and you should use the approach you find the simplest.

To illustrate one advantage of the rationalized form, consider the numerical computation of a decimal approximation for $\frac{\sqrt{2}}{\sqrt{3}}$. If we approach this problem directly, we obtain $\frac{\sqrt{2}}{\sqrt{3}} \approx \frac{1.414}{1.732}$ and arrive at a problem in long division. If we first rationalize the denominator, we obtain $\frac{\sqrt{2}}{\sqrt{3}} = \frac{\sqrt{6}}{3} \approx \frac{2.449}{3}$ and arrive at a simple division process, in fact, one which can be done mentally.

EXERCISES 8.6

■ *Simplify.*

Sample Problems

a. $\frac{\sqrt{12}}{\sqrt{3}}$

$\sqrt{\frac{12}{3}}$

$\sqrt{4}$

Ans. 2

b. $\frac{2\sqrt{30}}{\sqrt{6}}$

$2\sqrt{\frac{30}{6}}$

Ans. $2\sqrt{5}$

c. $\frac{3\sqrt{6a}}{\sqrt{2a}}$

$3\sqrt{\frac{6a}{2a}}$

Ans. $3\sqrt{3}$

1. $\frac{\sqrt{18}}{\sqrt{2}}$ **2.** $\frac{\sqrt{8}}{\sqrt{2}}$ **3.** $\frac{\sqrt{75}}{\sqrt{3}}$ **4.** $\frac{\sqrt{80}}{\sqrt{5}}$

5. $\frac{\sqrt{27a}}{\sqrt{3a}}$ **6.** $\frac{\sqrt{28a}}{\sqrt{7a}}$ **7.** $\frac{\sqrt{8a}}{\sqrt{2}}$ **8.** $\frac{\sqrt{12a}}{\sqrt{3}}$

9. $\frac{\sqrt{15b}}{\sqrt{5}}$ **10.** $\frac{\sqrt{21b}}{\sqrt{7b}}$ **11.** $\frac{\sqrt{ab}}{\sqrt{a}}$ **12.** $\frac{\sqrt{7abc}}{\sqrt{bc}}$

13. $\frac{2\sqrt{14bc}}{\sqrt{2c}}$ **14.** $\frac{3\sqrt{8a}}{\sqrt{2}}$ **15.** $\frac{\sqrt{2}\sqrt{3}}{\sqrt{6}}$ **16.** $\frac{\sqrt{6}\sqrt{8}}{\sqrt{3}}$

17. $\frac{\sqrt{a}\sqrt{ab}}{\sqrt{b}}$ **18.** $\frac{\sqrt{3a}\sqrt{6a}}{\sqrt{2}}$ **19.** $\frac{\sqrt{ab}\sqrt{3b}}{\sqrt{a}}$ **20.** $\frac{\sqrt{3}\sqrt{10}}{\sqrt{6}}$

■ *Rationalize each denominator.*

Sample Problems

a. $\frac{\sqrt{3}}{\sqrt{a}}$ or $\frac{\sqrt{3}}{\sqrt{a}}$

$\frac{\sqrt{3}\sqrt{a}}{\sqrt{a}\sqrt{a}}$ $\sqrt{\frac{3 \cdot a}{a \cdot a}}$

$\frac{\sqrt{3a}}{\sqrt{a^2}}$ $\frac{\sqrt{3a}}{\sqrt{a^2}}$

Ans. $\frac{\sqrt{3a}}{a}$

b. $\sqrt{\frac{1}{2}}$ or $\sqrt{\frac{1}{2}}$

$\sqrt{\frac{1 \cdot 2}{2 \cdot 2}}$ $\frac{\sqrt{1}\sqrt{2}}{\sqrt{2}\sqrt{2}}$

$\frac{\sqrt{2}}{\sqrt{4}}$ $\frac{\sqrt{2}}{\sqrt{4}}$

Ans. $\frac{\sqrt{2}}{2}$

21. $\frac{5}{\sqrt{2}}$ **22.** $\frac{5}{\sqrt{3}}$ **23.** $\frac{2}{\sqrt{x}}$ **24.** $\frac{5}{\sqrt{x}}$

25. $\frac{a}{\sqrt{b}}$ **26.** $\frac{x}{\sqrt{y}}$ **27.** $\sqrt{\frac{1}{3}}$ **28.** $\sqrt{\frac{1}{5}}$

29. $\sqrt{\frac{2}{a}}$ **30.** $\sqrt{\frac{2a}{b}}$ **31.** $\sqrt{\frac{3a}{b}}$ **32.** $\sqrt{\frac{5b}{a}}$

Sample Problems

a. $\frac{\sqrt{12}}{\sqrt{3x}}$ or $\frac{\sqrt{12}}{\sqrt{3x}}$

$\sqrt{\frac{12}{3x}}$ $\frac{\sqrt{12}\sqrt{3x}}{\sqrt{3x}\sqrt{3x}}$

$\sqrt{\frac{4 \cdot x}{x \cdot x}}$ $\frac{\sqrt{36x}}{\sqrt{9x^2}}$

b. $\frac{2\sqrt{5}}{\sqrt{8}}$ or $\frac{2\sqrt{5}}{\sqrt{8}}$

$2\sqrt{\frac{5 \cdot 2}{8 \cdot 2}}$ $\frac{2\sqrt{5}\sqrt{2}}{\sqrt{8}\sqrt{2}}$

$2\sqrt{\frac{10}{16}}$ $\frac{2\sqrt{10}}{\sqrt{16}}$

(Continued)

$$\frac{\sqrt{4}\,\sqrt{x}}{\sqrt{x^2}} \quad \text{or} \quad \frac{6\sqrt{x}}{3x} \qquad\qquad \frac{2\sqrt{10}}{4} \quad \text{or} \quad \frac{2\sqrt{10}}{4}$$

Ans. $\frac{2\sqrt{x}}{x}$ *Ans.* $\frac{\sqrt{10}}{2}$

33. $\frac{\sqrt{18}}{\sqrt{2x}}$ **34.** $\frac{\sqrt{8}}{\sqrt{2y}}$ **35.** $\frac{\sqrt{75}}{\sqrt{3y}}$ **36.** $\frac{\sqrt{80}}{\sqrt{5x}}$

37. $\frac{a\sqrt{2}}{\sqrt{a}}$ **38.** $\frac{b\sqrt{3}}{\sqrt{b}}$ **39.** $\frac{4\sqrt{3x}}{\sqrt{8}}$ **40.** $\frac{9\sqrt{5x}}{\sqrt{27}}$

41. $\frac{a\sqrt{32}}{\sqrt{2a}}$ **42.** $\frac{b\sqrt{21}}{\sqrt{3b}}$ **43.** $\frac{4y\sqrt{3x}}{\sqrt{4y}}$ **44.** $\frac{9x\sqrt{2y}}{\sqrt{27x}}$

Sample Problems

a. $\sqrt{\frac{20}{3}}$ or $\sqrt{\frac{20}{3}}$

$$\frac{\sqrt{4}\,\sqrt{5}}{\sqrt{3}} \qquad \sqrt{\frac{4\cdot 5\cdot 3}{3\cdot 3}}$$

$$\frac{2\sqrt{5}\,\sqrt{3}}{\sqrt{3}\,\sqrt{3}} \qquad \frac{\sqrt{4}\,\sqrt{15}}{\sqrt{9}}$$

Ans. $\frac{2\sqrt{15}}{3}$

b. $\sqrt{\frac{4}{3x}}$ or $\sqrt{\frac{4}{3x}}$

$$\frac{2}{\sqrt{3x}} \qquad \sqrt{\frac{4\cdot 3x}{3x\cdot 3x}}$$

$$\frac{2\sqrt{3x}}{\sqrt{3x}\,\sqrt{3x}} \qquad \frac{\sqrt{4}\,\sqrt{3x}}{\sqrt{9x^2}}$$

Ans. $\frac{2\sqrt{3x}}{3x}$

45. $\sqrt{\frac{8}{3}}$ **46.** $\sqrt{\frac{18}{5}}$ **47.** $\sqrt{\frac{9}{2}}$ **48.** $\sqrt{\frac{12}{7}}$

49. $\sqrt{\frac{72}{5}}$ **50.** $\sqrt{\frac{98}{3}}$ **51.** $\sqrt{\frac{50}{2x}}$ **52.** $\sqrt{\frac{75}{3y}}$

53. $\sqrt{\frac{24}{3x}}$ **54.** $\sqrt{\frac{32}{4y}}$ **55.** $\sqrt{\frac{x^3}{xy}}$ **56.** $\sqrt{\frac{y^5}{xy^2}}$

■ *Rationalize each denominator and find a decimal approximation (round off answers to two decimal places).*

Sample Problems

a. $\sqrt{\frac{5}{3}}$

$$\frac{\sqrt{5}\,\sqrt{3}}{\sqrt{3}\,\sqrt{3}}$$

b. $4\sqrt{\frac{1}{3}}$

$$4\sqrt{\frac{1\cdot 3}{3\cdot 3}}$$

$\frac{\sqrt{15}}{3}$ ($\sqrt{15} \approx 3.873$)

$\frac{\sqrt{15}}{3} \approx \frac{3.873}{3} = 1.291$

Ans. 1.29

$\frac{4}{3}\sqrt{3}$ ($\sqrt{3} \approx 1.732$)

$\frac{4}{3}\sqrt{3} \approx \frac{4}{3}(1.732) = 2.309$

Ans. 2.31

57. $\sqrt{\frac{1}{7}}$ **58.** $\sqrt{\frac{3}{5}}$ **59.** $\frac{3}{\sqrt{2}}$ **60.** $\frac{2}{\sqrt{3}}$

61. $\frac{3}{\sqrt{5}}$ **62.** $\frac{2}{\sqrt{6}}$ **63.** $3\sqrt{\frac{1}{3}}$ **64.** $5\sqrt{\frac{1}{5}}$

CHAPTER REVIEW

1. Which of the following are irrational numbers?

$$\sqrt{6}, -\sqrt{9}, \sqrt{4}, \sqrt{\frac{9}{25}}, -\sqrt{\frac{2}{3}}$$

2. Using the table of square roots, find a decimal approximation for:

a. $\sqrt{93}$ b. $2\sqrt{47}$. c. $\frac{1}{4}\sqrt{32}$.

3. Locate the numbers 8, $\sqrt{80}$, and $\frac{19}{2}$ on a line graph.

■ *Simplify each expression in Exercises 4–8.*

4. a. $\sqrt{72}$ b. $-\sqrt{90}$ c. $\sqrt{175}$

5. a. $\sqrt{y^4}$ b. $\sqrt{x^{15}}$ c. $\sqrt{x^3y^7}$

6. a. $x\sqrt{27}$ b. $3\sqrt{3x^2}$ c. $\frac{1}{3}\sqrt{27x^2}$

7. a. $\sqrt{7} + 3\sqrt{7}$
b. $4\sqrt{6} - 3\sqrt{6} + \sqrt{6}$
c. $7\sqrt{2} - 3\sqrt{2} + \sqrt{2}$

8. a. $2\sqrt{12} - 4\sqrt{27}$
b. $\sqrt{9x} - \sqrt{4x}$
c. $\sqrt{x^2y} - 3\sqrt{4x^2y} + 7x\sqrt{y}$

9. Express without parentheses.

a. $3(\sqrt{y} - 6)$ b. $y(\sqrt{xy} - 2y)$ c. $\sqrt{3}(\sqrt{2} - \sqrt{6})$

10. Simplify radicals where possible and factor.

a. $2 - \sqrt{8}$ b. $\sqrt{27} - 3\sqrt{5}$ c. $3x^2 - \sqrt{5x^4}$

■ *Simplify each expression.*

11. a. $\frac{3 - 2\sqrt{27}}{3}$ b. $\frac{6 - 3\sqrt{12}}{3}$ c. $\frac{4 - \sqrt{32}}{4}$

12. a. $\frac{3}{2} - \frac{\sqrt{3}}{2}$ b. $\frac{\sqrt{15}}{5} + \frac{2\sqrt{15}}{5}$ c. $\frac{2}{3} - \frac{\sqrt{3}}{6}$

13. a. $\frac{\sqrt{5}}{3} - \frac{3}{2}$ b. $\frac{2\sqrt{3}}{5} - 1$ c. $\frac{3\sqrt{6}}{4} + 2$

14. a. $\sqrt{5}\ \sqrt{3}$ b. $\sqrt{16}\ \sqrt{32}$ c. $\sqrt{3x^2}\ \sqrt{6x}$

15. a. $\sqrt{5}\ \sqrt{2}\ \sqrt{15}$ b. $(2\sqrt{6})(\sqrt{3})(\sqrt{2})$ c. $(x\sqrt{3})(\sqrt{4x})$

16. a. $\sqrt{3}(2 - \sqrt{2})$ b. $\sqrt{7}(\sqrt{2} - \sqrt{14})$ c. $\sqrt{8}(\sqrt{2} - \sqrt{3})$

17. a. $(2 - \sqrt{3})(2 + \sqrt{3})$
b. $(\sqrt{5} - \sqrt{7})(\sqrt{5} + \sqrt{7})$
c. $(2 - \sqrt{3})(3 - 2\sqrt{3})$

18. a. $\frac{\sqrt{27}}{\sqrt{3}}$ b. $\frac{\sqrt{12a}}{\sqrt{2}}$ c. $\frac{\sqrt{3a}\ \ \sqrt{ab}}{\sqrt{b}}$

19. a. $\sqrt{\frac{3}{5}}$ b. $\sqrt{\frac{12}{5}}$ c. $\sqrt{\frac{3}{2x}}$

20. a. $\frac{\sqrt{85}}{\sqrt{5x}}$ b. $\frac{\sqrt{3y}}{\sqrt{2x^2}}$ c. $\frac{4\sqrt{7x}}{\sqrt{2x}}$

Cumulative Review

1. A positive number has ___?___ square roots.

2. If $a = -4$, $b = 2$, and $c = 0$, find the value of $\frac{2a - 4}{3 - bc}$.

3. Multiply: $(-3ab)(2c)(b^2)$.

4. Represent $\frac{a - b}{2} + \frac{a - 3b}{3}$ as a single fraction.

5. Reduce to lowest terms: $\frac{3a^2x - 3ax}{24a^2x^2 + 60ax^2}$.

6. Divide: $(x^4 + 3x^3 - x^2)$ by (x^2).

7. Factor: $4x^2 - 12x + 9$.

8. Solve: $3(x - 2) - 7 = 5(2x + 5) - 3$.

9. Solve: $\frac{y - 6}{5} = \frac{12y + 3}{5} - 3y + 3$.

10. The sum of two numbers is 146. If n represents the smaller of the two numbers, represent the larger number in terms of n.

11. Write in terms of x, the number of cents in x quarters.

12. Solve the system: $\frac{x}{2} - \frac{y}{3} = 1$

$$x - \frac{2y}{3} = 2.$$

13. Solve for y: $2by = 9d - 5by$.

14. Solve: $2x^2 = 5x + 3$.

15. The sum of two numbers is 154. If the larger number is eight less than twice the smaller, what are the numbers?

16. Simplify: $\sqrt{63x^7}$.

17. Simplify: $\sqrt{18} - 2\sqrt{5} - \sqrt{20} + \sqrt{50}$.

18. Simplify: $(2 - \sqrt{3})(2 + \sqrt{3})$.

19. Simplify: $\frac{\sqrt{4x}\ \ \sqrt{9x}}{\sqrt{6x}}$.

20. Simplify: $\frac{3\sqrt{2y}}{\sqrt{3y}}$.

9

SOLUTION OF QUADRATIC EQUATIONS BY OTHER METHODS

In Chapter 7 you learned to solve, by factoring methods, quadratic equations whose solutions are rational numbers. Having studied irrational numbers in Chapter 8, you are now ready to examine additional methods of solving quadratic equations.

9.1 EXTRACTION OF ROOTS

We can solve an equation of the form

$$x^2 - a = 0$$

by writing it in the form

$$x^2 = a,$$

and observing that for a greater than or equal to zero, x is a number which, when multiplied by itself, yields a. Therefore, by the definition of a square root, x must be a square root of a. Since a has two square roots, we have $\sqrt{a}$ and $-\sqrt{a}$ as solutions for the equation. That is,

$$(\sqrt{a})^2 = \sqrt{a}\,\sqrt{a} = a,$$

and

$$(-\sqrt{a})^2 = (-\sqrt{a})(-\sqrt{a}) = a.$$

We sometimes use the double sign notation to represent both roots, that is, $x = \pm\sqrt{a}$. For example, if

$$x^2 - 5 = 0,$$

we have

$$x^2 = 5,$$

from which

$$x = \pm\sqrt{5}.$$

This method of solving equations is sometimes referred to as solving by **extraction of roots,** since in a sense, we have simply equated the square roots of the members of the equation.

Quadratic equations of the form

$$(x + k)^2 = d$$

can also be solved by extraction of roots. For example, from

$$(x - 2)^2 = 9$$

we obtain

$$x - 2 = \pm 3,$$
$$x = \pm 3 + 2,$$

from which

$$x = +3 + 2 = 5$$

or

$$x = -3 + 2 = -1.$$

EXERCISES 9.1

■ *Solve.*

Sample Problem

$3y^2 = 48$	
	Write with y^2 as left-hand member.
$y^2 = 16$	
	Extract square root of each member.
$y = \pm\sqrt{16}$	
	Simplify.
$y = \pm 4$	

Ans. $y = 4$; $y = -4$

Check.

$3(4)^2 = 48$	$3(-4)^2 = 48$
$3(16) = 48$	$3(16) = 48$
$48 = 48$	$48 = 48$

1. $x^2 = 4$ **2.** $y^2 = 9$ **3.** $x^2 - 16 = 0$

4. $x^2 - 25 = 0$ **5.** $98 = 2x^2$ **6.** $12 = 3x^2$

7. $x^2 - 3 = 0$ **8.** $7 - z^2 = 0$ **9.** $y^2 - 10 = 0$

10. $3x^2 - 15 = 0$ **11.** $4x^2 - 24 = 0$ **12.** $7x^2 = 42$

13. $12 = z^2$ **14.** $24 = b^2$ **15.** $x^2 - 18 = 0$

16. $3x^2 - 24 = 0$ **17.** $3x^2 - 54 = 0$ **18.** $5x^2 - 100 = 0$

Sample Problem

$$5x^2 - 3 = x^2 + 9$$

Write with x^2 as left-hand member.

$$4x^2 = 12$$
$$x^2 = 3$$

Extract square root of each member.

$$x = \pm\sqrt{3}$$

Ans. $x = \sqrt{3}; \quad x = -\sqrt{3}$

19. $2x^2 - 4 = x^2$ **20.** $6x^2 + 3 = 4x^2 + 11$ **21.** $3 = x^2 - 2$

22. $4t^2 - 16 = 16$ **23.** $4y^2 - 10 = y^2 - 10$ **24.** $s^2 - 2 = 4 - s^2$

Sample Problem

$$\frac{1}{3}x^2 - \frac{3}{4} = 0$$

Multiply each member by L.C.D. 12.

$$\overset{4}{(\cancel{12})}\frac{1}{\cancel{3}}x^2 - \overset{3}{(\cancel{12})}\frac{3}{\cancel{4}} = (12)0$$

Write with x^2 as left-hand member.

$$4x^2 - 9 = 0$$
$$4x^2 = 9$$
$$x^2 = \frac{9}{4}$$

Extract square root of each member.

$$x = \pm\frac{3}{2}$$

Ans. $x = \frac{3}{2}; \quad x = \frac{-3}{2}$

25. $\frac{1}{4}x^2 = 5$ **26.** $\frac{2}{3}y^2 - 4 = 0$ **27.** $\frac{2}{3}x^2 = 6$

28. $\frac{2x^2}{3} - 4 = \frac{x^2}{3}$ **29.** $\frac{5x^2}{2} - 4 = 2x^2$ **30.** $\frac{1}{2}x^2 - 4 = \frac{3}{2}$

■ *Solve each of the following equations for the indicated variable.*

Sample Problem

$$K = \frac{v^2}{64}, \quad \text{for } v$$

Multiply each member by 64.

$$(64)K = (\cancel{64})\frac{v^2}{\cancel{64}}$$

$$64K = v^2$$

$$v^2 = 64K$$

Extract the square root of each member and simplify.

$$v = \pm\sqrt{64K}$$

$$v = \pm\sqrt{64}\,\sqrt{K}$$

$$v = \pm 8\sqrt{K}$$

Ans. $v = 8\sqrt{K}$; $v = -8\sqrt{K}$

31. $x^2 - a = 0$, for x

32. $b = x^2$, for x

33. $\frac{x^2}{3} - b^2a^3 = 0$, for x

34. $\frac{ax^2}{2} - b = 0$, for x

35. $\frac{2y^2}{5} = \frac{b}{3}$, for y

36. $\frac{y^2}{2} + a = \frac{2a}{3} + 2y^2$, for y

37. $s = \frac{1}{2}gt^2$, for t

38. $V = \frac{1}{3}\pi r^2 h$, for r

39. $A = 4\pi r^2$, for r

40. $C = bh^2r$, for h

41. $I = \frac{3k}{d^2}$, for d

42. $F = \frac{k}{d^2}$, for d

■ *Given $x^2 + y^2 = z^2$ and values for two of the variables. Find values for the third variable.*

Sample Problem

$$y = \quad , \quad z =$$

Substitute 4 for y and 5 for z in $x^2 + y^2 = z^2$.

$$x^2 + (4)^2 = (5)^2$$

Simplify.

$$x^2 + 16 = 25$$

(Continued)

Write with x^2 as left-hand member.

$x^2 = 9$

Extract square root of each member.

$x = \pm 3$

Ans. $x = 3$; $x = -3$

43. $x = 15, y = 20$

44. $x = 10, y = 24$

45. $x = 6, z = 10$

46. $y = 9, z = 41$

47. $x = 5, y = 5$

48. $x = 11, z = 61$

■ *Solve for x.*

Sample Problem

$(x + 3)^2 = 25$

Extract square root of each member.

$x + 3 = \pm 5$

Solve resulting first-degree equations.

$x = \pm 5 - 3$

$x = +5 - 3$; $x = -5 - 3$

Ans. $x = 2$; $x = -8$

49. $(x - 1)^2 = 4$

50. $(x + 3)^2 = 9$

51. $(x - 2)^2 = 25$

52. $(x + 1)^2 = 36$

53. $(x - 5)^2 = 1$

54. $(x + 7)^2 = 1$

55. $(x - a)^2 = 25$

56. $(x + b)^2 = 4$

57. $(x - 3)^2 = a^2$

58. $(x + 5)^2 = b^2$

59. $(x - a)^2 = b^2$

60. $(x + a)^2 = b^2$

Sample Problem

$(x - 2)^2 = 20$

Extract square root of each member.

$x - 2 = \pm\sqrt{20}$

Solve resulting first-degree equations.

$x - 2 = \pm 2\sqrt{5}$

$x = 2 \pm 2\sqrt{5}$

Ans. $x = 2 + 2\sqrt{5}$; $x = 2 - 2\sqrt{5}$

61. $(x + 3)^2 = 2$

62. $(x - 2)^2 = 3$

63. $(x + 5)^2 = 5$

64. $(x - 6)^2 = 7$

65. $(x + 10)^2 = 8$

66. $(x - 1)^2 = 12$

67. $(x - 5)^2 = a$

68. $(x + 2)^2 = a$

69. $(x + 1)^2 = b$

70. $(x-7)^2 = b$ **71.** $(x-b)^2 = a$ **72.** $(x+b)^2 = a$

9.2 COMPLETING THE SQUARE

Thus far, the methods we have used to solve quadratic equations apply to special cases only. Let us now develop a method which is applicable to any quadratic equation.

First you should recall that

$$(x+q)^2 = x^2 + 2qx + q^2,$$

from which you can see that if the square of a binomial is expressed as a trinomial, the last term, q^2, must be the square of one half the coefficient of x; that is, q^2 is the square of one half of $2q$. For example, the value for k^2, such that the expression

$$x^2 + 6x + k^2$$

is equivalent to the square of the binomial

$$(x+k)^2,$$

is 9, the square of one half of 6, the coefficient of x. Thus,

$$x^2 + 6x + 9 = (x+3)^2.$$

Using this procedure, any quadratic equation can be written in the form $(x+k)^2 = d$, which we learned to solve in the preceding section. As an example, consider

$$x^2 - 6x - 7 = 0.$$

We first write the equation equivalently in the form

$$x^2 - 6x \qquad = 7.$$

Now, by adding 9 [the square of $\frac{1}{2}(-6)$] to each member, we have

$$x^2 - 6x + 9 = 7 + 9$$

or, equivalently,

$$(x-3)^2 = 16,$$

which we can solve by extraction of roots. In the event the coefficient on the second-degree term is not 1, we must divide each term in the equation by that coefficient before proceeding. Thus,

$$2x^2 - 3x - 9 = 0$$

should be written

$$x^2 - \frac{3}{2}x - \frac{9}{2} = 0$$

before proceeding further. The solution of this equation is completed in the sample problem on page 233. The foregoing method of solution is referred to as solving an equation by **completing the square.** We can summarize the procedure as follows:

1. Write the equation in standard form.
2. If the coefficient of the second-degree term is different from 1, divide each term in the equation by this coefficient.
3. Write the equation with the constant term in the right-hand member.
4. Add to each member the square of one half the coefficient of the first-degree term.
5. Rewrite the equation with the left-hand member expressed as a perfect square.
6. Solve by extraction of roots.

EXERCISES 9.2

■ *Solve by completing the square.*

Sample problem

$$x^2 - x - 2 = 0$$

Rewrite equation with constant term in the right-hand member.

$$x^2 - x \quad = 2$$

Square one half the coefficient of x and add to each member; $[\frac{1}{2}(-1)]^2 = \frac{1}{4}$

$$x^2 - x + \frac{1}{4} = 2 + \frac{1}{4}$$

Rewrite left-hand member as a perfect square.

$$\left(x - \frac{1}{2}\right)^2 = \frac{9}{4}$$

Extract square root of each member.

$$x - \frac{1}{2} = \pm\frac{3}{2}$$

Solve resulting first-degree equations.

$$x = \frac{1}{2} \pm \frac{3}{2}$$

$$x = \frac{1}{2} + \frac{3}{2}; \quad x = \frac{1}{2} - \frac{3}{2}$$

Ans. $x = 2; \quad x = -1$

1. $x^2 + 4x - 12 = 0$ **2.** $x^2 - 2x - 15 = 0$ **3.** $z^2 - 2z + 1 = 0$

4. $y^2 - y - 6 = 0$ **5.** $y^2 + y - 20 = 0$ **6.** $x^2 - x - 20 = 0$

7. $x^2 + 3x + 2 = 0$ **8.** $u^2 + 5u + 6 = 0$ **9.** $z^2 - 3z - 4 = 0$

10. $y^2 + 9y + 20 = 0$ **11.** $r^2 - 3r - 10 = 0$ **12.** $p^2 - 5p + 4 = 0$

13. $x^2 - 2x - 1 = 0$ **14.** $y^2 + 4y = 4$ **15.** $z^2 = 3z + 3$

16. $s^2 + 1 = -3s$ **17.** $t^2 = 3 - t$ **18.** $-x^2 - 6x = 1$

Sample Problem

$$2x^2 - 3x - 9 = 0$$

Divide each term by 2, the coefficient of x^2, and rewrite with constant in right-hand member.

$$x^2 - \frac{3}{2}x = \frac{9}{2}$$

Square one half the coefficient of x and add to each member.

$$x^2 - \frac{3}{2}x + \frac{9}{16} = \frac{9}{2} + \frac{9}{16}$$

Rewrite left-hand member as a perfect square.

$$\left(x - \frac{3}{4}\right)^2 = \frac{81}{16}$$

Extract square root of each member.

$$x - \frac{3}{4} = \pm\frac{9}{4}$$

Solve resulting first-degree equations.

$$x = \frac{3}{4} \pm \frac{9}{4}$$

$$x = \frac{3}{4} + \frac{9}{4}; \quad x = \frac{3}{4} - \frac{9}{4}$$

Ans. $x = 3; \quad x = -\frac{3}{2}$

19. $4x^2 + 4x - 3 = 0$ **20.** $4y^2 - 4y = 3$ **21.** $2x^2 = 2 - 3x$

22. $6z^2 + 6 = 13z$ **23.** $2t^2 - t - 15 = 0$ **24.** $1 - r = 6r^2$

9.3 QUADRATIC FORMULA

Solving the general quadratic equation

$$ax^2 + bx + c = 0, \quad a \neq 0$$

by completing the square, we can obtain a formula expressing the solutions of the equation in terms of the coefficients a, b, and c. We can then solve any quadratic equation by simply substituting the numerical coefficients of the terms in the formula and evaluating the result.

Completing the square in the general quadratic equation is accomplished as follows:

$$\begin{aligned}
ax^2 + bx + c &= 0 \\
ax^2 + bx &= -c \\
x^2 + \frac{b}{a}x &= \frac{-c}{a} \\
x^2 + \frac{b}{a}x + \frac{b^2}{4a^2} &= \frac{-c}{a} + \frac{b^2}{4a^2} \\
\left(x + \frac{b}{2a}\right)^2 &= \frac{-c(4a)}{a(4a)} + \frac{b^2}{4a^2} \\
\left(x + \frac{b}{2a}\right)^2 &= \frac{b^2 - 4ac}{4a^2} \\
x + \frac{b}{2a} &= \pm\sqrt{\frac{b^2 - 4ac}{4a^2}} \\
x &= \frac{-b}{2a} \pm \frac{\sqrt{b^2 - 4ac}}{2a} \\
\boldsymbol{x} &= \boldsymbol{\frac{-b \pm \sqrt{b^2 - 4ac}}{2a}}
\end{aligned}$$

The last equation is called the **quadratic formula.** Since this formula was developed from the quadratic equation in standard form, any quadratic equation should be written in standard form before attempting to determine values for a, b, and c to substitute in the formula. Furthermore, the sign on the coefficient must be substituted with the coefficient. For example, the equation

$$-x + 3x^2 = 4,$$

is first written as

$$3x^2 - x - 4 = 0,$$

from which

$$a = 3, \quad b = -1, \quad \text{and} \quad c = -4.$$

In the event that a quadratic equation has fractional coefficients, it is generally advantageous to clear the equation of fractions before proceeding further. For example, if

$$\frac{2}{3} - \frac{1}{2}x = -2x^2,$$

we have

$$(\overset{2}{\not{6}})\frac{2}{\not{3}} - (\overset{3}{\not{6}})\frac{1}{\not{2}}x = (6)(-2x^2),$$

or

$$4 - 3x = -12x^2,$$

from which

$$12x^2 - 3x + 4 = 0,$$

and

$$a = 12, \quad b = -3, \quad \text{and} \quad c = 4.$$

In actual practice, the quadratic formula should be used only when easier methods (factoring or extraction of roots) fail. Many of the following exercises are easier to solve by other methods, but they are included to indicate the complete generality of the quadratic formula. These exercises should be solved by use of the formula.

EXERCISES 9.3

■ *In Exercises 1–20, indicate the values for a, b, and c to be substituted in the quadratic formula.*

Sample Problems

a. $x^2 = x + 2$

$x^2 - x - 2 = 0$

Ans. $a = 1, b = -1, c = -2$

b. $2x^2 = x$

Write in standard form.

$2x^2 - x = 0$

Ans. $a = 2, b = -1, c = 0$

1. $x^2 - 3x + 2 = 0$

2. $y^2 + 5y + 4 = 0$

3. $x^2 - x - 30 = 0$

4. $y^2 + 3y - 4 = 0$

5. $x^2 - 2x = 0$

6. $y^2 = 5y$

7. $4y^2 - 3 = 0$

8. $2y^2 - 1 = 0$

9. $2x^2 = 7x - 6$

10. $6x^2 + x = 1$

11. $6x^2 = 5x - 1$

12. $3x^2 - 5 = 0$

13. $y^2 + 4 = 8y$

14. $x^2 = 7x$

Sample Problem

$$\frac{x}{3} = 4 - \frac{x^2}{2}$$

Multiply each term by L.C.D. 6.

$$\overset{2}{(\not{6})}\frac{x}{\not{3}} = (6)4 - \overset{3}{(\not{6})}\frac{x^2}{\not{2}}$$

$$2x = 24 - 3x^2$$

Write in standard form.

$$3x^2 + 2x - 24 = 0$$

Ans. $a = 3,\ b = 2,\ c = -24$

15. $x^2 = x + \frac{1}{2}$ **16.** $x^2 = \frac{15}{4} - x$ **17.** $2x^2 - 1 + \frac{7}{3}x = 0$

18. $y^2 + 1 = \frac{13}{6}y$ **19.** $\frac{9}{4}y^2 + \frac{3}{2}y - 2 = 0$ **20.** $\frac{x^2}{3} = \frac{x}{2} + \frac{3}{2}$

■ *Solve by use of the quadratic formula.*

Sample Problem

$$x^2 - x - 6 = 0$$

$$x = \frac{-b \pm \sqrt{b^2 - 4ac}}{2a}$$

Substitute 1 for a, -1 for b, and -6 for c.

$$x = \frac{-(-1) \pm \sqrt{(-1)^2 - 4(1)(-6)}}{2(1)}$$

Perform indicated operations.

$$x = \frac{1 \pm \sqrt{1 + 24}}{2}$$

Simplify.

$$x = \frac{1 \pm \sqrt{25}}{2}$$

$$x = \frac{1 + 5}{2};\quad x = \frac{1 - 5}{2}$$

Ans. $x = 3;\quad x = -2$

Check. $(3)^2 - (3) - 6 = 0$ $\quad$ $(-2)^2 - (-2) - 6 = 0$

$9 - 3 - 6 = 0$ $\quad$ $4 + 2 - 6 = 0$

$0 = 0$ $\quad$ $0 = 0$

21. $x^2 - 3x + 2 = 0$ **22.** $y^2 + 5y + 4 = 0$

23. $z^2 - 4z - 12 = 0$

24. $x^2 - x - 30 = 0$

25. $x^2 + 2x - 15 = 0$

26. $y^2 + 3y - 4 = 0$

27. $x^2 + 3x - 1 = 0$

28. $y^2 + 5y + 5 = 0$

29. $y^2 - 3y - 2 = 0$

30. $x^2 + x - 1 = 0$

Sample Problem

$$x^2 - 9 = 0$$

$$x = \frac{-b \pm \sqrt{b^2 - 4ac}}{2a}$$

Substitute 1 for a, 0 for b, and -9 for c.

$$x = \frac{-(0) \pm \sqrt{(0)^2 - 4(1)(-9)}}{2(1)}$$

Perform indicated operations.

$$x = \frac{\pm\sqrt{36}}{2}$$

Simplify.

$$x = \frac{+6}{2}; \quad x = \frac{-6}{2}$$

Ans. $x = 3; \quad x = -3$

31. $x^2 - 2x = 0$ (*Hint.* $c = 0$)

32. $x^2 - 4 = 0$ (*Hint.* $b = 0$)

33. $y^2 = 5y$

34. $z^2 = 9$

35. $7x = x^2$

36. $16 = y^2$

37. $z^2 - 3z = 0$

38. $x^2 = 1$

39. $4y^2 - 3 = 0$

40. $2y^2 - 1 = 0$

Sample Problem

$$2x^2 = 2 - 3x$$

Write in standard form.

$$2x^2 + 3x - 2 = 0$$

$$x = \frac{-b \pm \sqrt{b^2 - 4ac}}{2a}$$

Substitute 2 for a, 3 for b, and -2 for c.

$$x = \frac{-(3) \pm \sqrt{(3)^2 - 4(2)(-2)}}{2(2)}$$

Perform indicated operations.

$$x = \frac{-3 \pm \sqrt{9 + 16}}{4}$$

$$x = \frac{-3 \pm \sqrt{25}}{4}$$

(*Continued*)

Simplify.

$$x = \frac{-3+5}{4}; \quad x = \frac{-3-5}{4}$$

Ans. $x = \frac{1}{2}; \quad x = -2$

41. $2x^2 = 7x - 6$ **42.** $5 = 6y - y^2$ **43.** $6x^2 + x = 1$

44. $-z = 3 - 2z^2$ **45.** $6x^2 - 13x - 5 = 0$ **46.** $6x^2 = 5x - 1$

47. $x^2 = 2x + 1$ **48.** $x^2 = 2x + 4$ **49.** $y^2 - 4y - 2 = 0$

50. $z^2 + 4 = 8z$ **51.** $2x^2 - x - 1 = 0$ **52.** $3x^2 - 5 = 0$

Sample Problem

$$\frac{x^2}{3} = \frac{1}{3} - \frac{x}{2}$$

Multiply by L.C.D. 6.

$$\overset{2}{(\cancel{6})}\frac{x^2}{\cancel{3}} = \overset{2}{(\cancel{6})}\frac{1}{\cancel{3}} - \overset{3}{(\cancel{6})}\frac{x}{\cancel{2}}$$

$$2x^2 = 2 - 3x$$

Write in standard form.

$$2x^2 + 3x - 2 = 0$$

Solve as shown in preceding sample problem.

53. $x^2 = \frac{15}{4} - x$ **54.** $2x^2 - 1 + \frac{7}{3}x = 0$ **55.** $y^2 + 1 = \frac{13}{6}y$

56. $\frac{9}{4}y^2 + \frac{3}{2}y - 2 = 0$ **57.** $\frac{1}{3}x^2 = \frac{1}{2}x + \frac{3}{2}$ **58.** $\frac{3}{5}x^2 - x - \frac{2}{5} = 0$

9.4 GRAPHING QUADRATIC EQUATIONS IN TWO VARIABLES

In Chapter 6, you learned how to graph first-degree equations in two variables. The same procedure can be used to graph second-degree equations of the form

$$y = ax^2 + bx + c.$$

Since the graph of a second-degree equation of this form is not a straight line but a curve, called a **parabola** (see examples in Figure 9.1), we have to plot more than two points to determine the graph. The significant features of a parabola include the maximum or minimum point of the curve and the points, if any, where the curve intersects the axes. We shall therefore choose first components for the ordered pairs used to graph the equation so that these features are displayed. The ability to determine the number of ordered pairs necessary and to choose the proper first components is acquired through experience.

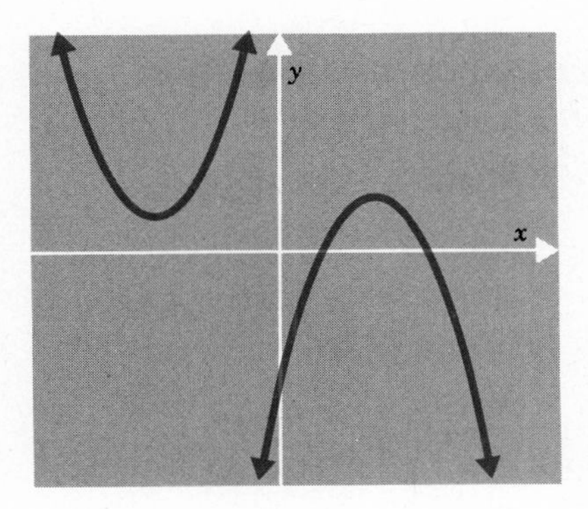

Figure 9.1

EXERCISES 9.4

■ *In Exercises 1-6, find a second component such that each of the ordered pairs satisfies the equation* $y = x^2 - 2x - 3$.

Sample Problem

$(-3, ?)$

Substitute -3 for x in $y = x^2 - 2x - 3$.

$y = (-3)^2 - 2(-3) - 3$

Simplify.

$y = 9 + 6 - 3$
$y = 12$

Ans. $(-3, 12)$

1. $(0, ?)$ **2.** $(1, ?)$ **3.** $(-1, ?)$ **4.** $(2, ?)$ **5.** $(-2, ?)$ **6.** $(3, ?)$

■ *In Exercises 7-12, find a second component such that each of the ordered pairs satisfies the equation* $y = x^2 + x - 2$.

7. $(0, ?)$ **8.** $(-1, ?)$ **9.** $(1, ?)$ **10.** $(2, ?)$ **11.** $(-2, ?)$ **12.** $(-3, ?)$

■ *In Exercises 13-18, find a second component such that each of the ordered pairs satisfies the equation* $y = x^2 - 7x + 12$.

13. (0, ?) **14.** (1, ?) **15.** (2, ?) **16.** (3, ?) **17.** (4, ?) **18.** (5, ?)

19. Graph the ordered pairs obtained in Exercises 1-6 and connect the points with a smooth curve.

20. Graph the ordered pairs obtained in Exercises 7-12 and connect the points with a smooth curve.

■ *In Exercises 21-28, graph each equation. For x components, use all integers between the given numbers.*

Sample Problem

$y = x^2 - 3x + 1$, $(-2 \text{ and } 5)$

Determine ordered pairs which are solutions to the equation.

$(-1,5)$, $(0,1)$, $(1,-1)$, $(2,-1)$, $(3,1)$, and $(4,5)$

Graph these ordered pairs and connect them with a smooth curve.

x	y
-1	5
0	1
1	-1
2	-1
3	1
4	5

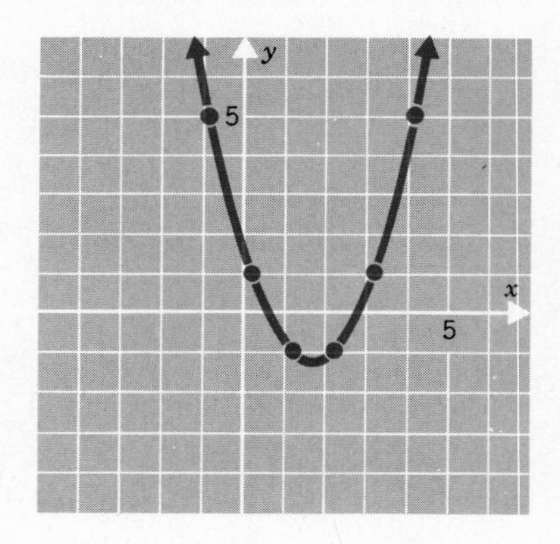

21. $y = x^2 - 2x$, $(-3 \text{ and } 5)$

22. $y = x^2 - 4$, $(-3 \text{ and } 3)$

23. $y = x^2 + 2x$, $(-5 \text{ and } 3)$

24. $y = x^2 + 1$, $(-3 \text{ and } 3)$

25. $9 - x^2 = y$, $(-4 \text{ and } 4)$

26. $3x - x^2 = y$, $(-2 \text{ and } 5)$

27. $y = x^2 - 5x + 4$, $(-2 \text{ and } 6)$

28. $y = x^2 + x - 6$, $(-4 \text{ and } 3)$

29. In each Exercise 21-24, estimate from the graph the values of x for which y is 0.

30. In each Exercise 25-28, estimate from the graph the values of x for which y is 0.

■ *Graph each equation.*

31. $y = x^2$ **32.** $y = x^2 + 3$ **33.** $y = x^2 - 3$

34. $y = 4 - x^2$ **35.** $x^2 - 2x - 3 = y$ **36.** $x^2 + 2x - 3 = y$

■ *Solve by graphing.*

Sample Problem

$x^2 - 2x - 3 = 0$

Write the equation in two variables $x^2 - 2x - 3 = y$ or $y = x^2 - 2x - 3$. Find appropriate ordered pairs which satisfy the equation:

$(-2,5)$, $(-1,0)$, $(0,-3)$, $(1,-4)$, $(2,-3)$, and $(3,0)$.

Complete the graph. The points at which the curve crosses the x-axis $(y = 0)$ correspond to values for x which satisfy $x^2 - 2x - 3 = 0$.

x	y
−2	5
−1	0
0	−3
1	−4
2	−3
3	0

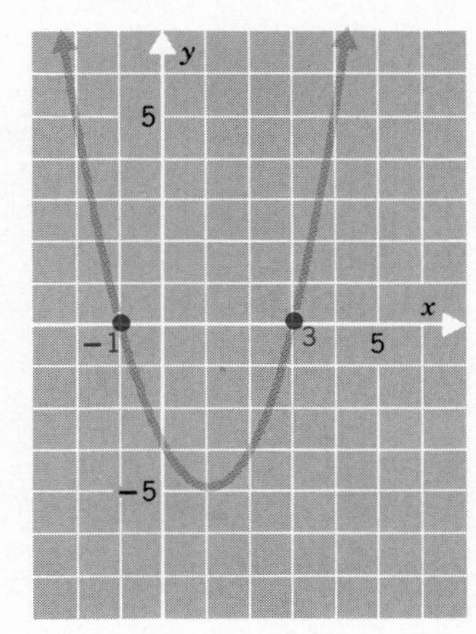

37. $x^2 - 4 = 0$ **38.** $x^2 + 3x - 4 = 0$ **39.** $x^2 + x - 12 = 0$

40. $x^2 - 2x = 0$ **41.** $x^2 - 2x + 1 - 0$ **42.** $x^2 - x - 6 = 0$

43. $2x^2 + 3x - 2 = 0$ **44.** $2x^2 - 5x + 3 = 0$ **45.** $2x^2 + x - 3 = 0$

46. $2x^2 = 3x$ **47.** $5x^2 + 3x = 2$ **48.** $5x^2 - x = 6$

9.5 THE PYTHAGOREAN THEOREM

A particularly useful application of quadratic equations is illustrated in the solution of problems involving the sides of a right triangle. The early Greeks proved that in any right triangle the sum of the squares of the lengths of the shorter sides, called **legs,** of a right triangle, is equal to the square of the length of the longest side, called the **hypotenuse.** Thus, in Figure 9.2,

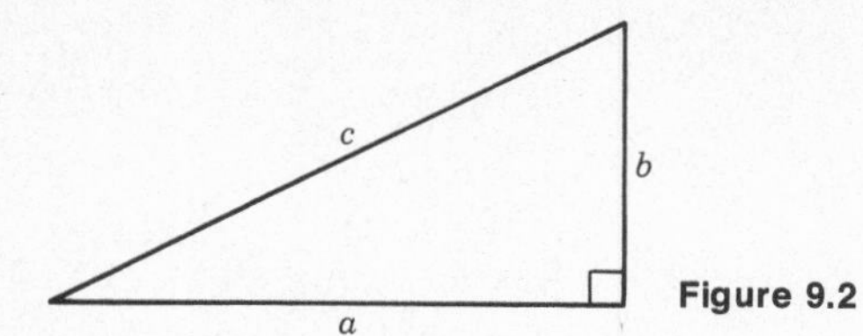

Figure 9.2

$$a^2 + b^2 = c^2.$$

This relationship is known as the Pythagorean theorem, in honor of the Greek mathematician Pythagoras.

If we apply this theorem to compute the length of the diagonal of a square with sides of length one unit, shown in Figure 9.3, we have the following:

$$c^2 = a^2 + b^2,$$
$$c^2 = (1)^2 + (1)^2,$$
$$c^2 = 2,$$
$$c = \pm\sqrt{2}.$$

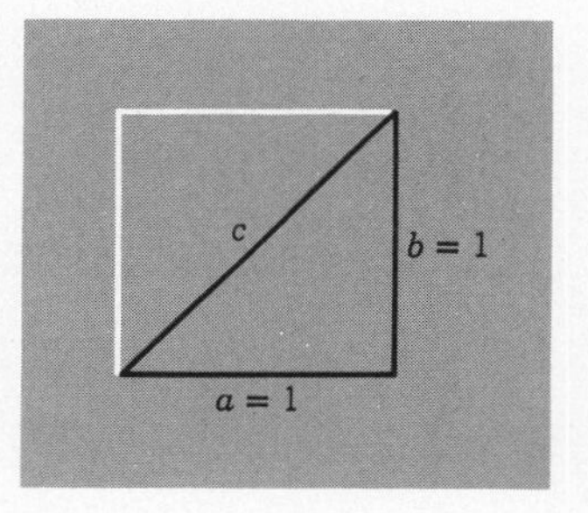

Figure 9.3

The number $-\sqrt{2}$, does not meet the physical conditions of length. Hence, the length of the diagonal is $\sqrt{2}$ units.

EXERCISES 9.5

■ *Solve.*

Sample Problem

Find the length of the diagonal of a rectangle whose length is 6 meters and whose width is 4 meters.

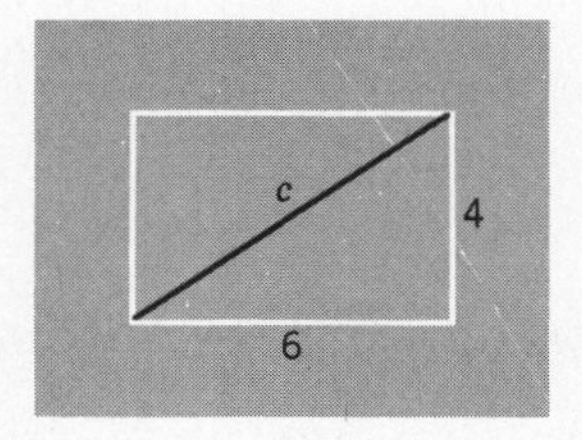

Sketch the rectangle.

Use the Pythagorean theorem.

$c^2 = a^2 + b^2$

Substitute 4 for a and 6 for b.

$c^2 = (4)^2 + (6)^2$
$c^2 = 16 + 36$
$c = \pm\sqrt{52}$

$c = \pm 2\sqrt{13}$ — $2\sqrt{13}$ is the only meaningful answer.

Ans. Length of diagonal is $2\sqrt{13}$ meters.

1. Find the length of the diagonal of a rectangle whose length is 4 centimeters and whose width is 3 centimeters.
2. Find the length of the diagonal of a rectangle whose length is 12 meters and whose width is 5 meters.
3. Find the length of the diagonal of a square whose side is 3 kilometers in length.
4. Find the length of the diagonal of a square whose side is 2 meters in length.
5. A baseball diamond is a square whose sides are 90 feet in length. Find the straight line distance from home plate to second base. (Use the table of square roots and find the length to the nearest foot.)
6. Find the length of the diagonal of a square whose side is a millimeters in length.

Sample Problem

Find the width of a rectangle whose length is 8 centimeters and whose diagonal is 10 centimeters long.

Sketch the rectangle.

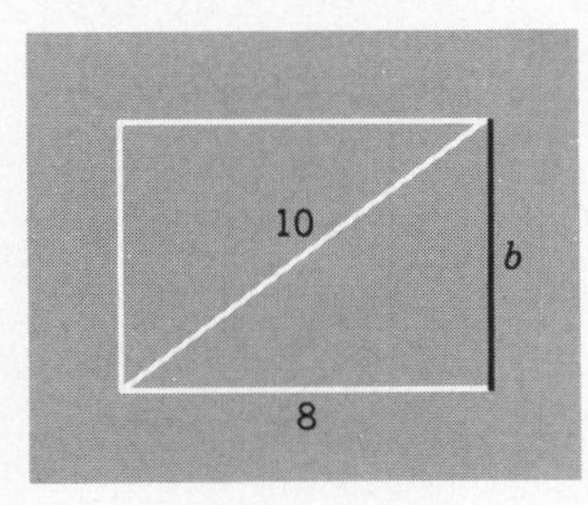

Substitute 8 for a a 10 for c in $a^2 + b^2 = c^2$.

$(8)^2 + b^2 = (10)^2$
$64 + b^2 = 100$
$b^2 = 100 - 64$
$b^2 = 36$
$b = \pm 6$

6 is the only meaningful answer.

Ans. Width is 6 centimeters.

7. Find the length of a rectangle whose width is 5 inches and whose diagonal is 13 inches long.
8. Find the length of a rectangle whos diagonal is 20 meters long and whose width is 12 meters.
9. Find the width of a rectangle whose diagonal is 11 yards long and whose length is 9 yards.
10. Find the width of a rectangle whose length is 5 feet and whose diagonal is 6 feet long.

Sample Problem

The length of a rectangle is 2 meters greater than the width. The diagonal is 10 meters long. Find dimensions of the rectangle.

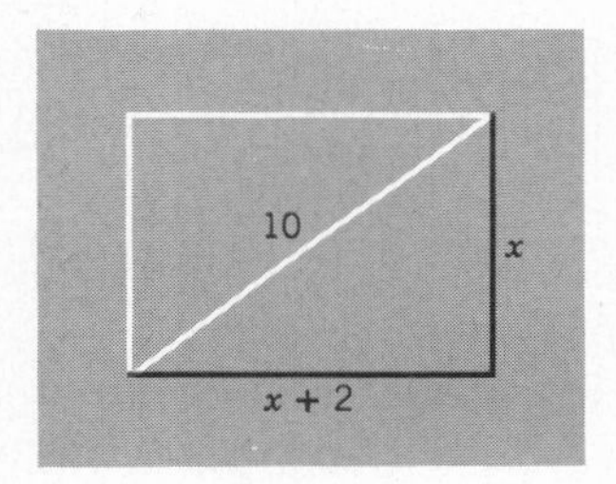

Sketch the rectangle.

Substitute x for a, $x + 2$ for b, and 10 for c in $a^2 + b^2 = c^2$.

$$(x)^2 + (x + 2)^2 = (10)^2$$

Solve resulting equation.

$$x^2 + x^2 + 4x + 4 = 100$$
$$2x^2 + 4x - 96 = 0$$
$$2(x - 6)(x + 8) = 0$$
$$x - 6 = 0 \qquad x + 8 = 0$$
$$x = 6 \qquad x = -8$$

6 is the only meaningful solution.

Ans. Width = 6 meters,
length = 8 meters.

11. The length of a rectangle is 3 meters greater than the width, and the diagonal is 15 meters in length. Find the dimensions of the rectangle.
12. The width of a rectangle is 7 centimeters less than the length, and the diagonal is 13 centimeters in length. Find the dimensions of the rectangle.
13. The width of a rectangle is 3 millimeters less than the length, and the square of the length of the diagonal is 29 millimeters. Find the dimensions of the rectangle.

14. The length of a rectangle is 5 inches greater than the width, and the square of the length of the diagonal is 73 inches. Find the dimensions of the rectangle.

15. The length of a rectangle is twice the width, and the diagonal is $3\sqrt{5}$ yards in length. Find the dimensions of the rectangle.

16. The length of a rectangle is three times the width, and the diagonal is $2\sqrt{10}$ feet in length. Find the dimensions of the rectangle.

Sample Problem

A 25-foot ladder is placed against a wall so that its foot is 7 feet from the foot of the wall. How far up the wall does the ladder extend?

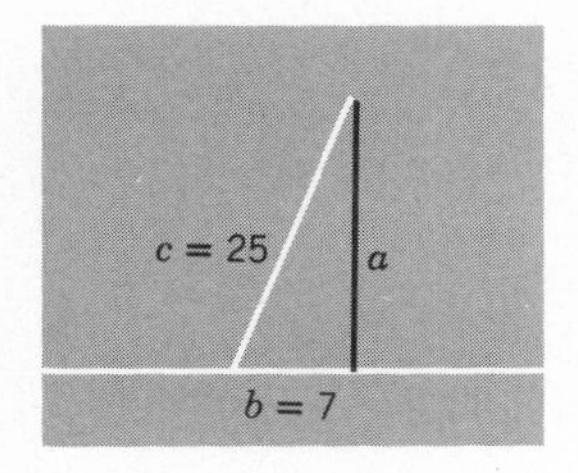

Sketch figure.

Substitute 7 for b and 25 for c in $a^2 + b^2 = c^2$.

$$a^2 + (7)^2 = (25)^2$$

Simplify.

$$a^2 + 49 = 625$$
$$a^2 = 576$$

Solve by extraction of roots.

$$a = \pm\sqrt{576}$$
$$a = \pm 24$$

24 is the only meaningful solution.

Ans. The ladder extends 24 feet up the wall.

17. How long must a wire be to stretch from the top of a 40-meter telephone pole to a point on the ground 30 meters from the foot of the pole?

18. How high on a building will a 25-foot ladder reach if its foot is 15 feet from the wall against which the ladder is to be placed?

19. If a 30-meter pine tree casts a shadow of 30 meters, how far is it from the tip of the shadow to the top of the tree?

20. One leg of a right triangle is 6 centimeters longer than the other, and the hypotenuse has a length of 6 centimeters less than twice that of the shorter leg. Find the lengths of the sides of the right triangle.

CHAPTER REVIEW

■ *Solve for x, y, or z by any method.*

1. *a.* $x^2 - 25 = 0$ *b.* $3x^2 - 27 = 0$

2. *a.* $6x^2 - 42 = 0$ *b.* $\frac{2y^2}{3} = 4$

3. *a.* $(z - 2)^2 = 9$ *b.* $(z + 3)^2 = 1$

4. *a.* $(x - 7)^2 = 16$ *b.* $(x - a)^2 = c^2$

5. *a.* $(x + 3)^2 = a$ *b.* $(y + a)^2 = 4$

6. *a.* $x^2 + 3x - 4 = 0$ *b.* $y^2 = 3y + 3$

7. *a.* $2x^2 + 4x + 2 = 0$ *b.* $y^2 - y = 2$

8. Solve by completing the square: $y^2 - 4y = 5$.

■ *In Exercises 9-11, solve each equation using the quadratic formula.*

9. $\frac{x^2}{4} = \frac{15}{4} - \frac{x}{2}$ **10.** $x^2 + 3x + 1 = 0$ **11.** $\frac{x^2}{4} + 1 = \frac{13}{12}x$

12. Graph: $y = x^2 + 3x$.

13. Graph: $y = x^2 + 3x - 4$.

14. Solve by graphing: $x^2 - x - 6 = 0$.

15. If $x^2 + y^2 = z^2$, and $x = 12$, $z = 20$, find y.

16. If $a^2 + b^2 = c^2$, and $a = 6$, $b = 6$, find c.

17. Find the length of the diagonal of a rectangle whose width is 7 meters and whose length is 9 meters.

18. Find the length of the diagonal of a square with sides 8 centimeters long.

19. A cable is to be stretched from a point 18 meters up a pole to a point on the ground 24 meters from the base of the pole. If the pole is mounted on level ground, what is the length of the stretched cable?

20. The length of a rectangle is 3 times the width and the diagonal is $5\sqrt{10}$ meters in length. Find the dimensions of the rectangle.

Cumulative Review

1. Write $24x^3y^2$ in completely factored form.

2. If $x = -2$ and $y = -4$, find the value of $\frac{x^2 + 2x - 4}{y}$.

3. For what value of x is $\frac{3x - 2}{x - 1}$ meaningless?

4. Simplify: $\frac{6a^3b^2}{3ab^2}$.

5. Solve for x: $3x - a = \frac{2ax - a^2}{a}$.

6. Solve the system: $3x + 4y = 11$

$2x - y = 0.$

7. Represent $\frac{3}{x + y} + \frac{4}{x^2 + xy}$ as a single fraction.

8. Find three ordered pairs that are solutions of $2x - y = 4$.

9. Simplify: $3x\sqrt{x^2y} + 2\sqrt{x^4y}$.

10. The sum of two numbers is a. If one of the numbers is x, the other is __?__.

11. How many solutions has the equation $2x^2 - 8 = 0$?

12. What conclusions can be drawn concerning a and b if $ab = 0$?

13. The sum of two numbers is 18. If one of the numbers is -9, the other is __?__.

14. One number is four times another. If their sum is 40, find the numbers.

15. The perimeter of a triangle is 30 inches. If the second side is 2 inches longer than the first, and the first is two thirds of the third, find the length of each side.

16. The first angle of a triangle is one-half of the second, and the third is equal to the sum of the other two, find the size of each angle.

17. The area of a triangle is 18 square centimeters. If the length of the base is four times the length of the altitude, find the base and altitude of the triangle.

18. The sum of two numbers is 16, and their product is 63. Find the numbers.

19. The length of a rectangle is twice the width; and its area is 242 square centimeters. Find its dimensions.

20. The base of a triangle is three times its altitude. If the triangle contains the same area as a rectangle whose length and width are 4 centimeters and 6 centimeters respectively, find the base and altitude of the triangle.

FINAL CUMULATIVE REVIEWS

REVIEW I

1. List all prime numbers between 70 and 100.
2. If a and b are natural numbers, which of the following statements are true for all values of a and b?

   $a + b = b + a$; $a - b = b - a$; $ab = ba$; $\frac{a}{b} = \frac{b}{a}$

3. Arrange the numbers 6, -2, -4, 5, 3, -1 in order, from smallest to largest.
4. Write an equation expressing the word sentence: "The volume (V) of a cylindrical silo is equal to π times the square of the radius (r) times the height (h)."
5. If a 30-meter rope is cut into two pieces and x represents one of the pieces, how could the second piece be represented in terms of x?
6. Simplify: $2a(a - 1) + a(a + 2) - a^2$.
7. Divide $(6x^2 + 3x - 1)$ by $3x$.
8. Side b of a triangle is 2 centimeters longer than a second side a. Side c is twice as long as side a. How long is each side of the triangle if its perimeter is 34 centimeters?
9. Simplify: $\frac{2ab - 4b^2}{a^2 - 4b^2}$.

10. Represent $\frac{3}{4} + \frac{3}{4a}$ using a single fraction.

11. Simplify: $\frac{3xy^2}{4a} \cdot \frac{2xa^2}{5y}$.

12. Solve for x: $\frac{2a}{b} = \frac{6}{4 - x}$.

13. Graph the following ordered pairs on a rectangular coordinate system. $(2, 6); (3, -4); (-5, -5); (-5, 0)$.

14. Solve the system: $y = x$
$3x + 2y = 5$.

15. Where does the graph of $3x + 4y = 12$ cross the x-axis?

16. Simplify: $\frac{\sqrt{4^4 y^3}}{8^2 y}$.

17. For what value(s) of y will the fraction $\frac{5(y - 2)}{y + 3}$ equal 0?

18. An approximate value of $\sqrt{3}$ is 1.73. Find an approximation for $\sqrt{12}$.

19. Simplify: $(\sqrt{2} - \sqrt{5})(\sqrt{2} + \sqrt{5})$.

20. Find two consecutive positive odd integers whose product is 195.

REVIEW II

1. Simplify: $\frac{5^2 + 5}{5} - \frac{3^2 + 3}{3}$.

2. Division by __?__ is meaningless.

3. State whether $-2a$ is a solution of $x - a = 4x + 5a$.

4. The product of two numbers is 24. If one of the numbers is b, represent the other number in terms of b.

5. Simplify: $(a + b) - (2a - b) + (a + 3b)$.

6. Solve: $7(a + 3) = a - 9$.

7. Two trucks deliver 47 tons of gravel to a building site. The larger truck carries 5 tons more than the smaller, and makes only 1 trip while the smaller truck makes 5 trips. What is the capacity of each truck?

8. Represent $\frac{a}{b} + \frac{b}{a}$ using a single fraction.

9. Express $\frac{3}{x + 3}$ as an equivalent fraction with denominator $x^2 - 9$.

10. Simplify: $\frac{2x}{3y} \div \frac{x^2}{y}$.

11. A statement that two ratios are equal is called a __?__.

12. Graph $2x + 3y = 6$.

13. Which equation has a straight line for a graph?
$y = x^2 - 1$; $x + y = 1$; $x = y^2 - 1$.

14. Solve the system: $2y - x = 7$
$y + 2x = 1$.

15. Solve: $(x - 3)^2 = 25$.

16. Which of the following numbers are irrational?
$\sqrt{4}$, $\sqrt{5}$, $\sqrt{6}$, $\sqrt{7}$, $\sqrt{8}$, $\sqrt{9}$, $\sqrt{10}$.

17. Simplify: $\sqrt{160x^3y^3z}$.

18. Simplify: $\frac{8 - \sqrt{32}}{8}$.

19. Solve: $x^2 + 5x + 2 = 0$.

20. If y varies directly as x, and $y = 12$ when $x = 5$, find x when $y = 20$.

REVIEW III

1. Express 360 in completely factored form.
2. If $a = -2$, $b = 1$, $c = 3$, find the value of $a^2bc - abc^2$.
3. Simplify: $a(a + 1) - 2(a^2 + a) - a$.
4. State whether -2 is a solution of $4x + 4 = 2x - 4$.
5. What is the area of a square whose perimeter is 100 millimeters?
6. Factor completely: $abc - ab$.
7. Factor completely: $6x^2 - 3x - 9$.
8. In a collection of coins, there are 6 more dimes than nickels. If n represents the number of nickels, represent the total value of the collection (in cents) in terms of n.
9. Represent $\frac{a - 3b}{5} - \frac{a + 3b}{5}$ using a single fraction.
10. Represent $\frac{1}{2x} + \frac{1}{3x}$ using a single fraction.
11. Divide $(x^2 - 6x - 16)$ by $(x + 2)$.
12. Simplify: $\frac{ab}{a^2 - b^2} \div \frac{ab}{2a - 2b}$.

13. The denominator of a certain fraction is four more than the numerator and the fraction is equivalent to $5/6$. Find the numerator of the fraction.

14. If 228 bricks are required for 4 meters of a wall, how many bricks will be required for 10 meters?

15. First-degree equations are called __?__ equations because their graphs are straight lines.

16. Solve for y in terms of x: $x = \frac{3y - 2}{4}$.

17. What value would be substituted for b in the quadratic formula when solving the equation $x^2 + x - 5 = 0$?

18. Simplify: $\sqrt{36a^6b^4c^5}$.

19. Simplify: $4\sqrt{a} + \sqrt{4a} - \sqrt{9a}$.

20. How far from the root of a vertical pole will a 14-meter wire reach if the other end of the wire is tied to the pole at a height of 6 meters?

REVIEW IV

1. Graph the first ten prime numbers on a line graph.

2. If r and s represent two numbers: (a) What is their sum? (b) What is their product?

3. If $4a$ and $-2b$ are two factors of $24ab$, what is the third factor?

4. If $a = 3$, $b = 2$, $c = -1$, find the value of $\frac{ab}{c^2} - \frac{a + c}{b}$.

5. Find three consecutive even integers whose sum is -78.

6. Simplify: $(x - b)(x - 2b) - (x^2 + 2b^2)$.

7. Factor completely: $2x^2 - 8$.

8. At a baseball game 260 tickets were sold. Adults paid 80 cents for their tickets, and children paid 30 cents each. If the total receipts for the game were $160, how many tickets of each kind were sold?

9. Simplify: $\frac{2x - 2}{2}$.

10. Represent $\frac{3}{x + 2} - \frac{1}{(x + 2)^2}$ using a single fraction.

11. Divide $12x^2$ by $\frac{3x}{4}$.

12. Solve: $\frac{13}{x} = 1 + \frac{4}{x}$.

13. If $1/4$ inch on a map represents 8 miles, how many miles does 6 inches represent?

14. Only __?__ points are needed to determine the graph of a first degree equation.

15. What is the value of s when r equals 6 if $2r + 6s = 20$?

16. Supply the missing components so that the ordered pairs (0,), (2,), and (−3,) satisfy the equation $3x - y = 6$.

17. Divide $\frac{a^2\sqrt{3}}{4}$ by $\frac{3a}{\sqrt{3}}$.

18. Solve: $8r(r - 1) = -6r$.

19. Where does the graph of $y = x^2 + x - 30$ intersect the x-axis?

20. If the perimeter of a rectangle is 56 centimeters and its area is 192 square centimeters, what are the dimensions of the rectangle?

REVIEW V

1. Simplify: $\frac{6 + 2^2}{5} - \frac{3^3 - 5^2}{2}$.

2. The signed whole numbers together with zero are called __?__.

3. Solve: $\frac{3x}{2} - 9 = 6$.

4. Factor: $y^2 - 12y + 11$.

5. Simplify: $(x + 3b)(x - 3b) - (x^2 - b^2)$.

6. The length of a rectangle is 8 feet more than its width w. Represent two-thirds of the length in terms of w.

7. Represent $-\frac{1 - a}{3}$ as a positive fraction with a positive denominator.

8. Divide $(x^3 - 2x^2 + x)$ by x.

9. Divide $\frac{2\pi rh}{6 + h}$ by $\frac{2\pi r}{h}$.

10. A plane travels 875 miles in t hours. Represent the distance the plane can travel in 1 hour in terms of t.

11. Solve: $\frac{x - 9}{2} = \frac{x + 12}{9}$.

12. In a proportion, the product of the __?__ equals the product of the __?__.

13. Graph the set of all points for which $x = 5$ in a rectangular coordinate system.
14. Solve the system: $y = 2x + 5$
$$x - 3y = -20.$$
15. For what value(s) of a will the product $(a + 3)(a - 7)$ equal zero?
16. Graph $-\sqrt{4}$, $\sqrt{7}$, $\sqrt{21}$, $\sqrt{29}$ on a line graph.
17. Simplify the product of $\sqrt{3}$ and $\sqrt{27}$.
18. Solve: $x^2 + 5x + 2 = 0$.
19. Solve: $x^2 - 2 + \frac{7}{3}x = 0$.
20. If y varies directly as x, and $y = 6$ when $x = 24$, find y when $x = 30$.

REVIEW VI

1. If $x = -1$, $y = -2$, $z = -3$, find the value of $x^3 - y^3 + z^3$.
2. Simplify: $\frac{3^3 - 2^2 - 1^2}{2}$.
3. Represent $\frac{3x^2 - x^2}{x} - \frac{x^3}{x^2}$ using a single fraction.
4. The temperature drops 23° from a reading of 6°. What is the new temperature?
5. Simplify: $\frac{a^2 - 3a - 4}{a^2 - 1} \cdot \frac{a + 2}{a - 4}$.
6. At a recent election, the winning candidate received 62 votes more than his opponent. If there were 3626 votes cast in all, how many did each candidate receive?
7. Simplify: $3(y + 3)^2 - (18y + 27)$.
8. Factor completely: $a^2 - 25b^2$.
9. One typist averages 8 words-per-minute fewer than a second typist. If the rate of the second typist is r words per minute, represent the output of the first typist in 5 minutes in terms of r.
10. Find the lowest common denominator of $\frac{1}{2}$, $\frac{1}{3}$, $\frac{1}{4}$, $\frac{1}{5}$, and $\frac{1}{6}$.
11. Simplify: $\dfrac{3 + \frac{2}{3}}{1 - \frac{1}{3}}$.

12. Represent $\frac{x-1}{6} - \frac{2x+5}{3}$ using a single fraction.

13. Solve: $\frac{7}{8} = \frac{21}{y+2}$.

14. Solve: $\frac{3}{5}x + \frac{3}{10} = x - \frac{1}{2}$.

15. Solve for x: $\frac{a+c}{x} = \frac{2}{b}$.

16. An approximate value for $\sqrt{2}$ is 1.41, find an approximation for $\sqrt{72}$.

17. Solve the system $5x - 3y = -1$ using graphical methods.

$$3x + 3y = 9$$

18. Solve: $3w(2w+3) = 0$.

19. Solve: $(x-1)(x+3) = 1$.

20. Given $a^2 + b^2 = c^2$. Find b, if $a = 5$ and $c = 13$.

REVIEW VII

1. What is the numerical coefficient of $-x^3$?

2. Solve: $\frac{x^2}{2} - x = \frac{5}{2}$.

3. Simplify: $(a + b - 2c) - (3a - b + 2c) + (2a + 2b - c)$.

4. Which of the following equations are true for all values of x?

   $x^3 \cdot x^2 = x^5$; $\quad x^3 \cdot x^2 = x^6$; $\quad 3x^2 = 9x^2$

5. Factor completely: $4y^2 + 16y + 15$.

6. A 22-meter cable is divided into 2 parts. If y represents the longer piece, represent six times the shorter piece in terms of y.

7. Represent $\frac{4}{3y} + \frac{7}{3y} - \frac{2}{3y}$ using a single fraction.

8. Solve for x: $\frac{a}{x} = \frac{b}{2c}$.

9. Simplify: $\dfrac{1 - \frac{a}{b}}{1 + \frac{a}{b}}$.

10. A car travels 90 miles in the same time that a slower car travels 60 miles. If the first car goes 10 miles per hour faster than the second, find the rate of each.

11. Find second components for the ordered pairs $(3,\quad)$, $(-2,\quad)$, $(0,\quad)$, and $(6,\quad)$ so that each ordered pair satisfies $y = 3x + 4$.

12. Solve the system $2x + 3y = -5$ by graphical methods.
$$2y + 8 = x$$

13. Solve the system $x - 2y = 7$ by algebraic methods.
$$2x + y = 4$$

14. Solve: $14x^2 = 28x$.

15. Solve: $\frac{4}{x-2} - \frac{7}{x-3} = \frac{2}{15}$.

16. What term must be added to the expression $x^2 - 8x$ in order to make the expression a perfect square?

17. Simplify: $\frac{2 + 2\sqrt{2}}{2}$.

18. Simplify: $\frac{\sqrt{12a^2b^3}}{ab}$.

19. Solve for x: $b^2x^2 - c = 0$.

20. A man rowed 9 miles downstream and back again in 6 hours. The rate of the current was 2 miles per hour. Find the rate of the boat in still water.

REVIEW VIII

1. Simplify: $\frac{-3^2 + (-3)^2 - 2^3}{3}$.

2. For what value of x is the expression $\frac{x+2}{x-5}$ meaningless?

3. If n is an odd integer, represent the next three odd integers in terms of n.

4. Factor completely: $a^2 - 6ab + 8b^2$.

5. Simplify: $(x - 2y)^2 - (x^2 + 4y^2)$.

6. Where should a 64-meter cable be cut so that twice the length of the longer piece equals five times that of the shorter?

7. Simplify: $\frac{x^2 - x - 20}{x^2 - 7x + 10} \cdot \frac{x^2 + 9x + 18}{x^2 + 7x + 12}$.

8. Divide $\frac{ab}{a^2 - b^2}$ by $\frac{ab}{2a - 2b}$.

9. Simplify: $\dfrac{\dfrac{1}{a} - \dfrac{1}{b}}{\dfrac{1}{ab}}$.

10. Divide $(x^2 + 3x - 7)$ by $(x - 5)$.

11. Solve the system $\begin{aligned} 3y - x - 1 &= 0 \\ y + 6x + 6 &= 0 \end{aligned}$ algebraically.

12. Solve the system $\begin{aligned} y - 2x - 5 &= 0 \\ x - 3y + 20 &= 0 \end{aligned}$ by graphical methods.

13. Two packages weighed together total 146 kilograms. One of the packages weighs 12 kilograms more than the other. How much does each package weigh?

14. Solve: $x^2 - 5x - 14 = 0$.

15. Simplify: $\sqrt{200x^3y^2}$.

16. Simplify: $\sqrt{125} + 2\sqrt{5}$.

17. Graph: $y = x^2 + x - 6$.

18. Simplify: $5\sqrt{\dfrac{2}{5}}$.

19. Solve: $2x^2 - 2x = 7$.

20. The length of a rectangle is 4 centimeters less than twice its width, and the area is 240 square centimeters. Find the dimensions of the rectangle.

REVIEW IX

1. Simplify: $\dfrac{4^2 + 2}{2} - \dfrac{3^3 - 7}{4}$.

2. If $a = 1$, $b = -2$, $c = 2$, find the value of $abc - b^2 + c^2$.

3. Simplify: $2a(a - b) - b(a + b) - (a^2 - b^2)$.

4. Find three consecutive even integers whose sum is 78.

5. Factor completely: $2x^2 - 24x + 22$.

6. The difference of two numbers is 28. If n represents the smaller number, represent the larger in terms of n.

7. Solve: $2(a - 1) = a + 3$.

8. Represent $\dfrac{2}{x} - \dfrac{3}{y}$ using a single fraction.

Divide $\frac{3b}{4a}$ by $\frac{12b^2}{a}$.

. Divide $(2x^2 + 3x - 1)$ by $(x + 2)$.

. Solve for x: $\frac{2b - 2a}{x} = 2c$.

2. Graph: $x - 3y = 6$.

13. Graph: $y = -2x^2 + 3$.

14. Solve the system $3y - 5x = 1$ by algebraic methods.
$$x + y = 3$$

15. The value of a collection of coins is \$3.15. There are three more dimes than nickels and two more quarters than dimes. How many of each kind of coin is there in the collection?

16. Simplify the product of $\sqrt{6x}$ and $\sqrt{15x}$.

17. Simplify: $\frac{2 + 3\sqrt{12}}{2}$.

18. An approximate value for $\sqrt{2}$ is 1.41. Find an approximation for $\sqrt{18}$.

19. A plane flies directly east for 12 kilometers, turns and flies south for 16 kilometers. How far is the plane from its starting point?

20. Where does the graph of $y = 3x^2 + 4x - 1$ intersect the y-axis?

REVIEW X

1. Graph the prime numbers between 20 and 40 on a line graph.
2. Express $240x^2y$ in completely factored form.
3. Arrange the numbers $5, -2, 3, -5, 1, 7, -1$ in order, from smallest to largest.
4. Simplify: $(a - b) - 2(a + b) - (a + 2b)$.
5. Simplify: $(a - b)(a + b) - (a - b)^2$.
6. Factor completely: $3x^2 - 18xy + 24y^2$.
7. Simplify: $\frac{2a}{6a^2 + 8a}$.
8. Represent $\frac{a + 3b}{7} - \frac{a + 2b}{7}$ using a single fraction.
9. Represent $\frac{1}{a - 3} - \frac{2}{(a - 3)^2}$ using a single fraction.

10. Solve: $\frac{3}{5}x = x - \frac{4}{5}$.

11. Graph: $y = x$.

12. Solve the system $\begin{aligned} 5x + 2y - 8 &= 0 \\ 3x - 7y - 13 &= 0 \end{aligned}$ by algebraic methods.

13. Simplify: $3\sqrt{3} - \sqrt{12} + \sqrt{75}$.

14. Simplify: $\frac{\sqrt{18} - \sqrt{27}}{3}$.

15. Where does the graph of $y = x^2 + 4x + 3$ intersect the x-axis?

16. Find two consecutive integers such that twice the second less one half the first is 14.

17. In a right triangle, if two thirds of one acute angle is added to one half of the second acute angle, the result is 50°. Find each angle.

18. A man gave one third of his money to one son, one fourth to another son, and had $250 left. How much did he have to start with?

19. A furniture dealer sold a desk and a chair for $640. If the desk sold for $40 more than four times the chair, what was the price of each?

20. The hypotenuse of a right triangle is 20 meters long. If one of the remaining sides is 16 meters long, how long is the third side?

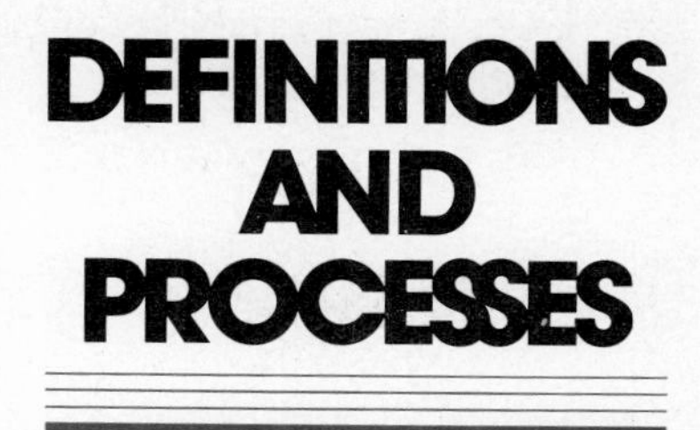

CHAPTER 1

[1.2] ■ The order in which the addends of a sum are considered does not change the sum.

$$a + b = b + a$$

■ The way in which the addends of a sum of three terms are grouped for addition does not change the sum.

$$(a + b) + c = a + (b + c)$$

■ $a - b$ is the number d, so that $b + d = a$.

■ The order in which factors are multiplied has no effect on the product.

$$ab = ba$$

■ The way in which factors are grouped has no effect on the product.

$$(ab)c = a(bc)$$

■ $\frac{a}{b}$ is the number q, so that $b \cdot q = a$.

■ Division by 0 is undefined.

[1.3] ■ The symbol a^n denotes the product of n factors a.

$$a^n = aaa \cdot \cdots \cdot a \text{ (n factors)}$$

[1.7] ■ To add like terms, add their numerical coefficients.

■ Distributive law.

$$ba + ca = (b + c)a$$

[1.8] ■ To subtract like terms, subtract their numerical coefficients.

[1.9] ■ The product of two powers with the same base equals a power with the same base and with an exponent equal to the sum of the exponents of the powers.

$$a^m \cdot a^n = a^{m+n}$$

[1.10] ■ The quotient of two powers with the same base equals a power with the same base with an exponent equal to the difference of the exponents of the powers.

$$\frac{a^m}{a^n} = a^{m-n}$$

CHAPTER 2

[2.1] ■ The number whose graph lies on the left is less than the number whose graph lies on the right.

[2.2] ■ To add two numbers with like signs:
Add the absolute values of the numbers and prefix the common sign of the numbers to the sum.
With unlike signs:
Find the nonnegative difference of the absolute values of the numbers and prefix the sign of the number with the larger absolute value to the difference.
The sum of any integer a and 0 equals a.

■ In expressions involving only addition, parentheses that are preceded by a (+) sign may be dropped; each term within the parentheses retains its original sign.

[2.4] ■ To subtract an integer b from an integer a, change the sign of b and add the two according to the rules for adding integers.

$$a - b = a + (-b)$$

■ In expressions involving only addition and subtraction, parentheses preceded by a $(-)$ sign may be dropped, provided that the sign of each term inside the parentheses is changed.

[2.6] ■ To find the product of two integers, multiply the absolute values of the integers. If the factors have like signs, the product is positive; if they have unlike signs, the product is negative. If one of the factors is 0 the product is zero.

[2.7] ■ To find the quotient of two signed numbers, find the quotient of the absolute values of the numbers. If the dividend and divisor have like signs, the quotient is positive; if they have unlike signs, the quotient is negative.

CHAPTER 3

[3.3] ■ If the same quantity is added to or subtracted from equal quantities, the resulting quantities are equal.

$$\text{If } a = b, \quad \text{then} \quad a + c = b + c.$$

■ Symmetric property of equality.

$$\text{If } a = b, \quad \text{then} \quad b = a.$$

[3.4] ■ If equal quantities are divided by the same (nonzero) quantity, the quotients are equal.

$$\text{If } a = b, \quad \text{then} \quad \frac{a}{c} = \frac{b}{c}, \quad c \neq 0.$$

[3.5] ■ If equal quantities are multiplied by the same quantity, their products are equal.

$$\text{If } a = b, \quad \text{then} \quad a \cdot c = b \cdot c.$$

CHAPTER 4

[4.1] ■ Distributive law

$$a(b+c) = ab + ac \quad \text{or} \quad (b+c)a = ba + ca$$

[4.7] ■ The difference of two squares, $a^2 - b^2$, equals the product of the sum $(a+b)$ and the difference $(a-b)$.

$$a^2 - b^2 = (a+b)(a-b)$$

CHAPTER 5

[5.1] ■ Any two of the three signs of a fraction may be changed without changing the value of the fraction.

$$\frac{a}{b} = \frac{-a}{-b} = -\frac{-a}{b} = -\frac{a}{-b} \quad \text{and} \quad \frac{-a}{b} = \frac{a}{-b} = -\frac{a}{b} = -\frac{-a}{-b}$$

[5.2] ■ If both the numerator and the denominator of a given fraction are divided by the same nonzero number, the resulting fraction is equivalent to the given fraction.

$$\frac{a}{b} = \frac{\frac{a}{c}}{\frac{b}{c}}, \quad c \neq 0$$

[5.5] ■ If both the numerator and denominator of a given fraction are multiplied by the same nonzero number, the resulting fraction is equivalent to the given fraction.

$$\frac{a}{b} = \frac{a}{b}\left(\frac{c}{c}\right) = \frac{a \cdot c}{b \cdot c}, \quad c \neq 0$$

[5.6] ■ The sum of two or more fractions with common denominators is a fraction with the same denominator and a numerator equal to the sum of the numerators of the original fractions.

$$\frac{a}{c} + \frac{b}{c} = \frac{a+b}{c}$$

[5.8] ■ The product of two fractions is a fraction whose numerator is the product of the numerators and whose denominator is the product of the denominators of the given fractions.

$$\frac{a}{b} \cdot \frac{c}{d} = \frac{ac}{bd}$$

[5.9] ■ The quotient of two fractions equals the product of the dividend times the reciprocal of the divisor.

$$\frac{a}{b} \div \frac{c}{d} = \frac{a}{b} \cdot \frac{d}{c}$$

[5.13] ■ In any proportion, the product of the extremes is equal to the product of the means.

$$\text{If } \frac{a}{b} = \frac{c}{d}, \quad \text{then} \quad ad = bc$$

CHAPTER 7

[7.1] ■ If the product of two or more factors is 0, at least one of the factors is 0.

$$ab = 0 \quad \text{if} \quad a = 0 \quad \text{or} \quad b = 0 \quad \text{or} \quad a \text{ and } b = 0$$

CHAPTER 8

[8.3] ■ The square root of a product is equal to the product of the square roots of its factors.

$$\sqrt{ab} = \sqrt{a}\sqrt{b}$$

[8.6] ■ The square root of a quotient is equal to the quotient of the square root of the numerator divided by the square root of the denominator.

$$\sqrt{\frac{a}{b}} = \frac{\sqrt{a}}{\sqrt{b}}, \quad b \neq 0$$

CHAPTER 9

[9.3] ■ Quadratic formula.

$$x = \frac{-b \pm \sqrt{b^2 - 4ac}}{2a}$$

GLOSSARY

Abscissa

The first component in an ordered pair. The distance of a point from the vertical axis in rectangular coordinates; to the right of the axis if the component is positive; to the left of the axis if the component is negative.

Absolute Value

$|x| = \begin{cases} x, \text{ if } x \text{ is greater than or equal to } 0. \\ -x, \text{ if } x \text{ is less than } 0. \end{cases}$

Addends

The numbers combined to form a sum.

Algebraic Expression

Any variable or numeral or meaningful combination thereof. $3xy - xy^2$, $\frac{x+y}{3}$, xyz, etc., are algebraic expressions.

Associative Law

a. Addition: $a + (b + c) = (a + b) + c$

b. Multiplication: $(ab)c = a(bc)$

Axis

A straight line in a plane used as a visual representation of the relative order of the real numbers. A line graph or number line.

Base (of a Power)

The number to which an exponent is attached. In the term x^4, x is the base to which the exponent 4 is attached.

Binomial

An algebraic expression consisting of two terms, $3a + 4b$ is a binomial.

Cartesian Coordinates

See Rectangular coordinates.

Coefficient

Any factor or group of factors in a product is the coefficient of the remaining factors. In $32ab$, 32 is the coefficient of ab; in ax, a is the coefficient of x.

Commutative Law

a. Addition: $a + b = b + a$

b. Multiplication: $ab = ba$

Complete Quadratic Equation

A quadratic equation in one variable containing a second-degree term, a first-degree term, and a nonzero constant term. $2x^2 + 3x - 4 = 0$ is a complete quadratic equation.

Complex Fraction

A fraction that contains other fractions in its numerator or denominator or both. $\dfrac{3 + \dfrac{1}{a}}{\dfrac{3}{a}}$ is a complex fraction.

Component (of an Ordered Pair)

Either number of an ordered pair.

Consecutive Integers

Integers that differ by 1. The numbers $-3, -2, -1, 0, 1, 2$, and 3 are consecutive integers.

Constant

A symbol representing a single number during a particular discussion.

Coordinates of a Point

A pair of numbers giving the position of the point with respect to an origin.

Degree of a Term (with One Variable)

The exponent of the variable. The term $2x^4$ is of degree 4.

Dependent Equations

A system of equations where every set of values which satisfies one of the equations satisfies them all.

Dependent Variable

A variable whose values are considered determined by the values of another variable.

Descending Powers

The arrangement of an expression so that each term is of higher degree in one of the variables than the next succeeding term. The algebraic expression $x^4 + x^3 - 2x^2 + x - 1$ is arranged in descending powers of x.

Difference

The result of subtracting one number from another.

Direct Variation

A relationship determined by the equation $y = kx$, where k is a constant.

Distributive Law

$a(b + c) = ab + ac$

Equation

An assertion that two expressions are names for the same number.

Equivalent Equations

Equations which have the same solutions. The equations $2x + 2 = 6$ and $2x = 4$ are equivalent, because 2 is the only solution of each.

Equivalent Expressions

Expressions that represent the same number for all values of any variables involved. The expressions, $2c + 3b + c$, $3c + 3b$, and $3(c + b)$ are equivalent, because for all values of b and c they represent the same number.

Even Natural Numbers

2, 4, 6, ...

Exponent

A number represented by a symbol placed to the right and above another symbol to indicate how many times the number represented by this latter symbol occurs as a factor in a product. In a^3, 3 is the exponent.

Factor

Any of a group of numbers that are multiplied together.

First-Degree Equation

An equation of degree 1; a linear equation.

Formula

A relationship between quantities expressed in symbols; an equation.

Graph

A geometric representation of a numerical relationship.

Incomplete Quadratic Equation

A quadratic equation where either the first-degree term or the constant term is missing. $x^2 - 2 = 0$ and $2x^2 + 3x = 0$ are incomplete quadratic equations.

Inconsistent Equations

Equations that have no common solution. Graphically, inconsistent linear equations (in two variables) appear as parallel lines.

Independent Variable

A variable considered free to assume any one of a given set of values.

Integer

The integers include the natural numbers, the negatives of the natural numbers, and 0.

Irrational Number

Any real number that is not rational. That is, any real number that is not the quotient of two integers. The numbers $\sqrt{2}$, $\sqrt{7}$, π, etc. are irrational.

Like Terms

Terms whose variable factors are identical. $22xy$ and $2xy$ are like terms.

Linear Equation

An equation of the first degree. The graph of a linear equation in two variables is a straight line.

Line Graph

A straight line used to represent the relative order of a set of numbers.

Literal Equations

An equation containing symbols for constants such as a, b, and c in addition to variables such as x, y, and z. The equation $2ax^2 + bx + c = 0$ is a literal equation.

Literal Number

A variable. A symbol representing any one of a set of numbers.

Lowest Common Denominator (L.C.D.)

The smallest natural number or the polynomial of least degree into which each of the denominators of a given set of fractions divide exactly.

Member of an Equation

The expression to the right of an equal sign constitutes the right-hand member of an equation, and that to the left of an equal sign constitutes the left-hand member. In $2x + 3 = x + 9$, the left-hand member is $2x + 3$ and the right-hand member is $x + 9$.

Monomial

An algebraic expression consisting of one term. The expression $2xy$ is a monomial.

Natural Number

A positive whole number, such as 1, 2, 3, 4, ...

Negative Number

A number less than 0.

Number Line

See line graph.

Numeral

A symbol representing a number. The symbols "2," "π," and "½" are numerals.

Numerical Evaluation

The act of finding the value of an expression. To evaluate $x + 4$ for $x = 3$, means to replace x with 3 and simplify the results (giving 7).

Odd Natural Numbers

1, 3, 5, ...

Ordered Pair

A pair of numbers in which the order in which the numbers are considered is important. An ordered pair is usually represented (x, y).

Ordinate

The second component in an ordered pair. The distance of a point from the horizontal axis in rectangular coordinates. The point is above the axis if the component is positive and below the axis if the component is negative.

Origin

The point on a line graph corresponding to 0. The point of intersection of coordinate axes.

Parentheses

Symbols, (   ), used to group factors or terms.

Polynomial

A special kind of algebraic expression. In this book, any term or sum of terms.

Positive Number

A number greater than 0.

Prime Factor

A factor that is a prime number.

Prime Number

Any natural number that has, as whole number factors, itself and one only. The number 1 is excluded from the set of primes.

Principal Square Root

The positive square root of a positive number.

Product

The result of the multiplication of two or more numbers.

Quadrant

One of the four regions into which a set of rectangular axes divides the plane.

Quadratic Equation

An equation of degree 2. $x^2 = 2$, $x^2 - 3x = 0$, and $x^2 + 2x - 1$ are quadratic equations.

Quadratic Formula

The formula $x = \dfrac{-b \pm \sqrt{b^2 - 4ac}}{2a}$ used to solve quadratic equations of the form $ax^2 + bx + c = 0, \quad a \neq 0$.

Quotient

The result of dividing one number by another.

Radical (of Order 2)

A symbol ($\sqrt{\ }$) indicating the positive square root of a number. Any expression under a radical sign is called a radicand.

Rational Number

The quotient of two integers, $\frac{a}{b}$, where b does not equal zero.

Real Number

Any number that is either a rational number or an irrational number.

Reciprocal

The reciprocal of a number is the quotient obtained by dividing the given number into 1. The reciprocal of 3 is $1/3$; the reciprocal of $3/4$ is $4/3$; etc. The number 0 has no reciprocal.

Rectangular Coordinates

Numbers specifying the distances of points from two perpendicular number lines.

Root of an Equation

A value for the variable that satisfies the equation; i.e., for which the equation is a true statement. A solution.

Simplify

To find an equivalent form for an expression that is simpler than the original.

Solution (of an Equation)

See Root of an equation.

Square Root

One of two equal factors of a number. Since $3 \cdot 3 = 9$, the number 3 is a square root of 9. Also, since $(-3)(-3) = 9$, the number -3 is a square root of 9.

Standard Form (for a Fraction)

A positive fraction with positive denominator, such as, $\frac{-a}{b}$ or $\frac{a}{b}$.

Standard Form (for an Equation)

An equation with the left-hand member arranged in descending powers of the variable and the right-hand member 0, such as $3x^2 - 2x + 1 = 0$.

Standard Form (for a System of Equations)

The equations arranged with like terms in order in the left-hand members. For example,

$$2x + 3y = 7$$
$$-x + 5y = 3.$$

Sum

The result of the addition of two or more numbers.

Symmetric Property of Equality

If $a = b$, then $b = a$. Thus, if $2x = y + 3$, then $y + 3 = 2x$.

Systems of Equations

A set of two or more equations considered together.

Term

Any part of an algebraic expression separated from other parts by plus or minus signs.

Trinomial

An algebraic expression consisting of 3 terms. $2x + 3y + 2$ is a trinomial.

Unlike Terms

Terms that differ in their variable factors. $23xy$ and $4x$ are unlike terms; $3x^2$ and $3x^3$ are unlike terms; etc.

Variable

A symbol representing any one of a given set of numbers.

Variation

See "Direct Variation."

Whole Number

A natural number or 0. One of the numbers 0, 1, 2, 3, ...

ODD-NUMBERED ANSWERS

EXERCISES 1.1 (page 3)

1. 7 **3.** 37 **5.** 2, 3, 5, 7 **7.** 23, 29
9. 41, 43, 47, 53, 59

11.
6 7 8 9 10 11 12 13 14
0 5 10 15

13.
3 6 9 12 15 18
0 5 10 15 20

15.

17.
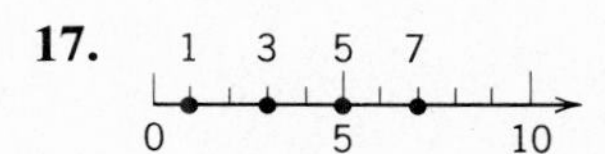

19.
3 6 9 12
0 5 10

21.
61 67 71 73 79
60 65 70 75 80

23.
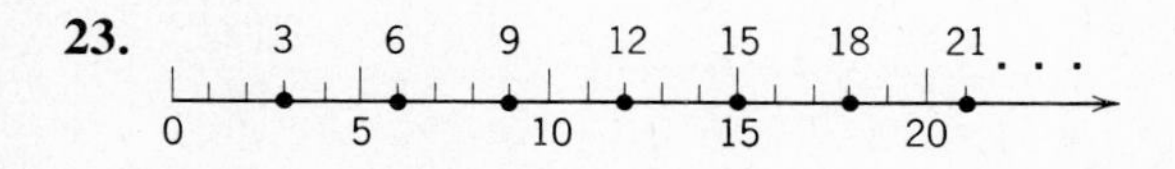

25. 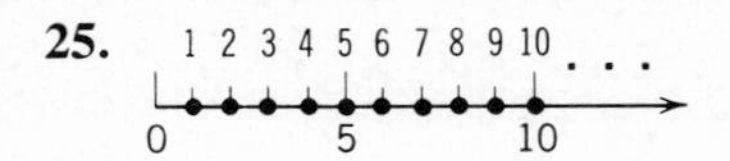

EXERCISES 1.2 (page 8)

1. a. Addition: increased by, add, more than, exceeded by, sum
b. Subtraction: take away, less than, difference, decreased by, diminished by, subtract
c. Multiplication: times, multiply, product
d. Division: divide, quotient

3. $3 + 5$ **5.** $6 \cdot 9$ **7.** $9 - 5$ **9.** xy
11. $a - b$ **13.** $5 + s$ **15.** $s - t$ **17.** $5 - y$
19. $h - 5$ **21.** Pr **23.** $l + w$ **25.** $5 + c$
27. $2(5 + y)$ **29.** $5 + 4x$ **31.** $y - (4 + x)$ **33.** $b(c + d)$
35. $(p + q) - 4y$ **37.** $\frac{r + s}{6z}$

EXERCISES 1.3 (page 10)

1. 4^2 **3.** x^3 **5.** ab^2c^3 **7.** $3 \cdot 4^2xy^2$
9. $(3x)^2$ **11.** $(x + 3)^2$ **13.** $x^2y + xy^2$ **15.** $2ab^3 - b^2$
17. $(2x)^2 - x^2$ **19.** $x^2 + y^2 - z^3$ **21.** $2^3x^2 + 3^2y + 4y^3$
23. $(2x)^3 - (x - 2)^2$ **25.** $2 \cdot 2 \cdot 2$ **27.** $xxxx$
29. $xxyyyy$ **31.** $3 \cdot 5abbc$ **33.** $(2x)(2x)(2x)(2y)(2y)(2y)$
35. $5(3a)(3a)(4b)(4b)$ **37.** $(a - 4)(a - 4)(a - 4)$
39. $yyy(3y + 4)(3y + 4)$ **41.** $xxyyy(x - y)(x - y)$

EXERCISES 1.4 (page 12)

1. 7 **3.** 18 **5.** 2 **7.** 14 **9.** 24 **11.** 45
13. 8 **15.** 40 **17.** 16 **19.** 3 **21.** 1 **23.** 0
25. 14 **27.** 7 **29.** 5 **31.** 0 **33.** 1 **35.** 90

EXERCISES 1.5 (page 14)

1. 45 **3.** 12 **5.** 18 **7.** 14 **9.** 18
11. 27 **13.** 33 **15.** 135 **17.** 7 **19.** 8
21. 18 **23.** 12 **25.** 2 **27.** 15 **29.** 2
31. 3 **33.** 6 **35.** 18 **37.** 18 **39.** 36
41. 5 **43.** 18 **45.** 36 **47.** 5 **49.** 10
51. 34 **53.** $W = 12{,}000$ **55.** $V = 76$
57. $A = 50{,}000$ **59.** $A = 30$

EXERCISES 1.6 (page 17)

1. Binomial, $3y^3$, 3; $2y$, 2
3. Binomial, $2x^3$, 2; x, 1
5. Monomial, $4y^3$, 4
7. Monomial, $7x^2$, 7
9. Binomial, y^5, 1; $2x^3$, 2
11. Binomial, $6x^3$, 6; $2y^2$, 2
13. Binomial, $2x^4$, 2; 2, 2
15. Trinomial, $3x^2$, 3; $3y$, 3; $4z$, 4
17. Binomial, x^3, 1; $4x$, 4
19. Binomial, $3xy^2$, 3; y, 1
21. Monomial, 3, 3
23. Binomial, $7xyz$, 7; $3x$, 3
25. Binomial, $3x^2y$, 3; x^2y^2, 1
27. Monomial, $4x^2yz^3$, 4
29. $4y^2$, 2
31. $2x^4$, 4; x, 1
33. x^2, 2; $3x$, 1
35. $2x^2$, 2; x, 1; $4x$, 1
37. y^4, 4; $2y^3$, 3; y, 1
39. z^5, 5; $2z^2$, 2; z, 1
41. Second degree
43. Fourth degree
45. Fifth degree

EXERCISES 1.7 (page 20)

1. $7x$
3. $6a$
5. $8x^2$
7. $5x^2 + 5x$
9. $8a^2 + a$
11. $3b^3 + 2b^2 + b$
13. $6xy$
15. $9xy^3$
17. $6x^2y + 2xy^2$
19. $2x + 7xy + 4y$
21. $4x + 4y + z$
23. $x^2y + 2xy + 2xy^2$
25. $3x^3 + 4x^2 + 5x$
27. $3z^2 + 2z + 8x + 7y$
29. $5x^2y + x^2 + 2xy + 3x + 2y$
31. $3x + 5ax + 3y$
33. $8s^2 + 2t^2 + 2st + r^2$
35. $2a + 5b + 2c + 3ab$
37. $2x^2yz + 4xy^2z + 3xyz^2$
39. $13x^2yz + 4xyz + 13x^3yz$

EXERCISES 1.8 (page 22)

1. $3y^3$
3. $6x^2$
5. $5x^2$
7. $6a^2$
9. $3a^2$
11. $7a^3$
13. $5a^2 + 3a$
15. $6b^2 + 4b$
17. $9x - 3y$
19. $6x^2y + xy^2$
21. $4ab^2 + a^2b$
23. $7a^2b - ab^2$
25. $7x^2y + xy^2$
27. $8x^3yz + 5x^2yz + 2xyz$
29. $22xz^2 + 17x^2z$

EXERCISES 1.9 (page 23)

1. x^7
3. y^5
5. $10x^5$
7. $4a^6$
9. $12b^5$
11. $12c^6$
13. $8x^3$
15. y^7
17. $14a^3b^4$
19. a^3b^3
21. $a^5b^3c^3$
23. $6a^2b^5c^4$
25. x^3
27. $3x^5$
29. $2y^3$
31. $3a^2b^2 - b^2$
33. $2y^4 - 6y^5$
35. $5a^3 - a^2$
37. $2ab + 3b$
39. $7a^3b$

EXERCISES 1.10 (page 26)

1. 5 **3.** 8 **5.** 0 **7.** Undefined
9. x **11.** x^2 **13.** xy^2 **15.** x^3y
17. $2xy^4$ **19.** $3xy^3$ **21.** a^3b^4 **23.** $2b^3$
25. 0 **27.** ab **29.** $3a^6bc$ **31.** Undefined
33. $6x$ **35.** $2x$ **37.** $9x^3$ **39.** $3x^2y$
41. $4b^2$ **43.** c **45.** $4x$ **47.** $5 + 8y$
49. Undefined

CHAPTER 1 REVIEW (page 28)

1. 11, 13, 17, 19, 23

2. 4 8 12 16 20 24
0 5 10 15 20 25 30

3. a. $6x$
b. $\frac{4+y}{6}$
c. $y(3+x)$

4. a. $2^2a^2b^3$
b. xy^2z^3
c. 3^2c^2d

5. a. $2 \cdot 3xyy$
b. $aaabb$
c. $3 \cdot 3 \cdot 3cdd$

6. a. 13
b. 8
c. 5

7. a. 3
b. 8
c. 2

8. a. 50
b. 26
c. 52

9. a. $6xy + 2y$
b. $5a^2 - 3a$
c. $2r + 4s$

10. a. x^3y^3
b. $12ab^4$
c. r^4s^3

11. a. $4x^2 - 2x^3$
b. $ab^3 - b^2$
c. r^2s^2

12. a. $2ab$
b. 1
c. $3xy$

13. a. 1
b. $7y$
c. $13a$

14. a. $x + y$
b. xy
c. $\frac{x}{y}$

15. Terms **16.** Binomials **17.** Numerical coefficient
18. 1 **19.** Third degree **20.** Fourth degree

EXERCISES 2.1 (page 31)

1. −5 −3 −1 0 2 5
−5 0 5

3. 6 **5.** 0 **7.** −2 **9.** −20
11. −9 **13.** −12 **15.** −5 **17.** −8

19. 1 2 3 4 5 6 7 8 9
−10 −5 0 5 10

21.

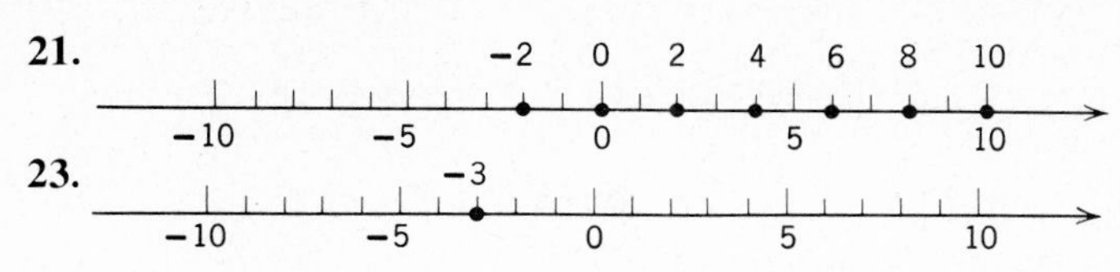

23.

25. 5 27. 8 29. −7 31. −3
33. 1 35. −4

EXERCISES 2.2 (page 34)

1. 9 3. −10 5. 5 7. −1
9. 7 11. 0 13. 8 15. −11
17. 6 19. −3 21. 5 23. 0
25. 7 27. −2 29. 3 31. 4
33. 0 35. −3 37. 0 39. 3

EXERCISES 2.3 (page 36)

1. $-x$ 3. $-2y$ 5. 0
7. $-hk$ 9. $-5cd$ 11. xy
13. $5x^2 - x$ 15. $3a^2 - 5a$ 17. $-x^2 + 5x$
19. $-x^3 + 3x^2 + 2x$ 21. $-2a^3 + 5a^2 + 1$ 23. $6x - y - z$
25. $4xy^2$ 27. $6x^2y - 7z + 3$ 29. $9m^2 - 5m + 3$
31. $-x^2 - 4x$ 33. $ab^2 - 4a^2bt + abt^2$ 35. $-5g^2 - 6ag$
37. $-3a^2bc + 4ab^2c$ 39. $7ab + 4c - 3d$ 41. $(-2) + x$
43. $2x + (-y)$ 45. $x + (-2y) + 3z$ 47. $y^2 + (-3y) + 2$

EXERCISES 2.4 (page 39)

1. 0 3. 3 5. −5 7. −5
9. −7 11. 2 13. 5 15. 7
17. 2 19. −2 21. 2 23. −4
25. −12 27. 1 29. 6 31. 0
33. 0 35. 1

EXERCISES 2.5 (page 41)

1. $-x$ 3. $-3x$ 5. $6y$
7. xy 9. $-5x^3y$ 11. $4y^2$
13. $6x$ 15. $-2g$ 17. 0
19. $5a + b$ 21. $-x^2 - x - 1$ 23. $-2x - 8y + 2z$
25. $-x^2 + 4x - 4$ 27. $-y^2 - 3y$ 29. $-z^2 - 4z$
31. 0 33. $-x^2y - 4xy + xy^2$ 35. $-3xy + 1 + x^2y^2$
37. $2xy^2 + xy - 2x$ 39. $4x + 2z$ 41. $2a$

43. $-x + 2y + 6z$ **45.** $3x$ **47.** $x + 7$
49. $2a + 10b$ **51.** $x - 4$ **53.** $y - x$
55. $y - (4 + x)$

EXERCISES 2.6 (page 44)

1. -15 **3.** 30 **5.** 0 **7.** -8 **9.** -16
11. -30 **13.** -18 **15.** -16 **17.** 0 **19.** -24
21. 12 **23.** 0 **25.** -4 **27.** -8 **29.** -18
31. 3 **33.** -4 **35.** -16 **37.** $6x^2y$ **39.** $-2x^3y$
41. x^3y^2 **43.** $6a^4$ **45.** $-a^2b^4c^2$ **47.** $6b^6$ **49.** $-x^3$
51. $-a^3$ **53.** $-x^3y^4$ **55.** $9x^2y^2$ **57.** $-2x^3y^2$ **59.** $6xyz^2$
61. $(-4)(3x);\quad -12x$ **63.** $(xy)(-y^2);\quad -xy^3$
65. $(-3xy)(x^2)(-y^2);\quad 3x^3y^3$ **67.** $-2(-x)^2\,(-x^2);\quad 2x^4$

EXERCISES 2.7 (page 47)

1. 4 **3.** -2 **5.** 0 **7.** Undefined
9. -6 **11.** -4 **13.** -3 **15.** xy^2
17. -1 **19.** $-4x^2$ **21.** $6xy^2$ **23.** $-2x$
25. x **27.** $-3y$ **29.** $-3x^2$ **31.** 1
33. $4xy$ **35.** $2hy^2$ **37.** -3 **39.** $-18y^2$
41. $x - 3$ **43.** $2x$ **45.** 0 **47.** 0
49. $-3x$ **51.** 0 **53.** 0 **55.** $5y$
57. $4ab$

EXERCISES 2.8 (page 50)

1. -6 **3.** 16 **5.** -4 **7.** 11
9. -40 **11.** 16 **13.** 6 **15.** -10
17. 0 **19.** 3 **21.** -2 **23.** -1
25. 0 **27.** -4 **29.** -8 **31.** 0
33. 3 **35.** -3 **37.** 5 **39.** 7
41. 0 **43.** -6 **45.** 0 **47.** -18
49. 5 **51.** 4 **53.** -16 **55.** 1
57. -18

CHAPTER 2 REVIEW (page 51)

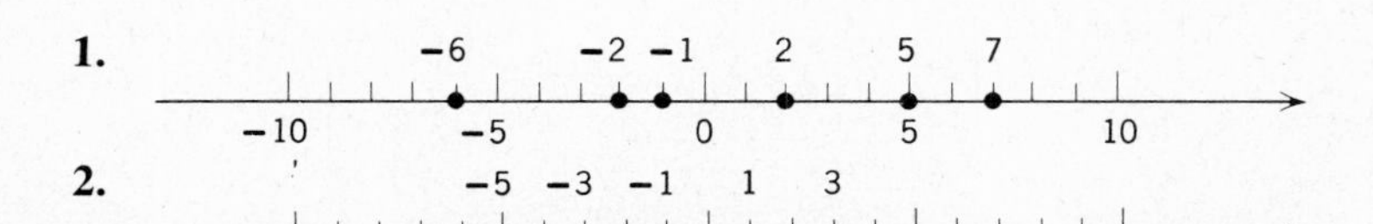

3. $-4, -3, 0, 2, 5$

4. 3

5. a. -2
b. -6
c. 1

6. a. -3
b. -3
c. -3

7. a. $12x$
b. $7x^2 + 2x$
c. $4ab^2$

8. a. -2
b. -4
c. 2

9. a. -7
b. x
c. x

10. a. $2x$
b. $8x^2 - xy - 3y^2$
c. $x^2y + 3xy + 3xy^2$

11. a. $x^2 - x - 1$
b. $x^2 - 5x + 13$
c. $-x^2 + 3y^2 - 2z^2$

12. a. $3y + z$
b. $a + 2b - 5c$
c. $-x$

13. a. $12x$
b. $6x$
c. 8

14. a. 5
b. -4
c. 8

15. a. $-y$
b. -1
c. x

16. a. 5
b. 0
c. $-2x$

17. a. $4x$
b. $-5x$
c. 1

18. a. 4
b. -3
c. 0

19. a. -2
b. 0
c. 3

20. a. -2
b. -7
c. 7

CHAPTER 2 CUMULATIVE REVIEW (page 52)

1. 19 23
15 20 25

2. -6 -5 -4 -3 -2 -1 0 1 2 3 4 5 6
-5 0 5

3. $2y^3$

4. $2 \cdot 2 \cdot 3 \cdot 3 \cdot xyyy$

5. 16

6. 5

7. 0

8. $3x^3 - 2y^3$

9. $-3x^4$

10. $-3n$

11. $2a - 4b + c$

12. $3b$

13. 0

14. 520 feet

15. 4

16. -5

17. $-(a - b + c)$

18. $-(a + b - c)$

19. 0

20. 0

EXERCISES 3.1 (page 55)

1. $3 + x = 15$

3. $2x + 4x = 42$

5. $8 + x = 2x$

7. $3 + x = 2x$

9. $2x - 5 = 3 + x$

11. $3x = x + 12$

13. $A = s^2$

15. $A = lw$

17. $V = lwh$

EXERCISES 3.2 (page 56)

1. Yes	**3.** Yes	**5.** Yes	**7.** No
9. Yes	**11.** Yes	**13.** No	**15.** Yes
17. Yes	**19.** Yes	**21.** Yes	**23.** No

EXERCISES 3.3 (page 58)

1. $x = 7$	**3.** $y = 4$	**5.** $z = 8$	**7.** $x = 3$
9. $z = 9$	**11.** 12	**13.** 5	**15.** 11
17. 5	**19.** 9	**21.** 4	**23.** 5
25. 7	**27.** -2	**29.** 2	**31.** 4
33. 3	**35.** -3	**37.** -5	**39.** 1
41. -2	**43.** 3	**45.** 2	**47.** 0
49. 4			

EXERCISES 3.4 (page 61)

1. $x = 3$	**3.** $z = -3$	**5.** $y = 2$	**7.** $x = 4$
9. $z = -5$	**11.** $y = 2$	**13.** 3	**15.** 4
17. -4	**19.** -6	**21.** 2	**23.** -1
25. 3	**27.** 1	**29.** 1	**31.** 0
33. 4	**35.** -4	**37.** 7	**39.** -4
41. 2	**43.** 4	**45.** 2	**47.** 3
49. 4			

EXERCISES 3.5 (page 63)

1. $x = 12$	**3.** $-20 = z$	**5.** $y = 18$	**7.** 16
9. 42	**11.** -12	**13.** -18	**15.** -10
17. 0	**19.** 20	**21.** 12	**23.** -20
25. -10	**27.** -36	**29.** 15	**31.** -12
33. -32	**35.** -3	**37.** 25	**39.** 10
41. 6	**43.** 15		

EXERCISES 3.6 (page 66)

1. 5	**3.** 3	**5.** -3	**7.** 4
9. 12	**11.** 2	**13.** -3	**15.** -10
17. -2	**19.** 3	**21.** 4	**23.** 3
25. 8	**27.** 3	**29.** 3	**31.** 2
33. 2	**35.** 6	**37.** 9	**39.** 0
41. 8	**43.** 24	**45.** -2	

EXERCISES 3.7 (page 67)

1. $x = a$ **3.** $x = a$ **5.** $x = 3a$ **7.** $y = \frac{b}{a}$

9. $y = \frac{3a}{b}$ **11.** $y = \frac{5}{3a}$ **13.** $x = \frac{a^2}{c}$ **15.** $x = 2$

17. $x = \frac{bc}{a}$ **19.** $t = \frac{d}{r}$ **21.** $l = \frac{v}{wh}$ **23.** $d = \frac{c}{\pi}$

25. $r = \frac{d}{t}$ **27.** $h = \frac{v}{lw}$ **29.** $r = \frac{I}{pt}$ **31.** $g = \frac{v - k}{t}$

33. $m = \frac{Fd^2}{kM}$ **35.** $x = -a$ **37.** $x = \frac{c^2}{2}$ **39.** $x = 2c$

EXERCISES 3.8 (page 69)

1. Let $x =$ the number; $x + 21 = 59$
3. Let $x =$ the number; $x + 15 = 53$
5. Let $x =$ the number of minutes; $x + 19 = 47$
7. Let $x =$ the number of minutes; $x + 9 = 30$
9. Let $x =$ the number of centimeters; $4x + 24 = 340$
11. Let $x =$ smaller integer; $x + 2 =$ second integer; $x + (x + 2) = 26$
13. Let $x =$ smaller integer; $x + 2 =$ second integer; $x + (x + 2) = 28$
15. Let $x =$ the shorter piece; $3x =$ the longer piece; $x + 3x = 112$
17. 7 **19.** 42, 44 **21.** $-12, -11, -10$ **23.** 7 kilometers
25. 94, 118 **27.** 2163, 2213

EXERCISES 3.9 (page 74)

1. 9 square inches
3. 120 square feet
5. 104 square inches
7. 144 square inches
9. 93.76 square inches
11. 624.76 square inches
13. 17 inches from one end
15. 18 centimeters
17. Width: 36 meters, length: 46 meters
19. 16 meters, 20 meters, 20 meters
21. 38°, 76°, 66°
23. Width: 7 feet, length: 21 feet
25. 6 feet
27. 70°
29. 50°, 60°, 70°

CHAPTER 3 REVIEW (page 77)

1. a. $3 + x = 2x - 2$
b. $12 - x = 2x$
c. $\frac{3x}{4} = 21 - 6$

2. a. 6
b. 2
c. -1

3. a. 5
b. 0
c. -1

4. a. -18
b. 9
c. -3

5. a. -2
b. 6
c. 3

8. a. $a = \dfrac{f}{m}$
b. $g = \dfrac{v - k}{t}$
c. $b = 2M - a$

9. $x + 2$
10. $x + 2$
11. $x + 1, x + 2, x + 3, x + 4$
12. $3x$
13. $x + 7, x - 7$
14. $x + 18, x - 12$
15. 12, 13, 14, 15
16. 3, 5, 7
17. 202, 229
18. 11 feet, 8 feet, 13 feet
19. 38 meters
20. 45°, 60°, 75°

CHAPTER 3 CUMULATIVE REVIEW (page 79)

1. 9
2. $3^2, \dfrac{4^2}{2}, \dfrac{4 + 2^3}{12}, 5^2 - 3^2$
3. Numerical coefficient or coefficient
4. Exponent
5. Positive
6. $|-5|$
7. Less than
8. a. x^5
b. x^6
9. a. 81
b. -81
10. 192 feet/second
11. 4 seconds
12. 6 hours
13. $-4y^2$
14. Before
15. Coefficients
16. Equivalent
17. 1
19. $\dfrac{b + 2}{2}$
20. $-14, -16, -18$

EXERCISES 4.1 (page 80)

1. $3x - 12$
3. $10y - 10$
5. $-2x - 8$
7. $10a^2 + 6a$
9. $-b^2 + 2b$
11. $x^2y + xy^2$
13. $-2x^3 - 3x^2y$
15. $x^3 - 2x^2 + x$
17. $-y^3 + y^2 - 2y$
19. $4x^5 - 12x^4 + 16x^3$
21. $-y^6 + y^4 - y^3$
23. $-x^3y - x^2y^2 - xy^3$
25. $-a$
27. $2ax + x + a$
29. $-ax + ay - 2y$
31. $-3ax$
33. $2x$
35. 6
37. $ax^2 - bx^2 + cx^2$
39. $-4abx - aby + 2ab + b$
41. $-a - c$
43. $a - 2b + c$
45. $-3x - 2y + z$
47. $-1 + 3x - x^2$
49. 0
51. $-2x$

EXERCISES 4.2 (page 83)

1. $3(x + 2)$
3. $2(x - 3y)$
5. $2y(y - 1)$
7. $y(ay + 1)$
9. $3y(3ay + 2)$
11. $3(y^2 - y + 1)$

13. $a(x + y - z)$
15. $x(x - 3 + y)$
17. $2y(2y^2 - y + 1)$
19. $6axy(x - 3y + 4)$
21. $-a(a + b)$
23. $-x(1 + x)$
25. $-b(ac + a + c)$
27. $-3y(2y^2 + y + 1)$
29. $-x(1 - x + x^2)$
31. $-xy^2(y^3 + y^2 - 1)$
33. $d = k(1 + at)$
35. $s = kr^2(h + 1)$
37. $V = 2ga^2(D - d)$
39. $A = r^2(a + b + c)$

EXERCISES 4.3 (page 84)

1. $x^2 + 7x + 12$
3. $y^2 - 2y - 3$
5. $a^2 + 7a + 10$
7. $b^2 - 2b - 8$
9. $x^2 + 9x + 8$
11. $y^2 - 8y + 7$
13. $a^2 + 8a + 16$
15. $b^2 - 25$
17. $x^2 + 2x + 1$
19. $y^2 - 1$
21. $4 - x^2$
23. $36 - y^2$
25. $x^2 + 8x + 16$
27. $x^2 - 14x + 49$
29. $x^2 - 2x + 1$
31. $x^2 + 4x + 4$
33. $x^2 - 4bx + 3b^2$
35. $x^2 + xy - 2y^2$
37. $x^2 + 4ax + 4a^2$
39. $a^2 - 2ab + b^2$
41. $y^2 - 36a^2$
43. $x^2 - t^2$
45. $2x^2 + 6x + 4$
47. $6y^2 + 60y + 150$
49. $6x^2 - 12x + 6$
51. $a^3 + 4a^2 - 5a$
53. $a^3 - 4a$
55. $xy^2 - 6xy + 9x$

EXERCISES 4.4 (page 88)

1. $(x + 3)(x + 2)$
3. $(y - 5)(y - 3)$
5. $(x - 4)(x + 3)$
7. $(y + 5)(y - 4)$
9. $(a + 7)(a - 5)$
11. $(b - 20)(b + 1)$
13. $(x + 6)(x + 5)$
15. $(y - 13)(y - 1)$
17. $(a - 10)(a + 5)$
19. $(b - 9)(b + 5)$
21. $(x - 45)(x - 1)$
23. $(y - 45)(y + 1)$
25. $(x + 2a)(x + 2a)$
27. $(a - 2b)(a - b)$
29. $(s + 3a)(s - 2a)$
31. $(ab - 2)(ab + 1)$
33. $2(x + 3)(x + 2)$
35. $y(y - 3)(y + 1)$
37. $5(c - 2)(c - 3)$
39. $4b(a + 6)(a - 3)$
41. $(7 + x)(3 - x)$
43. $(5 + y)(2 + y)$
45. $(9 + y)(7 - y)$
47. $(8 - z)(4 - z)$
49. $(8 - x)(1 - x)$
51. $(8 - y)(7 - y)$

EXERCISES 4.5 (page 90)

1. $2x^2 + 7x + 3$
3. $3y^2 + y - 2$
5. $6y^2 + 11y + 3$
7. $20x^2 + 7x - 6$
9. $4x^2 - 9$
11. $16y^2 - 9$
13. $4x^2 + 4x + 1$
15. $25x^2 + 20x + 4$
17. $16y^2 + 40y + 25$
19. $2x^2 + 3ax - 2a^2$
21. $3x^2 + 4ax + a^2$
23. $6x^2 + ax - a^2$
25. $9x^2 - 4y^2$
27. $16x^2 - 49y^2$
29. $x^2 - 4xy + 4y^2$
31. $9x^2 - 6xy + y^2$
33. $64x^2 + 48xy + 9y^2$
35. $4x^2 + 12xy + 9y^2$
37. $6x^2 - 16x - 6$
39. $12y^2 - 3$
41. $12x^2 - 60x + 75$
43. $2x^3 + x^2 - 10x$
45. $4x^3 - 4x^2 + x$
47. $9r^3 - r$

EXERCISES 4.6 (page 92)

1. $(3a + 1)(a + 1)$
3. $(2x - 1)(x - 1)$
5. $(3b - 1)(3b - 1)$
7. $(2x + 3)(x - 1)$
9. $(2x - 3)(x + 1)$
11. $(3a + 1)(2a - 1)$

13. $(2y-1)(2y-1)$ **15.** $(4y+1)(y-1)$ **17.** $(4a-3)(a-2)$
19. $(4a+5)(a-1)$ **21.** $(8x-5)(2x+1)$ **23.** $(16x+5)(x-1)$
25. $(3x+1)(3x-8)$ **27.** $(2y+3)(2y+5)$ **29.** $(2t+s)(t-3s)$
31. $(3x-a)(x-2a)$ **33.** $(4y+b)(y+b)$ **35.** $(2a+5b)(2a+3b)$
37. $2(3x+1)(x+1)$ **39.** $2(4y+1)(y-1)$ **41.** $9(2x-3)(x+1)$
43. $3y(3y+1)(3y-2)$ **45.** $3a(4b+a)(b+a)$
47. $2xy(5y+4x)(5y-2x)$

EXERCISES 4.7 (page 94)

1. $(x+3)(x-3)$ **3.** $(x+1)(x-1)$ **5.** $(x+z)(x-z)$
7. $(xy+4)(xy-4)$ **9.** $(ax+7b)(ax-7b)$ **11.** $(6+x)(6-x)$
13. $(2b+3)(2b-3)$ **15.** $(5x+4)(5x-4)$ **17.** $(3+2x)(3-2x)$
19. $(9+2x)(9-2x)$ **21.** $(2a+11b)(2a-11b)$ **23.** $(5y+7x)(5y-7x)$
25. $(7ax+12by)(7ax-12by)$ **27.** $(2xy+9)(2xy-9)$
29. $(6ab+1)(6ab-1)$ **31.** $5(x+1)(x-1)$
33. $3x(x+1)(x-1)$ **35.** $2(x+2y)(x-2y)$
37. $3(ab+2cd)(ab-2cd)$

EXERCISES 4.8 (page 96)

1. 7 **3.** -1 **5.** 3 **7.** -1
9. -2 **11.** 3 **13.** -5 **15.** 4
17. -5 **19.** 0 **21.** 2 **23.** -4
25. 3 **27.** 1 **29.** 4 **31.** 5
33. 1 **35.** 4

EXERCISES 4.9 (page 97)

1. a. $x+4$
b. $5x$
c. $5(x+4)$

3. $3(n+6)$

5. a. $27-n$
b. $3n$
c. $3(27-n)$

7. a. $n+16$
b. $5n$
c. $2(n+16)$

9. a. $10-x$
b. $2x$
c. $5(10-x)$

11. a. $x+4$
b. $25x$
c. $10(x+4)$

13. a. $n-3$
b. $n+2$
c. $25n$
d. $10(n-3)$
e. $n+2$

15. $5n+10(n+3)$ or $15n+30$

17. a. $x-4$
b. $7x$
c. $6(x-4)$

19. 5, 3 **21.** 22 and 26 gallons **23.** 8 feet from one end

25. 10 nickels, 13 dimes **27.** 9 pennies, 15 nickels, 3 dimes **29.** 16 grams

31. 40 kilograms **33.** 400 adults, 600 children **35.** 7 tons, 11 tons

CHAPTER 4 REVIEW (page 103)

1. a. $3x^3 + 3x^2$
b. $2xy^2 - 2x^2y$
c. $-x^2 + y - 1$

2. a. $2a - a^2$
b. $-ab + b^2$
c. $3ab + 3b^2 + 3bc$

3. a. $3a^2(1 - 2b)$
b. $2x(x^2 + 2x + 3)$
c. $-y^2(1 + y)$

4. a. $a^2(1 + b)$
b. $4(b - 1)$
c. $b(1 - b - b^2)$

5. a. $x^2 + x - 6$
b. $6a^2 - 17a + 12$
c. $4a^2 - 12a + 9$

6. a. $x^2 - ax - 2a^2$
b. $2x^2 + bx - b^2$
c. $4b^2 + 4b + 1$

7. a. $(x - 7)(x + 3)$
b. $(5a + 1)(2a + 3)$
c. $(2x - 3)(2x + 3)$

8. a. $(a - 7)(a - 3)$
b. $(3b + 1)(b + 1)$
c. $(2b - 1)(b + 2)$

9. a. $2(x + 3)(x + 4)$
b. $3(y + 10)(y - 2)$
c. $4x(x - 1)(x + 1)$

10. a. $(x - 2a)(x - a)$
b. $(x - a)(x + a)$
c. $2(2b - c)(b + 2c)$

11. a. 20
b. 2
c. −6

12. $A = 2kr(h + r)$; 747.32

13. $24 - x$ **14.** $10x$ **15.** $25(x + 3)$

16. $5(y - 2)$ **17.** $30(x + 4)$ **18.** 5, 11

19. $9\frac{5}{11}$ and $19\frac{5}{11}$ **20.** 15 dimes, 23 nickels

CHAPTER 4 CUMULATIVE REVIEW (page 104)

1.

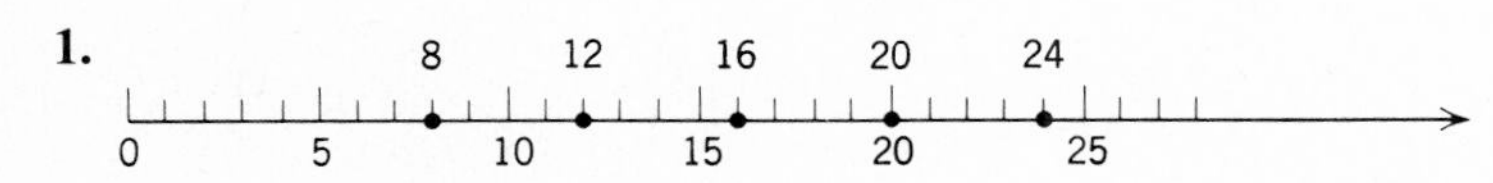

2. $3 \cdot 3 \cdot 3 \cdot 3xxxyyy$ **3.** 112 **4.** $-x^2 - 6x + 5$

5. $-x^2 - x - 3$ **6.** $2x^2 - 11x$ **7.** $x^2 - 10x + 6$

8. $x = -4b$ **10.** nc **11.** $(80 - y)^\circ$

12. $(y - 6)^\circ$ **13.** $(x - y)^\circ$ **14.** $47 - x$

15. 27, 29 **16.** 36, 37, 38 **17.** 12, 7

18. 3 quarters, 4 dimes, 6 nickels **19.** 7 centimeters, 21 centimeters, 7 centimeters

20. 20 hours and 30 hours

EXERCISES 5.1 (page 108)

1. $\frac{4}{5}$ **3.** $\frac{2x}{y}$ **5.** $\frac{5}{x - y}$ **7.** $\frac{5x}{x - 5}$

9. $\frac{1}{4}$ $\frac{3}{4}$
0 1

11. 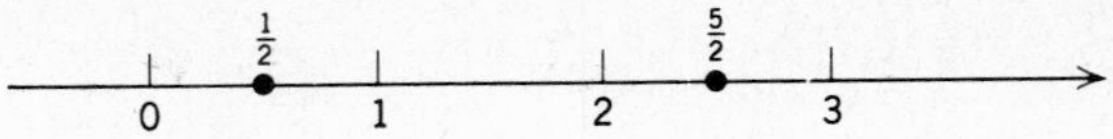

13. 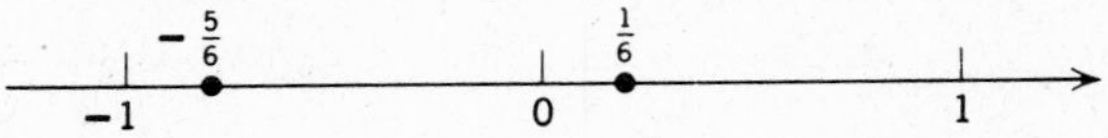

15. 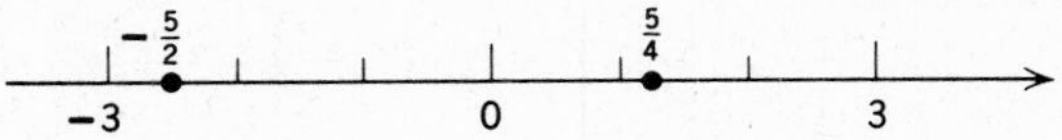

17. 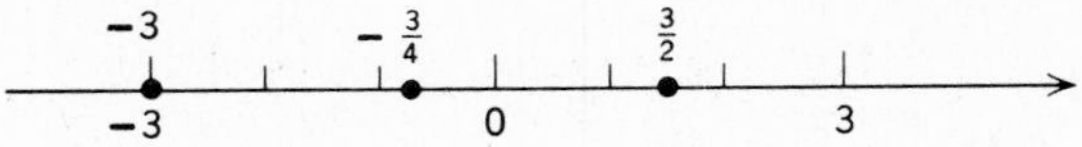

19. 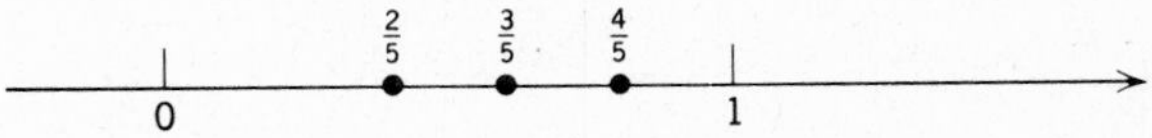

21. $\frac{3}{4}$ **23.** $\frac{2}{3}$ **25.** $\frac{-3}{5}$ **27.** $\frac{-a}{b}$

29. $\frac{-a}{b}$ **31.** $\frac{-x}{y}$ **33.** $\frac{7x}{8y}$ **35.** $-c$

37. $\frac{-(x+2)}{4}$ **39.** $\frac{5}{x+2}$

EXERCISES 5.2 (page 110)

1. $\frac{2}{3}$ **3.** $\frac{-6}{7}$ **5.** $\frac{8}{5}$ **7.** $\frac{1}{4x}$

9. $\frac{-2}{3y^4}$ **11.** $\frac{-1}{2x}$ **13.** x^2y **15.** $\frac{-1}{xy^2}$

17. $2x^2$ **19.** $\frac{3x^2}{2}$ **21.** $\frac{5bc^2}{4}$ **23.** $\frac{13a^2}{3c^2}$

25. $\frac{3}{4}$ **27.** $-4(x-y)$ **29.** 1 **31.** -2

33. $\frac{2}{x-a}$ **35.** $\frac{-1}{x+4}$ **37.** $\frac{1}{x+1}$ **39.** $\frac{1}{a-b}$

41. $\frac{a-b}{a+b}$ **43.** $\frac{a}{a+1}$ **45.** $\frac{x-2}{x-3}$ **47.** $\frac{a+3}{a-1}$

49. No **51.** No **53.** No **55.** No

57. Yes **59.** No

EXERCISES 5.3 (page 115)

1. $2x-1$ **3.** $y+2$ **5.** $x+3$

7. $3y^2-2y+1$ **9.** $2y^2-y+3$ **11.** $4y-x+1$

13. $3x^2 + 2x + \frac{-1}{3}$ **15.** $y + 2 + \frac{-1}{y}$ **17.** $3x^2 - 2 + \frac{-2}{3x^2}$

19. $y^2 - 3y + 2 + \frac{-1}{y}$ **21.** $y + 1 + \frac{1}{y}$ **23.** $x - 2 + \frac{3}{y}$

25. $x + 6$ **27.** $x + 1$ **29.** $x + 7$

31. $2x + 1$ **33.** $x + 3$ **35.** $2x + 3$

37. $x + 1 + \frac{-1}{x + 2}$ **39.** $x - 2 + \frac{1}{x + 5}$ **41.** $x - 6 + \frac{29}{x + 6}$

43. $2x - 1 + \frac{-1}{x + 1}$ **45.** $2x - 3 + \frac{-2}{2x + 1}$

EXERCISES 5.4 (page 118)

1. 12 **3.** 18 **5.** 84 **7.** 72
9. 24 **11.** 60 **13.** x^2y **15.** xyz
17. m^2n^3 **19.** $24x^2y$ **21.** $(x + y)(x - y)$
23. $x^2(x + 2)$ **25.** $(x + 4)(x - 1)^2$

EXERCISES 5.5 (page 120)

1. $\frac{4}{10}$ **3.** $\frac{12}{18}$ **5.** $\frac{66}{36}$

7. $\frac{98}{42}$ **9.** $\frac{10}{6x}$ **11.** $\frac{-12ab^2}{12b^3}$

13. $\frac{-3x^2y}{3y^3}$ **15.** $\frac{72}{36}$ **17.** $\frac{xy^2}{xy}$

19. $\frac{3x^4y}{3x^2y}$ **21.** $\frac{x + y}{2(x + y)}$ **23.** $\frac{-2a(a + 4)}{5(a + 4)}$

25. $\frac{2a(a + 3)}{a + 3}$ **27.** $\frac{3(x + y)}{(x - y)(x + y)}$ **29.** $\frac{-3(x + 1)}{(2x - 1)(x + 1)}$

31. $\frac{7a(b - 3)}{(b + 2)(b - 3)}$ **33.** $\frac{a^2}{a(a - 3)}$ **35.** $\frac{-3(x - y)}{(x + y)(x - y)}$

37. $\frac{y(y + 2)}{(y - 1)(y + 2)}$ **39.** $\frac{(x - 1)(x + 1)}{x^3 - 2x^2 + x}$ **41.** $\frac{3}{6}, \frac{2}{6}$

43. $\frac{5}{35}, \frac{21}{35}$ **45.** $\frac{-10}{24}, \frac{9}{24}$ **47.** $\frac{-3b}{3ab}, \frac{a}{3ab}$

49. $\frac{5a^2}{4ab^2}, \frac{-8b}{4ab^2}$ **51.** $\frac{-a^3}{a^2b^2}, \frac{2b}{a^2b^2}$ **53.** $\frac{-9}{3(x - a)}, \frac{2(x - a)}{3(x - a)}$

55. $\frac{3(a + b)}{ab(a + b)}, \frac{a^2b}{ab(a + b)}$

57. $\frac{-x(x + 4)}{(x + 1)(x + 2)(x + 4)}, \frac{2x(x + 2)}{(x + 1)(x + 2)(x + 4)}$

59. $\frac{3(x+3)}{(x-1)(x+1)(x+3)}, \frac{(x-2)(x-1)}{(x-1)(x+1)(x+3)}$

EXERCISES 5.6 (page 126)

1. $\frac{7}{9}$
3. $\frac{5x-3}{11}$
5. $\frac{3-x}{5}$
7. $\frac{4}{a}$
9. $\frac{1}{b}$
11. $\frac{4}{x}$
13. $\frac{x}{y}$
15. $\frac{x+4}{2}$
17. $\frac{2x+y}{3x}$
19. $\frac{x}{a}$
21. $x^2 + x$
23. $\frac{4x}{y}$
25. $\frac{x+6}{2}$
27. $\frac{a+3b}{a-b}$
29. $\frac{-a}{a+b}$
31. $\frac{3x+2y}{x+y}$
33. $\frac{1}{x+2y}$
35. $\frac{6a-b}{3}$
37. $3x + y$
39. $\frac{u}{2u-v}$
41. $\frac{x-1}{x+2}$
43. $\frac{4}{x-1}$
45. $\frac{1}{x-y}$
47. $\frac{x+5}{x}$

EXERCISES 5.7 (page 130)

1. $\frac{5}{8}$
3. $-\frac{1}{4}$
5. $\frac{21x}{8}$
7. $\frac{5+3a}{ax}$
9. $\frac{-x}{2y}$
11. $\frac{3x^2+2x-1}{x^3}$
13. $\frac{7}{12}$
15. $\frac{11}{20}$
17. $\frac{11x}{12}$
19. $\frac{-9y}{35}$
21. $\frac{2y+4x}{xy}$
23. $\frac{yz+xz+xy}{xyz}$
25. $\frac{-x-4}{6}$
27. $\frac{8y-5}{6}$
29. $\frac{11+2x}{6}$
31. $\frac{14x-5}{6x}$
33. $\frac{x-5y}{6x}$
35. $\frac{6a^2-5ab+6b^2}{12ab}$
37. $\frac{3}{2(x+y)}$
39. $\frac{-2}{3(x+1)}$
41. $\frac{9}{4(2a+b)}$
43. $\frac{2x^2}{(x-3)(x+3)}$
45. $\frac{38}{(3x-4)(5x+6)}$
47. $\frac{-3a-9}{(2a+1)(a-2)}$
49. $\frac{-8x}{(x+2)(x-2)}$
51. $\frac{-1}{(x+2)(x+3)}$
53. $\frac{3x^2-3xy+4y^2}{(x+y)(x-y)}$
55. $\frac{3}{(x+1)(x+1)(x-2)}$

57. $\dfrac{x^2 - x}{(x-2)(x+3)(x+5)}$ 59. $\dfrac{2x^2 + 13x + 6}{(x+1)(x+2)(x+2)}$

EXERCISES 5.8 (page 135)

1. $\dfrac{2x}{3}$ 3. $-\dfrac{2a}{5}$ 5. $\dfrac{3(a-b)}{4}$ 7. $-\dfrac{3(2x-y)}{5}$

9. $\dfrac{3}{7}x$ 11. $-\dfrac{5}{7}a$ 13. $\dfrac{5}{2}(a-b)$ 15. $\dfrac{1}{7}(x+y)$

17. $\dfrac{3}{8}$ 19. $\dfrac{81}{110}$ 21. $\dfrac{1}{3}$ 23. y

25. $\dfrac{4x^2}{5}$ 27. $4y$ 29. $\dfrac{-x}{y}$ 31. $\dfrac{49rt}{4}$

33. $\dfrac{-b}{az}$ 35. $\dfrac{3}{4}$ 37. 5 39. 1

41. $\dfrac{x-3}{x+7}$ 43. $\dfrac{3x-2}{3x+2}$ 45. $\dfrac{x+y}{x+3y}$ 47. $\dfrac{x-2}{2x(x+1)}$

49. 1 51. $\dfrac{y-3}{y-6}$

EXERCISES 5.9 (page 139)

1. $\dfrac{9}{10}$ 3. $\dfrac{10}{3}$ 5. $\dfrac{2}{25}$ 7. 1

9. $\dfrac{5}{16b^2}$ 11. $\dfrac{-x}{v}$ 13. $12y$ 15. $\dfrac{-x}{16y}$

17. $\dfrac{3bx}{ay}$ 19. $\dfrac{a}{2}$ 21. $\dfrac{5}{2}$ 23. $\dfrac{5}{6x}$

25. $\dfrac{2x+y}{x+2y}$ 27. $\dfrac{y-1}{y+7}$ 29. $\dfrac{(x-4)(x+3)}{(x-1)(2x+1)}$ 31. $\dfrac{1}{y-1}$

33. $\dfrac{2x-1}{x+3}$ 35. $\dfrac{(a+9)(a-7)(a-3)}{(a-1)(a+6)(a-8)}$

EXERCISES 5.10 (page 142)

1. $\dfrac{3}{2}$ 3. $\dfrac{5}{4}$ 5. $\dfrac{8}{7}$ 7. 6

9. $\dfrac{b}{3a}$ 11. $\dfrac{5}{3xy}$ 13. $\dfrac{1}{4}$ 15. $\dfrac{1}{4}$

17. $\dfrac{1}{14}$ 19. 5 21. $\dfrac{5}{3y^2}$ 23. $\dfrac{y^2-1}{y^2+1}$

25. $\frac{x}{4}$ **27.** $\frac{a}{11b}$

EXERCISES 5.11 (page 145)

1. 1 **3.** 6 **5.** 12 **7.** 2
9. 3 **11.** $\frac{1}{2}$ **13.** 15 **15.** 5
17. 6 **19.** 5 **21.** 4 **23.** 3
25. 3 **27.** 2 **29.** 15 **31.** $\frac{a+b}{2}$
33. -4 **35.** $\frac{4a-3b}{a}$ **37.** No solution **39.** $-2a-6$
41. No solution

EXERCISES 5.12 (page 148)

1. 5 **3.** 14, 15 **5.** 18 **7.** \$40
9. 150 kilograms **11.** 20° **13.** 72 **15.** Slower driver: 40 mph, faster driver: 60 mph
17. $1\frac{3}{4}$ miles per hour **19.** Slower man: 30 mph, faster man: 60 mph

EXERCISES 5.13 (page 152)

1. $\frac{2}{3}$ inches **3.** $\frac{5}{9}$ inches **5.** $\frac{2}{5}$ grams **7.** $\frac{2}{5}$
9. $\frac{3}{10}$ **11.** $\frac{8}{5}$ **13.** $\frac{8}{3}=\frac{24}{9}$ **15.** $\frac{21}{24}=\frac{7}{8}$
17. $\frac{15}{x}=\frac{10}{4}$ **19.** $\frac{6}{2}=\frac{x}{x+1}$ **21.** 15 **23.** 15
25. $\frac{3}{2}$ **27.** 4 **29.** -2 **31.** 4
33. 55.88 centimeters **35.** 15.25 miles
37. 16.36 kilograms **39.** 30.19 liters
41. 40 pounds **43.** 80 liters
45. 135 liters **47.** 825 bricks
49. \$2.24

CHAPTER 5 REVIEW (page 155)

1.

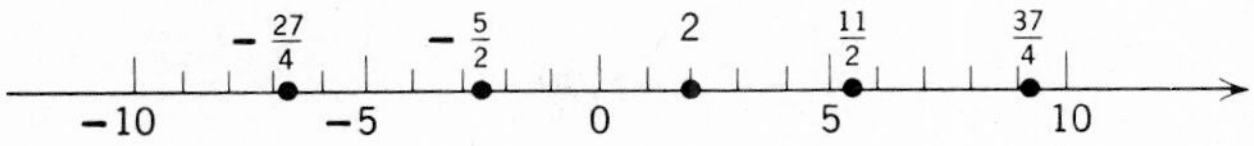

2. a. $\frac{-3}{x+y}$

b. $\frac{a}{x}$

c. $\frac{-(b-2)}{4}$

3. a. $\frac{2(x-3)}{3}$

b. $\frac{-(x^2+1)}{3}$

c. $\frac{-3(2x+y)}{4}$

4. a. $\frac{xz^2}{y}$

b. $\frac{1}{b+1}$

c. $\frac{1}{a-1}$

5. a. $\frac{6}{2(x-y)}$

b. $\frac{3(a+2)}{a^2+5a+6}$

c. $\frac{x(x-1)}{x^2-3x+2}$

6. a. $\frac{10}{15}, \frac{9}{15}$

b. $\frac{3y}{x^2y^2}, \frac{-2x}{x^2y^2}$

c. $\frac{a}{a^2-1}, \frac{3(a-1)}{a^2-1}$

7. a. $\frac{4}{5}$

b. $\frac{x}{y}$

c. $\frac{-5}{3}$

8. a. $\frac{7}{3x}$

b. $\frac{3s+2r}{rs}$

c. $\frac{2a-3b}{a^2b^2}$

9. a. $\frac{4a+2b}{(a-b)(a+b)}$

b. $\frac{a^2-a+1}{a(a-1)(a+1)}$

c. $\frac{6x+26}{(x+1)(x+5)(x-5)}$

10. a. $\frac{x^2}{6}$

b. $\frac{5(x-2)}{x}$

c. 1

11. a. $\frac{7s}{r}$

b. $\frac{(a-b)(a-1)}{a}$

c. x^3

12. a. $\frac{9}{4}$

b. $\frac{5}{7}$

c. $\frac{6}{11}$

13. a. $\frac{b-a}{b+2}$

b. $\frac{x^2(y-1)}{y^2(x-1)}$

c. $\frac{1+3y}{2y-3}$

14. a. 6

b. $\frac{-4}{3}$

c. 7

15. a. 3

b. 4

c. $\frac{-1}{2}$

16. a. 1

b. −1

c. 3

17. a. $\frac{2b}{3a}$

b. $\frac{-b}{9}$

c. 2

18. 2

19. 9

20. faster car: 30 mph,
slower car: 20 mph

CHAPTER 5 CUMULATIVE REVIEW (page 156)

1. $C = \frac{5F - 160}{9}$

3. Commutative

4. 13

5. 0

6. 375

7. $\frac{42}{13}$

8. -28

9. $x + 4$

10. $x(x-1)(x+1)$

11. $5b(b+1)(b+1)$

12. $x - 4$

13. $\frac{9}{5}$

14. 5

15. 6

16. $\frac{ab}{c}$

17. 160 kilometers

18. 30 nickels,
20 dimes

19. 8, 10, 12

20. $24

EXERCISES 6.1 (page 160)

1. a. 4
b. d, t
c. Increases

3. a. $(2,7)$
b. $(3,9)$
c. $(0,3)$
d. $(-2,-1)$
e. $(-3,-3)$
f. $(-6,-9)$

5. a. $(3,11)$
b. $(1,3)$
c. $(0,-1)$
d. $(-3,-13)$
e. $(-4,-17)$
f. $(-5,-21)$

7. a. $(2,6)$
b. $(1,3)$
c. $(0,0)$
d. $(-1,-3)$
e. $(-2,-6)$
f. $(-3,-9)$

9. a. $(4,13)$
b. $(1,7)$
c. $(0,5)$
d. $(-1,3)$
e. $(-2,1)$
f. $(-3,-1)$

11. a. $(5,-14)$
b. $(3,-10)$
c. $(0,-4)$
d. $(-1,-2)$
e. $(-3,2)$
f. $(-5,6)$

13. a. $\left(10, \frac{14}{3}\right)$
b. $\left(5, \frac{4}{3}\right)$
c. $(0, -2)$
d. $\left(-5, -\frac{16}{3}\right)$
e. $\left(-10, -\frac{26}{3}\right)$
f. $(-15, -12)$

15. a. $\left(7, -\frac{21}{2}\right)$
b. $\left(5, -\frac{13}{2}\right)$
c. $\left(0, \frac{7}{2}\right)$
d. $\left(-3, \frac{19}{2}\right)$
e. $\left(-5, \frac{27}{2}\right)$
f. $\left(-7, \frac{35}{2}\right)$

EXERCISES 6.2 (page 163)

1.

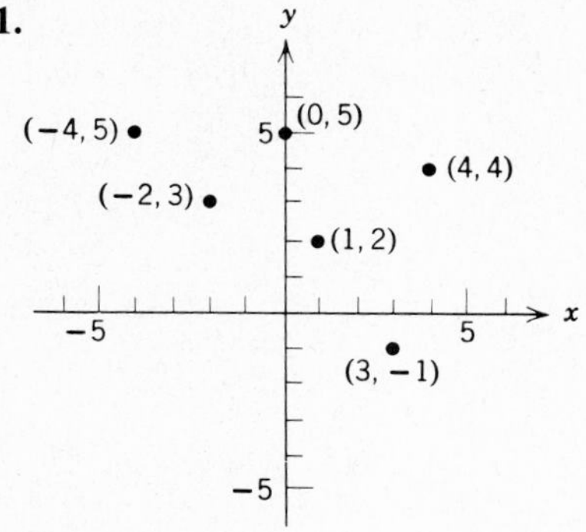

3.

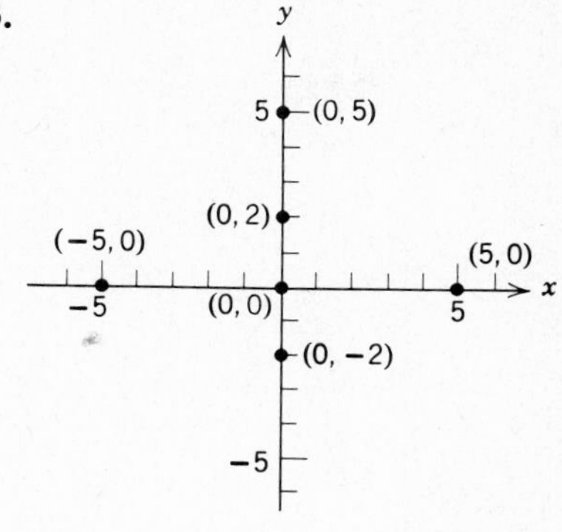

5.

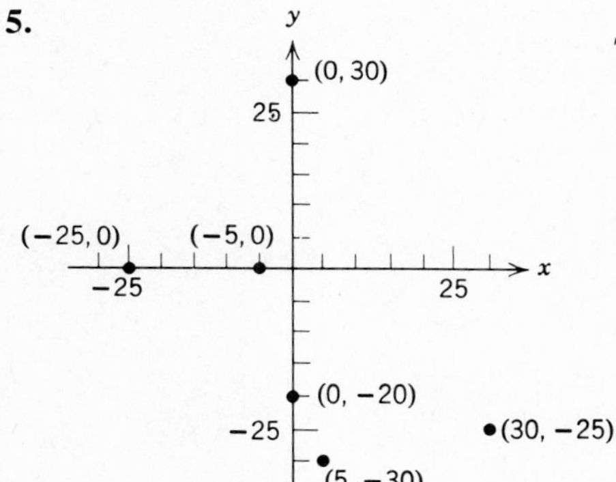

7.

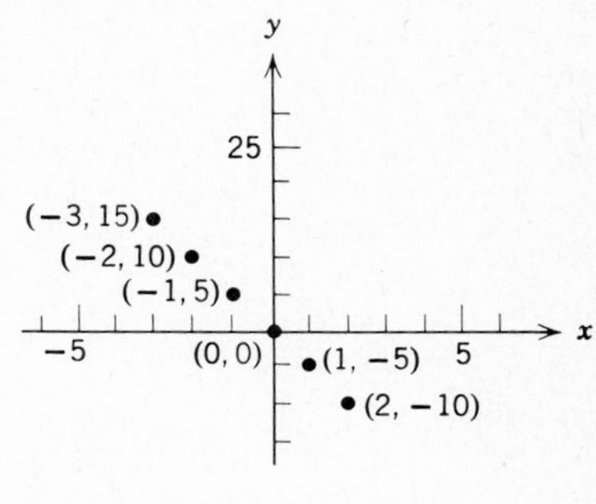

9. Result is a straight line

11. a. Ordinate
b. Abscissa

13. Origin

15. On a straight line bisecting the angles at the origin in the first and third quadrants

17.

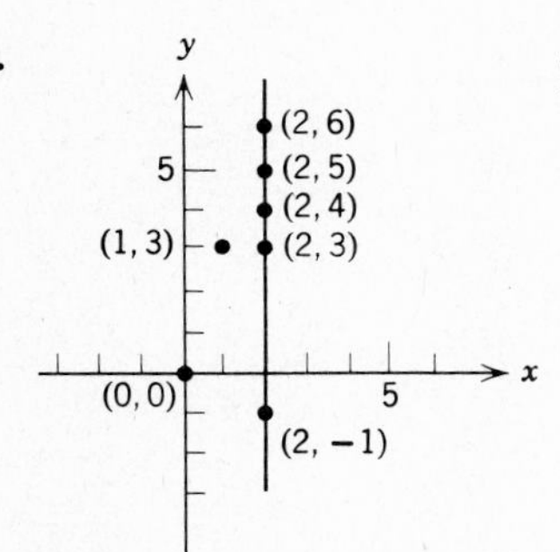

a. Yes
b. No
c. Yes
d. No

19. 2

EXERCISES 6.3 (page 165)

1.

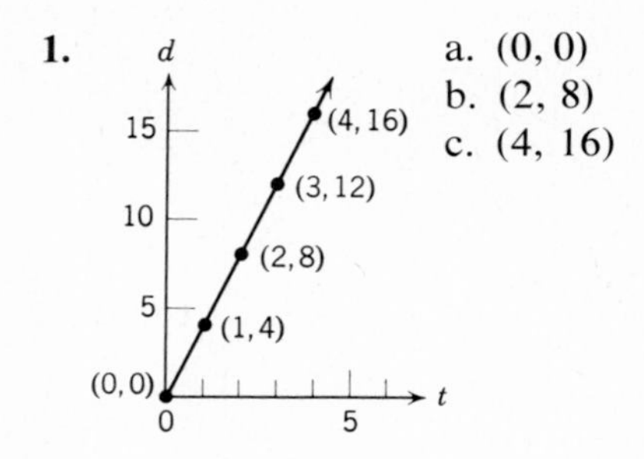

a. (0, 0)
b. (2, 8)
c. (4, 16)
d. On the line through these points
e. (1, 4), (3, 12)

3.

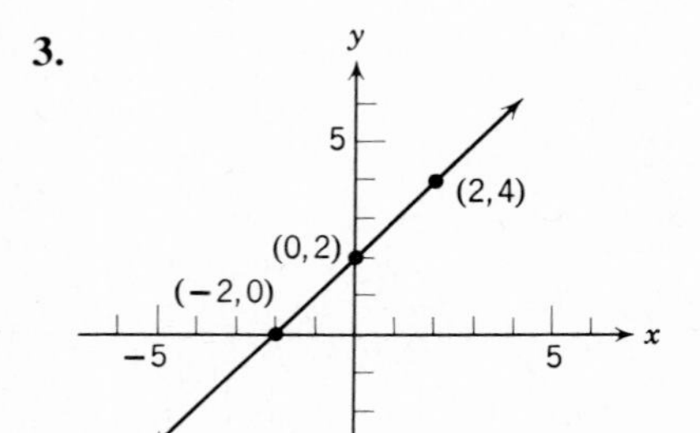

5.

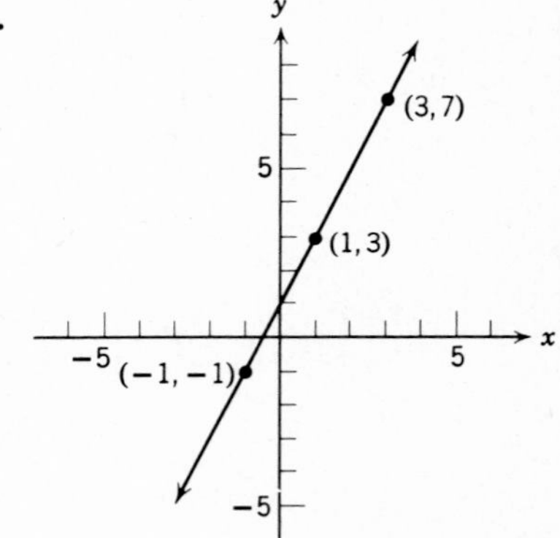

7.

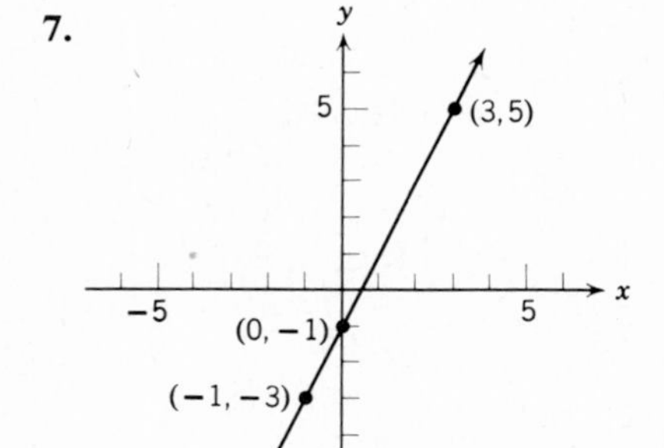

9.

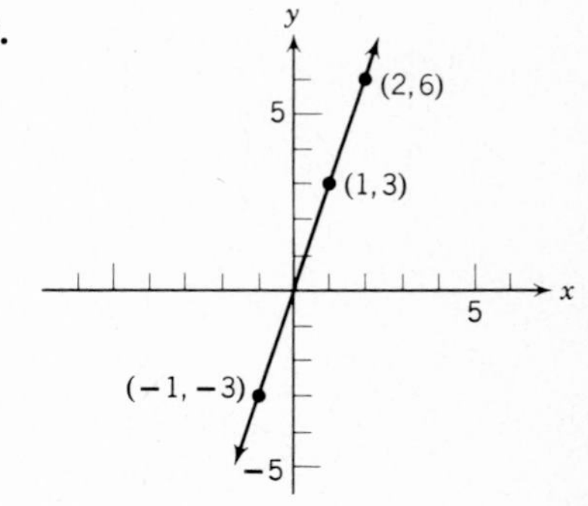

11.

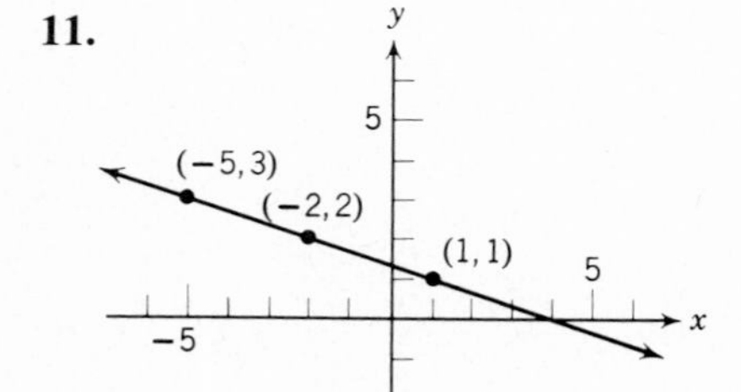

13.

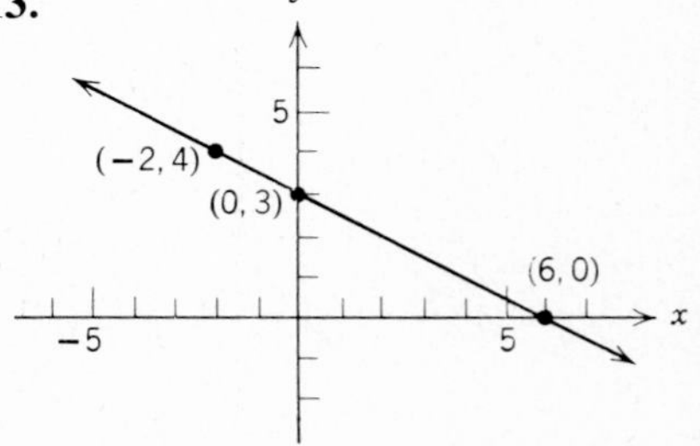

15.

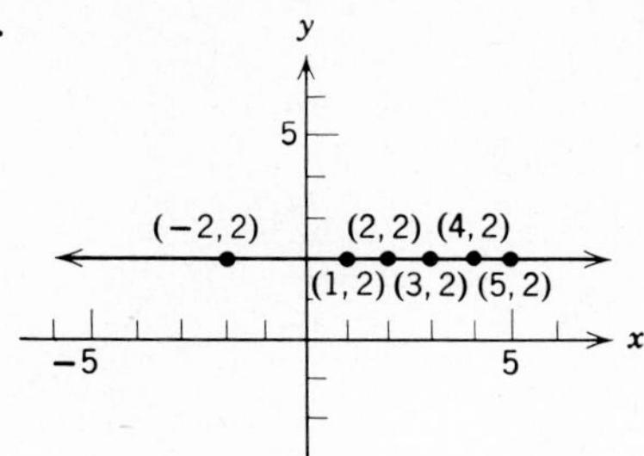

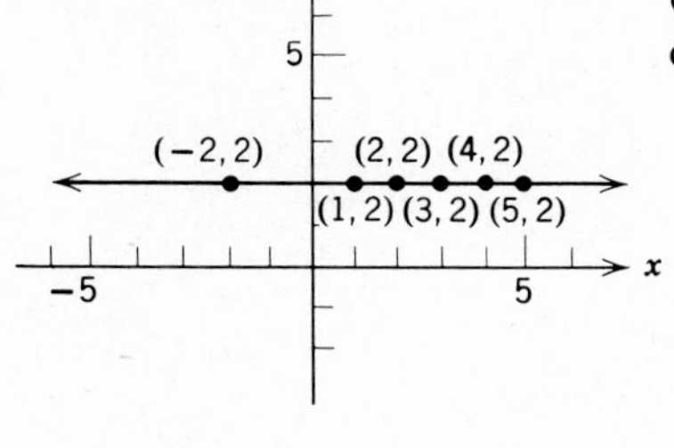

b. Yes
c. Yes
d. No
e. Yes
f. Yes

17.

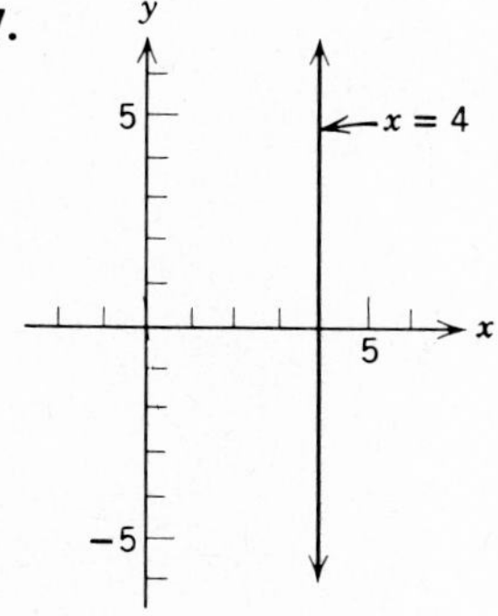

19.

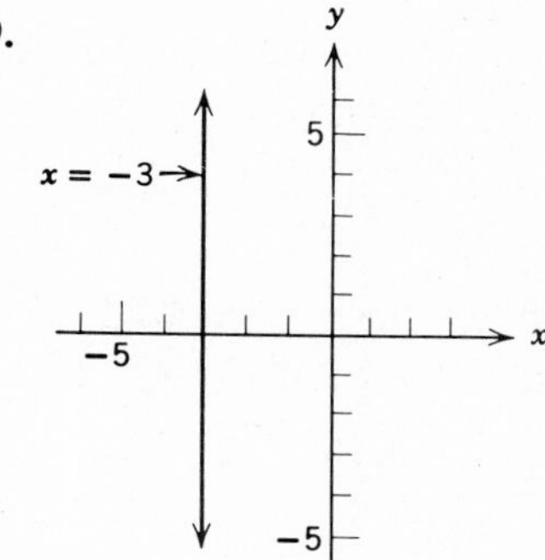

21.

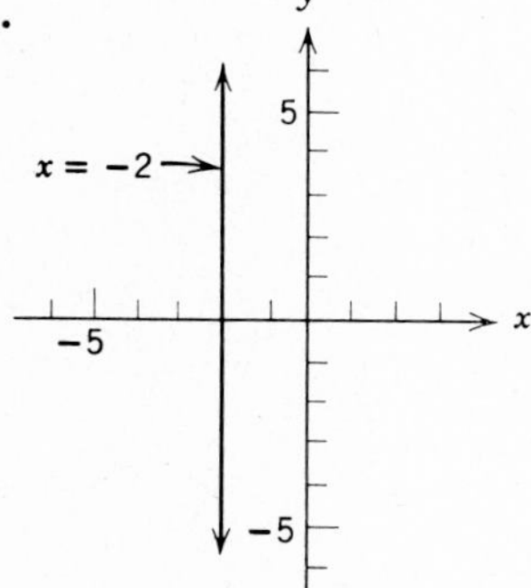

23.

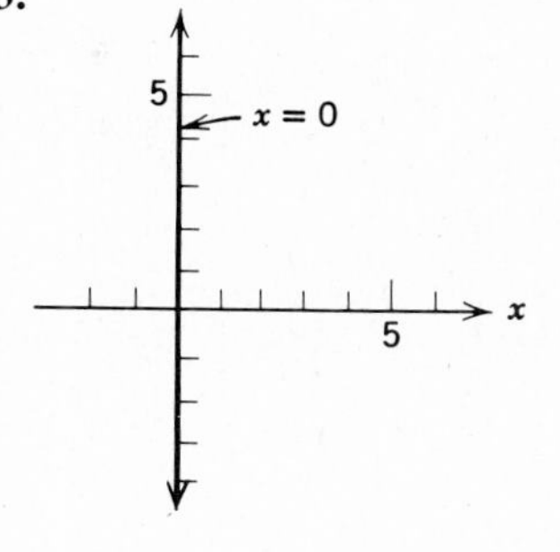

EXERCISES 6.4 (page 169)

1.

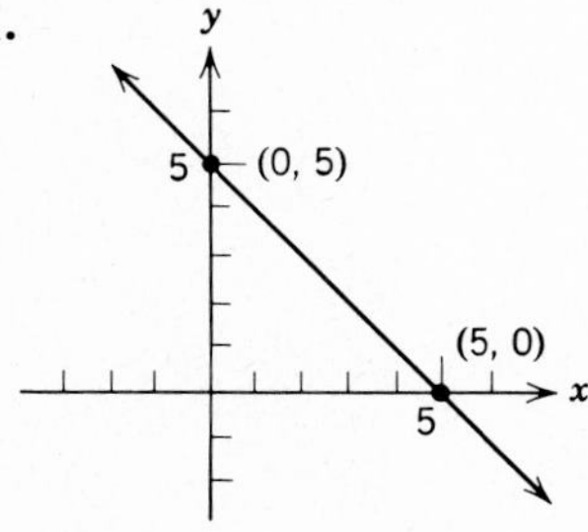

3.

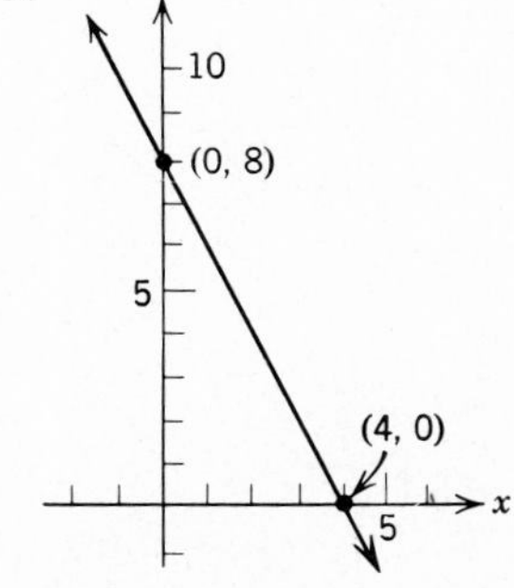

5.

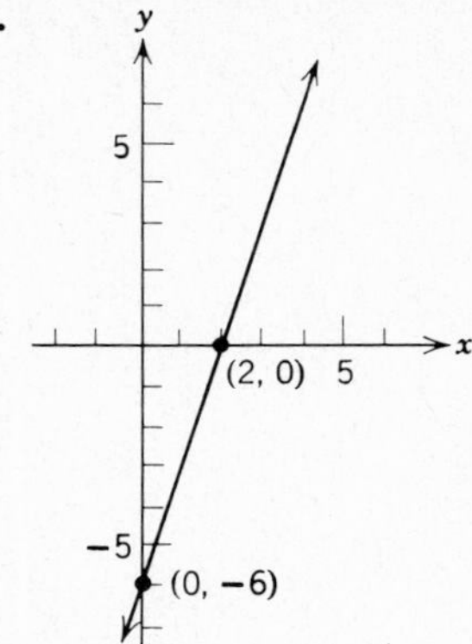

7.

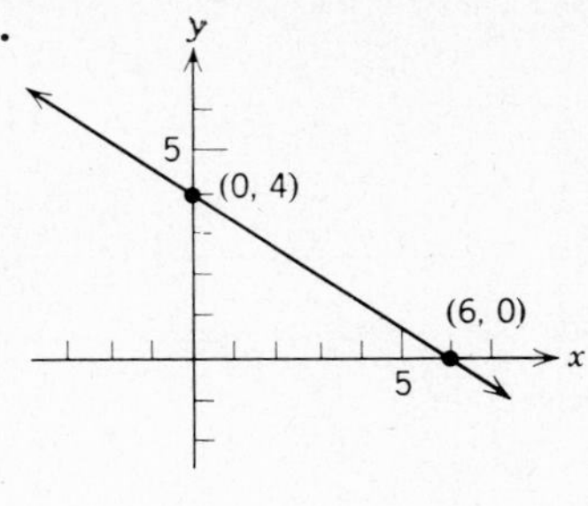

9.

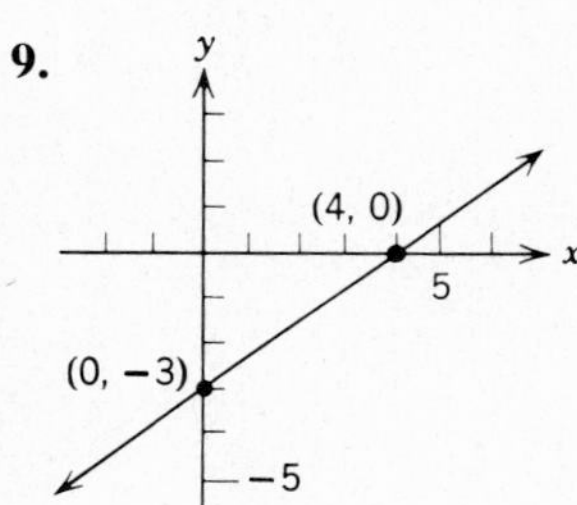

11.

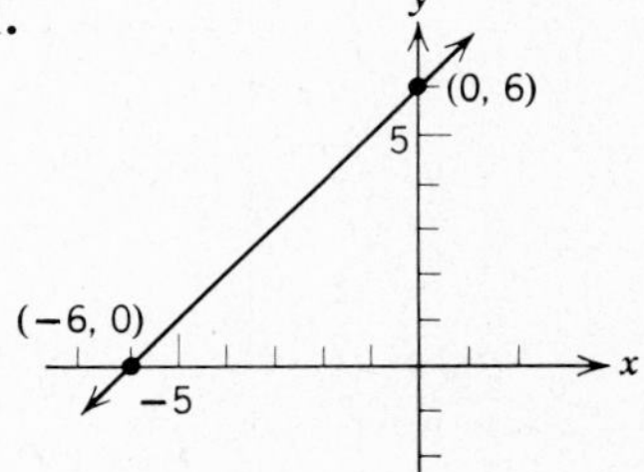

13.

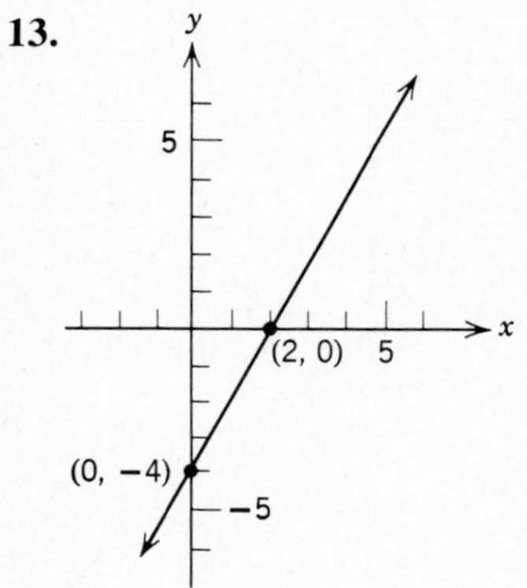

15.

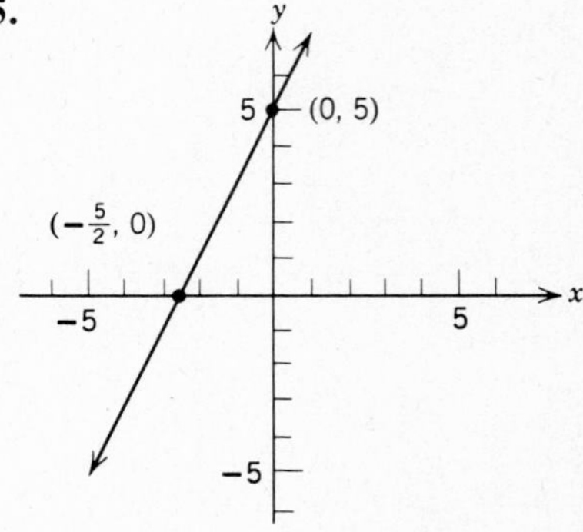

17.

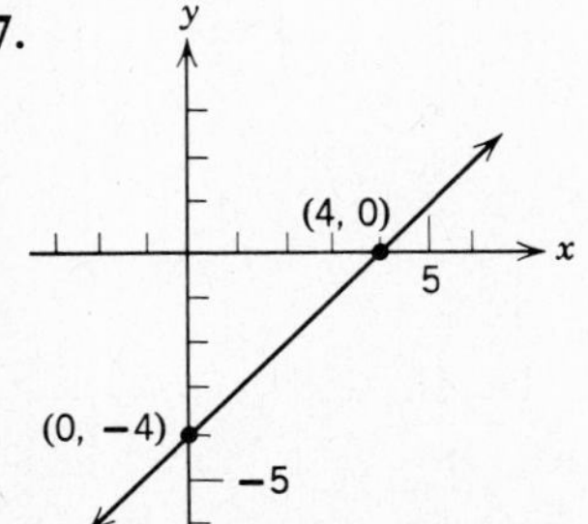

19.

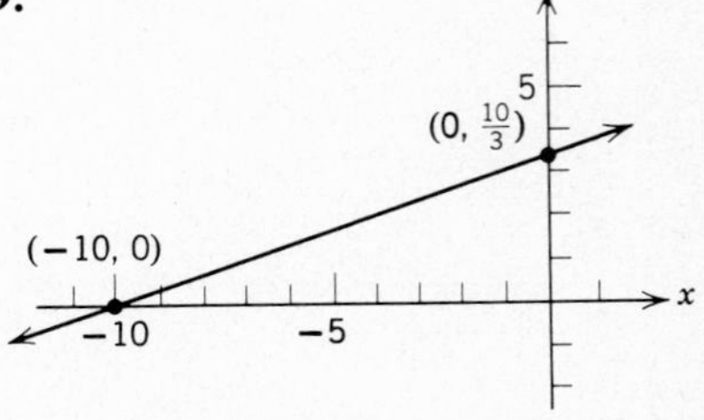

21.

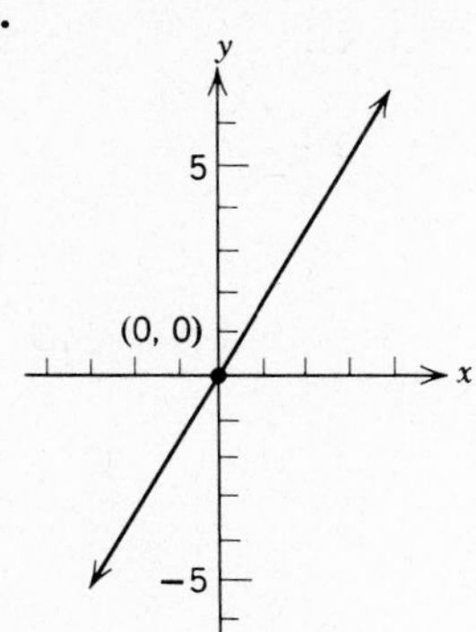

23.

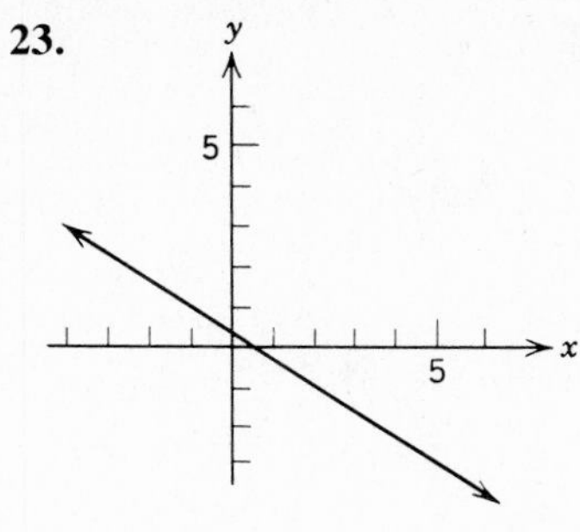

EXERCISES 6.5 (page 172)

1.

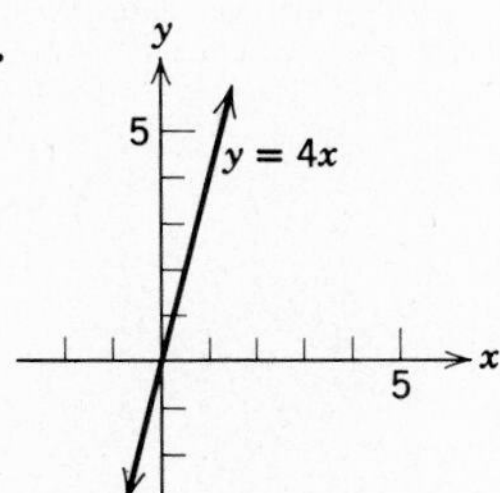

3.

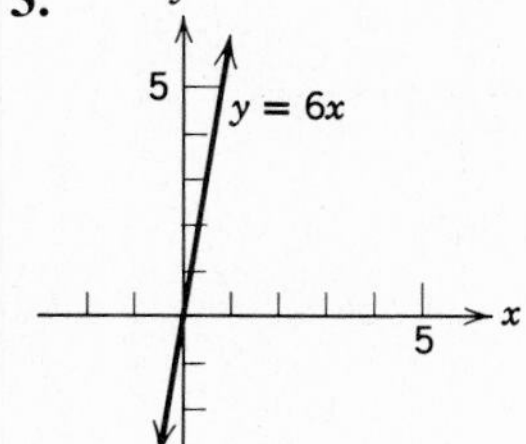

5.

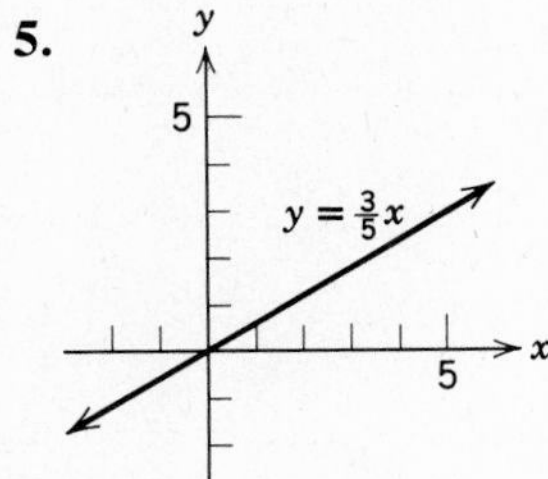

7.

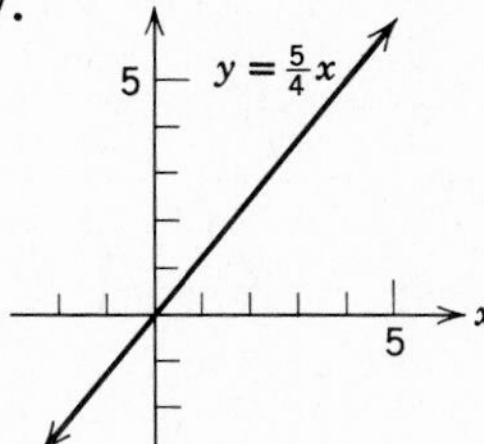

9.

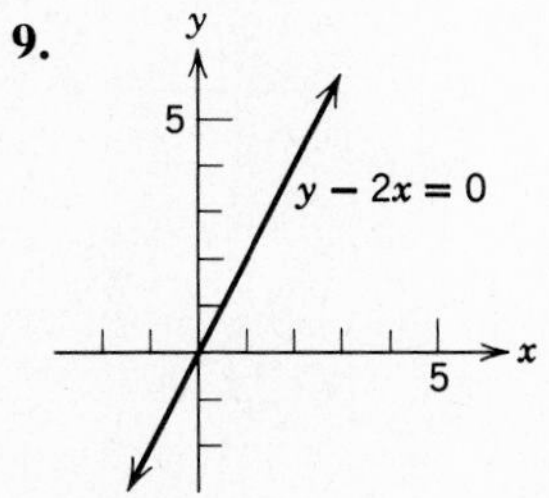

11.

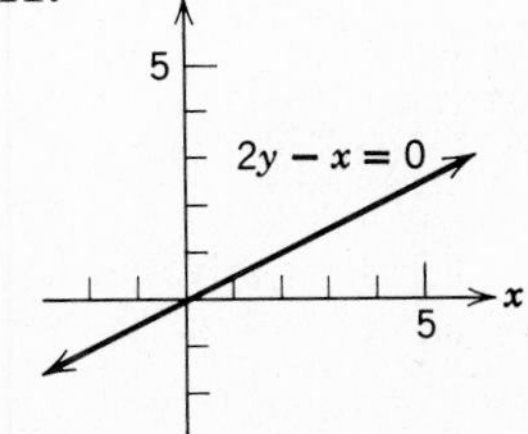

13. $y = 21$ **15.** $R = 63$ **17.** $d = 168$ *miles*

19. $E = 94.3$

EXERCISES 6.6 (page 175)

1. (2, 6)

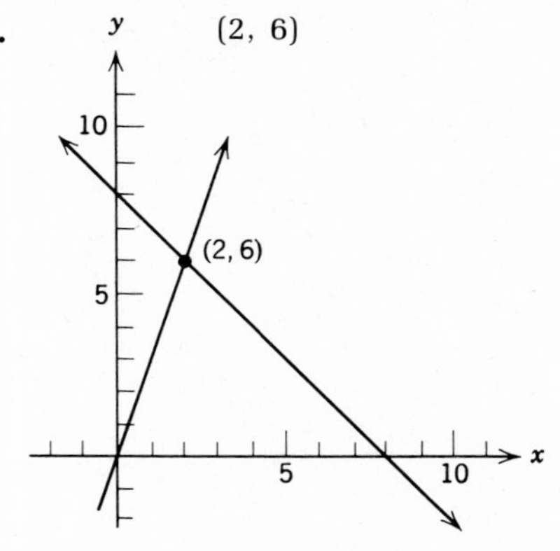

3. (1, 2)

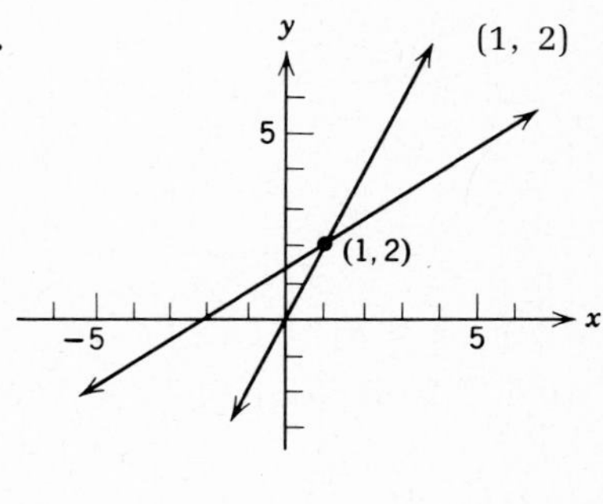

5. (−3, −1)

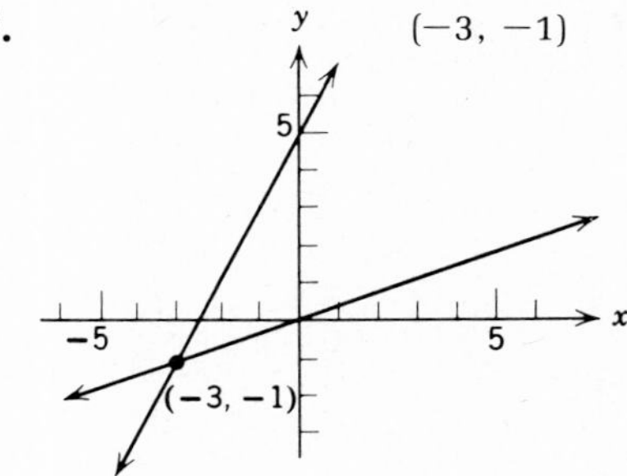

7. (2, 2)

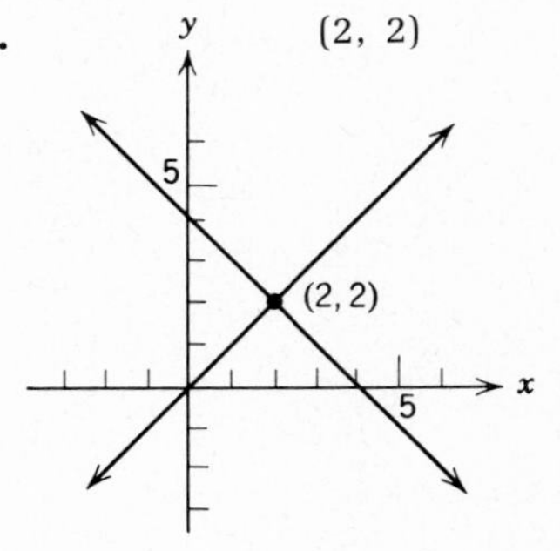

9. (4, 2)

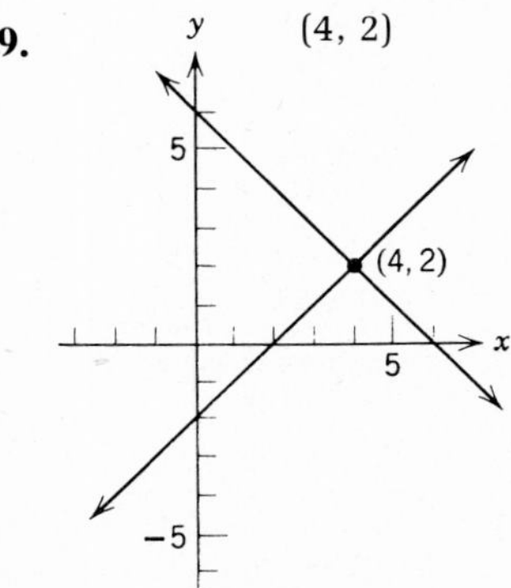

11. (−3, −2)

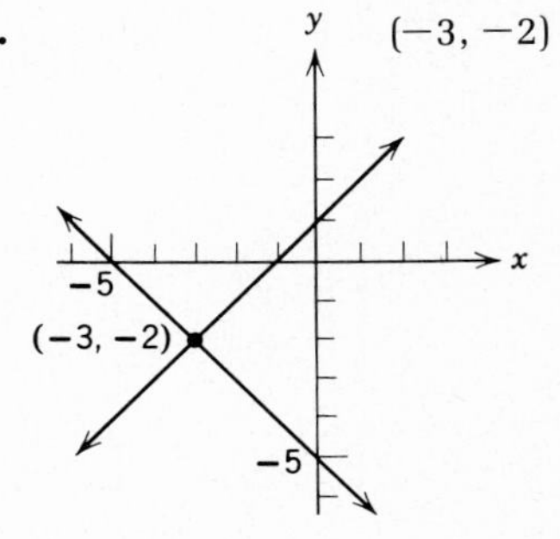

13. Dependent

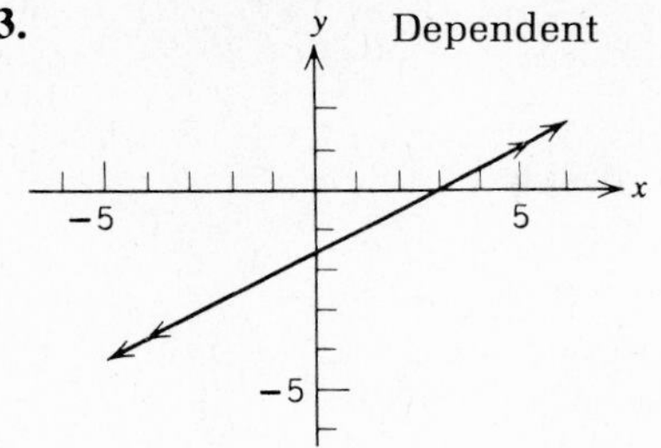

15. (1, 7)

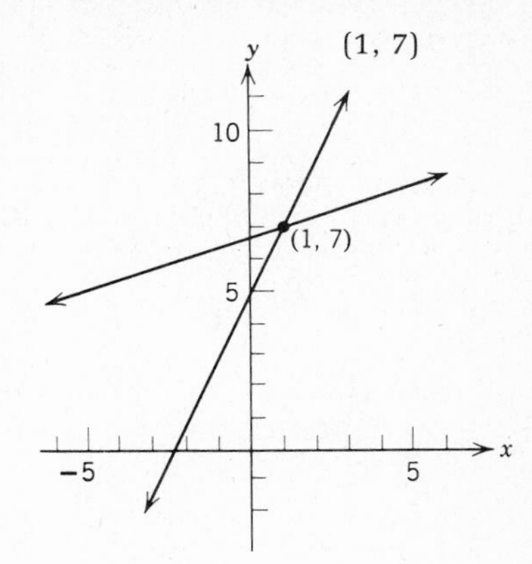

17. Inconsistent

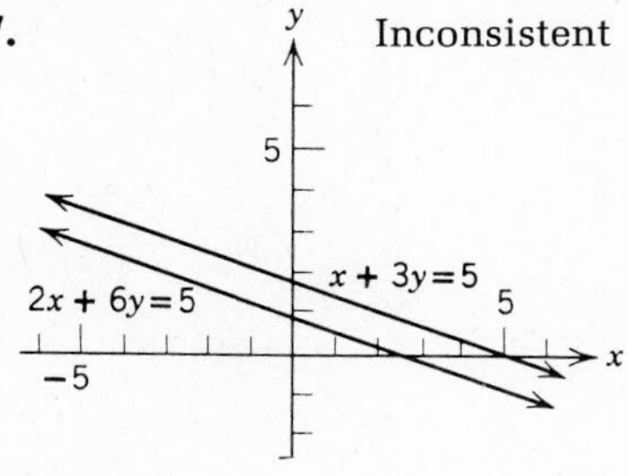

19. (−2, −1)

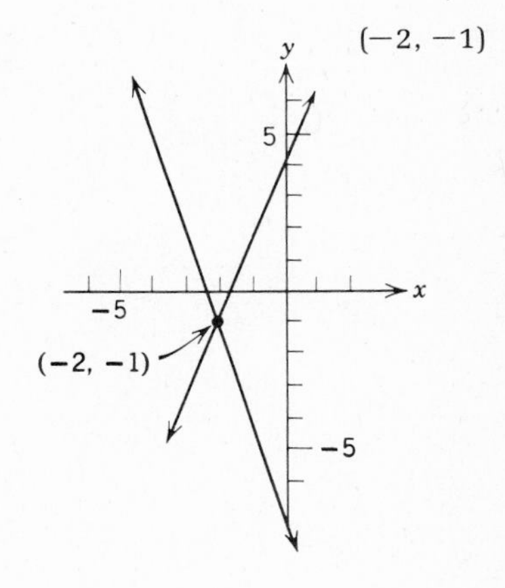

EXERCISES 6.7 (page 177)

1. (3, 2) **3.** (−1, 4) **5.** (6, 2) **7.** (1, 1)
9. (7, −2) **11.** (−1, 1) **13.** (1, 1) **15.** (5, 2)
17. (1, −1) **19.** (−4, 10) **21.** (5, −1) **23.** (4, 3)
25. (−4, 3) **27.** (−2, −1) **29.** (0, 2) **31.** (−4, 2)
33. (3, 1) **35.** (4, 3)

EXERCISES 6.8 (page 180)

1. (1, 2) **3.** (2, 2) **5.** (−2, −1) **7.** (−3, −2)
9. (5, −2) **11.** (8, 5) **13.** (1, 2) **15.** (10, −2)
17. (1, 1) **19.** (−1, −3) **21.** (−1, 1) **23.** (1, 0)
25. (1, −1) **27.** (4, 3) **29.** (4, 3) **31.** (−3, 2)
33. (2, 3) **35.** (−4, 10) **37.** (6, 2) **39.** Dependent

EXERCISES 6.9 (page 182)

1. 8, 17
3. 9 meters, 11 meters
5. 16 kilograms, 12 kilograms
7. Lot: \$3500, house: \$8500
9. Walls: 9 hours, trim: 15 hours
11. 13 kilometers
13. 60 flatcars
15. \$14 for doll, \$41 for train
17. 6, 18
19. Dimes: 20, quarters: 14
21. \$2160 at 4%, \$1440 at 6%

CHAPTER 6 REVIEW (page 185)

1. a. 0.05
   b. I, P
   c. Increases
2. $y = 2x - 4$
3. $y = \dfrac{3x + 6}{2}$
4. a. (3, 7)
   b. (−2, −3)
   c. (0, 1)
   d. (−½, 0)
5. a. (4, −1)
   b. (−2, −7)
   c. (0, −5)
   d. (−6, −11)

6.

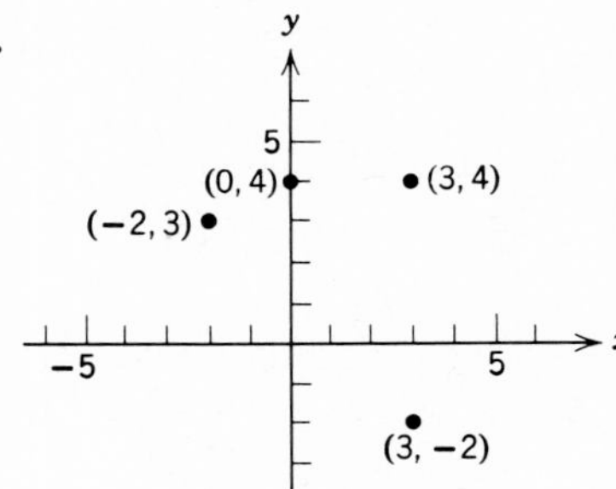

7.

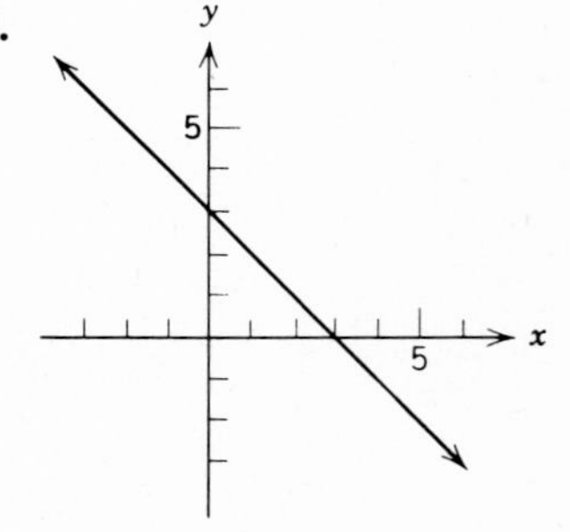

8.

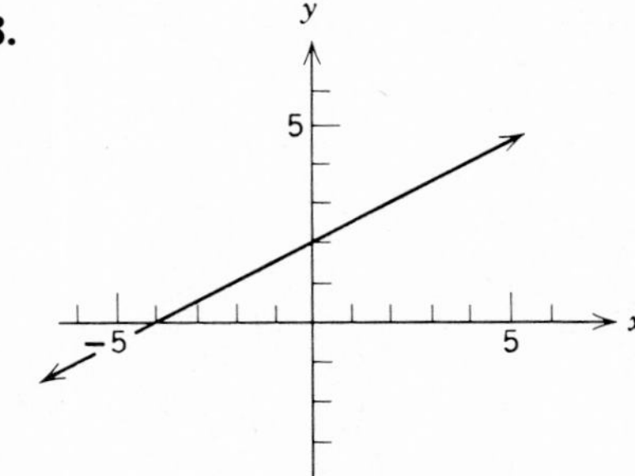

9.

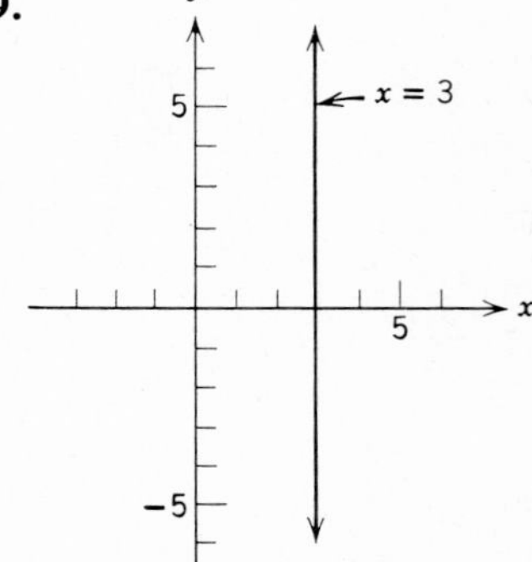

10.

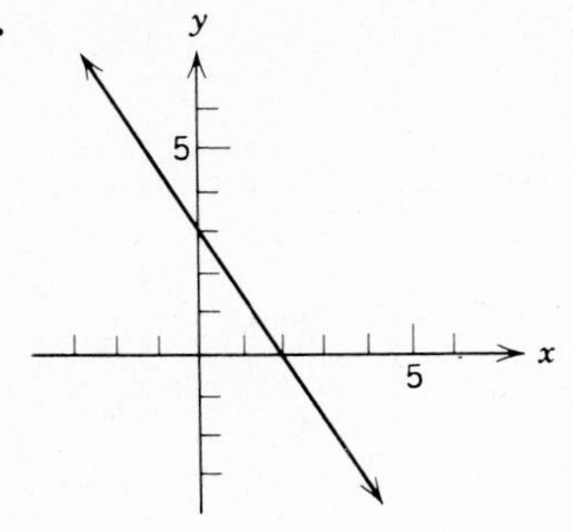

11.

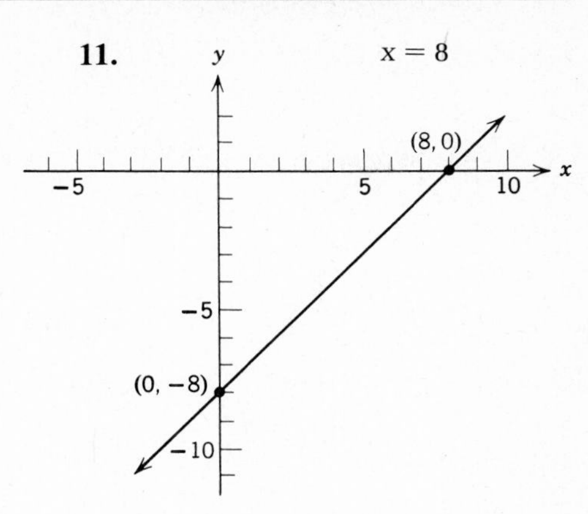

12. $y = -8$

13. $m = \dfrac{5}{2}$

14. $x = 13.2$

15.

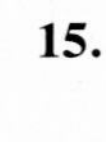

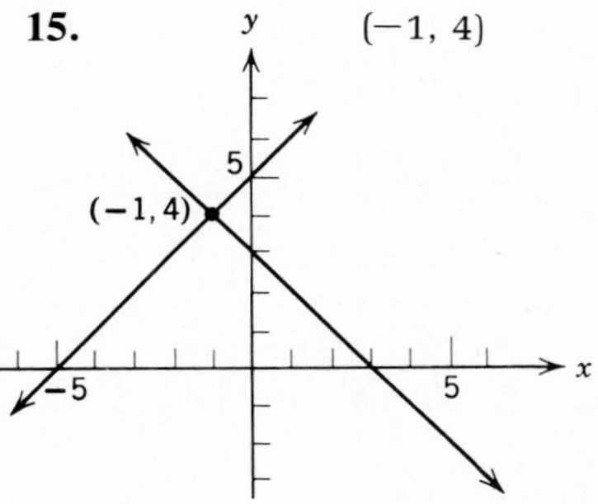

16.

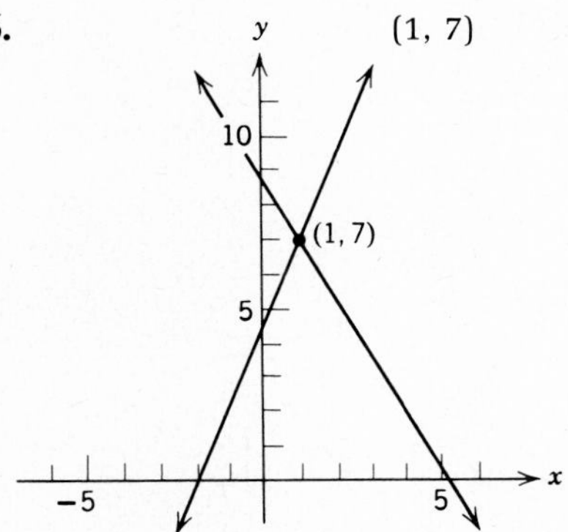

17. 32 pounds, 52 pounds

18. Quarters: 7, dimes: 18

19. $-16, -24$

20. $0, -2$

CHAPTER 6 CUMULATIVE REVIEW (page 186)

1.

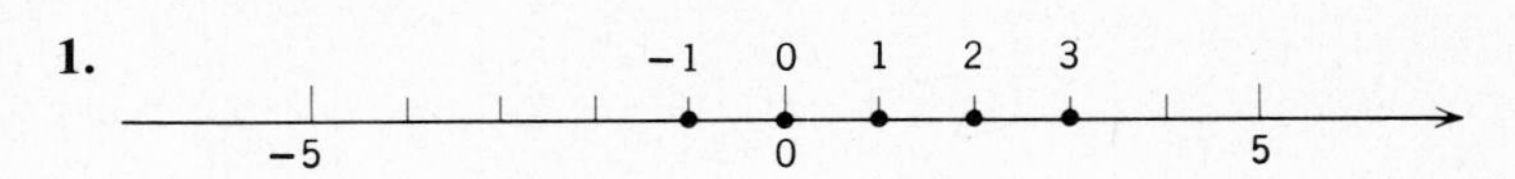

2. $\frac{1}{2}$ **3.** 7 **4.** 2064

5. $\frac{3}{2}$ **6.** $\frac{a+b}{a-b}$ **7.** $\frac{24a-5}{24}$

8. $\frac{-x+8}{(x+1)(x-2)}$ **9.** $\frac{3}{5}$ **10.** $\frac{3}{4}$

11. x **12.** $x-4$ **13.** 12

14. 1 **15.** 2 **16.** $ab-a$

17.

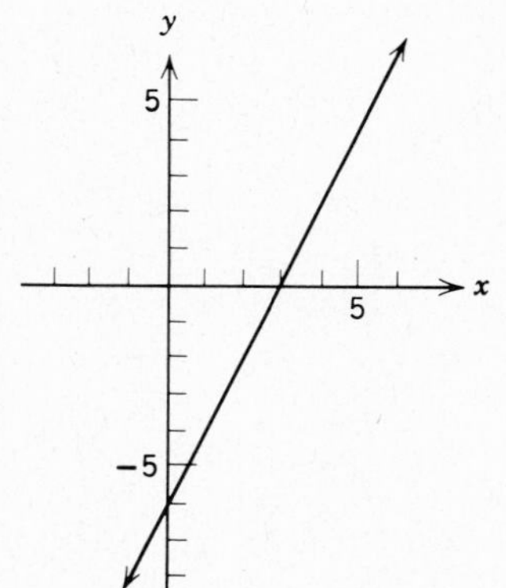

18. $(4, -1)$

19. 2, 26 **20.** 7, 8

REVIEW OF FACTORING (page 187)

1. $5(x+2y)$ **2.** $3(x^2+2x-1)$ **3.** $-2(x^2+2)$
4. $a(b-c)$ **5.** $a(bc+b-c)$ **6.** $xy(x-y+1)$
7. $(x+3)(x+2)$ **8.** $(y-3)^2$ **9.** $(y-8)(y+1)$
10. $(y+7)(y-5)$ **11.** $2(x+5)(x-2)$ **12.** $3(y+4)(y-1)$
13. $(2y-3)(y+1)$ **14.** $(3y-1)(2y+1)$ **15.** $(3x-2)(2x-3)$
16. $(3x+7)(x-5)$ **17.** $(3x+4)(2x-1)$ **18.** $(4y+1)(2y-1)$
19. $2(2y+1)(y+1)$ **20.** $3(2x+1)(x+3)$ **21.** $2(3x+1)(3x-8)$
22. $5(8x-5)(2x+1)$ **23.** $2(2x+1)(x-3)$ **24.** $3(3y-2)(2y+1)$
25. $(x+a)^2$ **26.** $(x-a)^2$ **27.** $(y+3b)^2$
28. $(y-2b)^2$ **29.** $(x+4a)^2$ **30.** $(y-5b)^2$
31. $(x+4)(x-4)$ **32.** $(y+6)(y-6)$ **33.** $(y+b)(y-b)$
34. $(x+c)(x-c)$ **35.** $(2x+5)(2x-5)$ **36.** $(3y+2c)(3y-2c)$

37. $12(y+2)(y-2)$ **38.** $2(5x+4)(5x-4)$

EXERCISES 7.1 (page 189)

1. 7 **3.** 2 **5.** $\frac{1}{3}$

7. -3 **9.** 3, 5 **11.** 0, -4

13. 3, -3 **15.** $-\frac{1}{3}, \frac{3}{2}$ **17.** $0, -\frac{4}{5}, \frac{7}{2}$

19. 2, 3 **21.** 0, 4 **23.** 0, -3

25. 2, -3 **27.** 1, -8 **29.** -2, 3

31. 0, 4 **33.** -4, 3 **35.** $\frac{3}{2}, -\frac{3}{4}$

37. $\frac{2}{3}, -\frac{2}{3}$ **39.** $0, -\frac{3}{2}$ **41.** 3, 2, 1

43. 0, -2, 1 **45.** 0, -4, 3 **47.** $0, -\frac{1}{2}, \frac{1}{2}$

49. $a, -a$ **51.** $-\frac{a}{2}, a$ **53.** $0, \frac{b}{4}$

55. $0, \frac{-3b}{2}, \frac{2b}{3}$ **57.** $0, b, -\frac{b}{8}$

EXERCISES 7.2 (page 192)

1. 0, -3 **3.** $0, \frac{5}{2}$ **5.** $0, \frac{9}{2}$

7. 0, 4 **9.** 1, -1 **11.** 2, -2

13. 4, -4 **15.** 8, -8 **17.** 3, -3

19. 3, -3 **21.** $\frac{1}{3}, -\frac{1}{3}$ **23.** $\frac{3}{2}, -\frac{3}{2}$

25. 4, -4 **27.** 0, -2 **29.** 0, -2

31. 0, 5 **33.** $a, -a$ **35.** $c, -c$

37. 0, 1 **39.** $0, \frac{c}{3}$ **41.** $0, \frac{1}{b}$

43. $2a, -2a$ **45.** $\frac{3b}{a}, -\frac{3b}{a}$

EXERCISES 7.3 (page 194)

1. 1, 2 **3.** -2, -2 **5.** -1, 4

7. 3, -7 **9.** 2, -7 **11.** -6, -6

13. 1, 15 **15.** $-1, 6$ **17.** $-1, 4$

19. $5, -6$ **21.** $-1, \frac{4}{3}$ **23.** $\frac{1}{3}, \frac{3}{2}$

25. $\frac{1}{2}, \frac{1}{2}$ **27.** $\frac{3}{2}, \frac{3}{2}$ **29.** $\frac{1}{3}, \frac{1}{3}$

31. $1, -\frac{7}{2}$ **33.** $-4, 5$ **35.** $\frac{3}{2}, -2$

37. $-\frac{3}{2}, -\frac{3}{2}$ **39.** $-\frac{3}{4}, -\frac{5}{2}$ **41.** $-5, 3$

43. $-1, -2$ **45.** $1, -3$ **47.** $\frac{1}{2}, 1$

49. $-\frac{1}{2}, 2$ **51.** $-1, 2$ **53.** $-2, 3$

55. 1, 1 **57.** $-\frac{1}{2}, 2$ **59.** $\frac{1}{3}, -\frac{3}{2}$

61. 1, 1 **63.** $-3, 5$ **65.** 2, 2

67. $-2, 5$ **69.** 10, 20 **71.** 2, 13

73. $5, -6$

EXERCISES 7.4 (page 199)

1. 0, 5 **3.** 8, 9 **5.** 4

7. 5 seconds **9.** 6 hours **11.** $\frac{3}{4}, \frac{4}{3}$

13. 5, 7 **15.** 8, 9 **17.** 10 mph

19. Rate going: 45 mph,
rate returning: 60 mph

CHAPTER 7 REVIEW (page 203)

1. a. $2, -5$ b. $0, -3$ **2.** a. 0, 2 b. 0, 2

3. a. $7, -7$ b. $3, -3$ **4.** a. $6, -6$ b. $\frac{3}{5}, -\frac{3}{5}$

5. a. $\frac{b^2}{a}, \frac{-b^2}{a}$ b. $0, ca^2$ **6.** a. $-1, 5$ b. $-2, 9$

7. a. $-1, 3$ b. 2, 2 **8.** a. $3, -4$ b. 4, 4

9. a. $-2, 8$ b. $2, -4$ **10.** a. $-1, 4$ b. $-2, 3$

11. a. $-3, 5$ b. $4, -6$ **12.** 5, 12

13. 5, 6 **14.** 5, 6

CHAPTER 7 CUMULATIVE REVIEW (page 203)

1. -1 **2.** 0 **3.** $-3, -\frac{5}{2}, -\frac{5}{3}, 0, \frac{3}{8}, \frac{3}{7}, 3$

4. $x(x-1)(x-2)$ **5.** $2x^2+5x-12$ **6.** $\dfrac{x+2}{3}$

7. $2x-6$ **8.** $\dfrac{3-2x}{x(x+1)}$ **9.** $\dfrac{3(x+3)}{x^2+2x-3}$

10. $\dfrac{3x+2}{2x+3}$ **11.** $(3, 3)$

12.

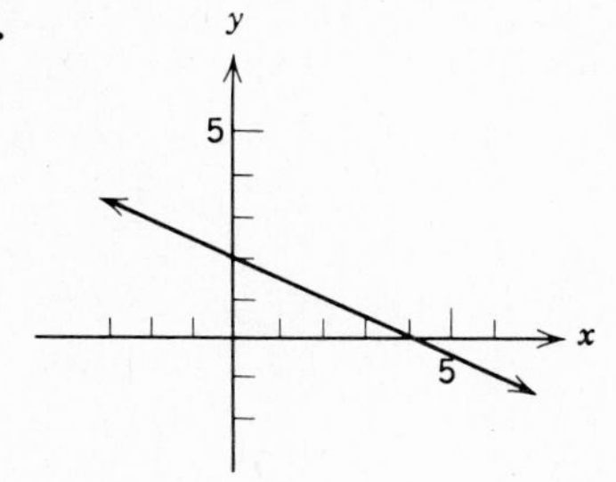

13. 44

14. 54.5 kilograms **15.** $\dfrac{20}{15}$ **16.** Lighter package: 55 kilograms, heavier package: 85 kilograms

17. $y=1$ or $y=0$ **18.** $x=\dfrac{3}{5}$ or $x=-\dfrac{3}{5}$

19. $x=3$ or $x=-1$ **20.** $T=31.04$

EXERCISES 8.1 (page 206)

1. 4 **3.** -9 **5.** ± 12

7. 3 **9.** $\dfrac{1}{6}$ **11.** $\pm\dfrac{2}{3}$

13. $\sqrt{9}$ **15.** $-\sqrt{49}$ **17.** $\sqrt{169}$

19. $-\sqrt{25}$ **21.** $\sqrt{\dfrac{1}{4}}$ **23.** $-\sqrt{\dfrac{49}{64}}$

25. x **27.** $2x$ **29.** $-ac^2$

31. $\pm 6a^3$ **33.** $11ab$ **35.** $-(a+b)$

37. $\dfrac{a}{b}$ **39.** $\pm\dfrac{3}{xy}$ **41.** $\sqrt{x^2}$

43. $\sqrt{x^2y^4}$ **45.** $\sqrt{16y^4}$ **47.** $-\sqrt{64x^4y^4}$

49. $-\sqrt{100x^{20}y^{20}}$ **51.** $\pm\sqrt{(2x+y)^2}$ **53.** $\sqrt{\dfrac{1}{4}}$

55. $\sqrt{\dfrac{9}{16}}$ **57.** $\pm\sqrt{\dfrac{16b^4}{25}}$ **59.** $\sqrt{\dfrac{1}{a^2}}$

61. $-\sqrt{\dfrac{9x^2}{y^2}}$ **63.** $\sqrt{\dfrac{(a+b)^2}{a^2}}$

EXERCISES 8.2 (page 208)

1. Rational; 6 **3.** Irrational **5.** Irrational
7. Rational; 5 **9.** Rational; 12 **11.** Irrational
13. Rational; $\frac{2}{3}$ **15.** Irrational **17.** Rational; 3
19. Irrational **21.** 7.55 **23.** 1.73
25. 2.24 **27.** −5.10 **29.** 3.46
31. −8.66 **33.** 1.41 **35.** −1.73
37. 2.73 **39.** 1.59 **41.** 12.94
43. −22.88 **45.** 2.91 **47.** 0.51
49. 0.32 **51.** 0.72

53.
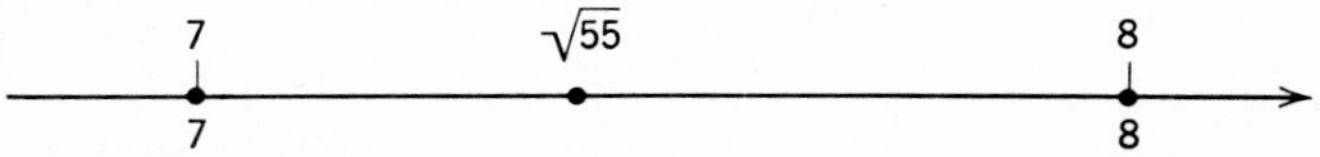

55.
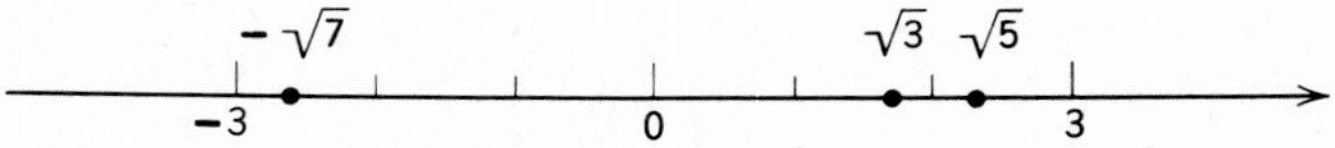

57.
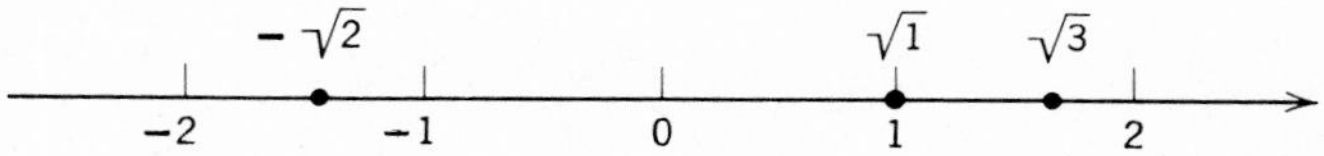

59.
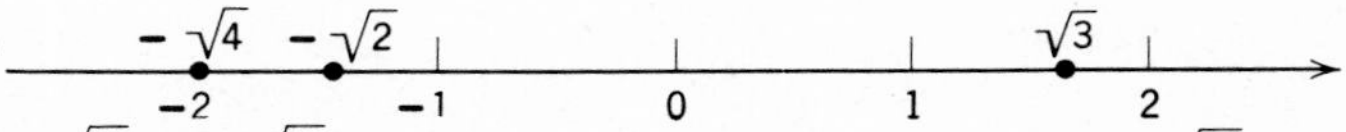

61.
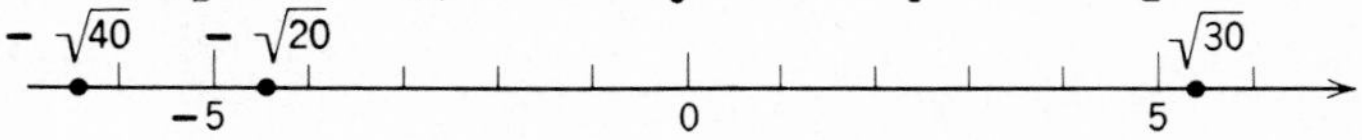

63.
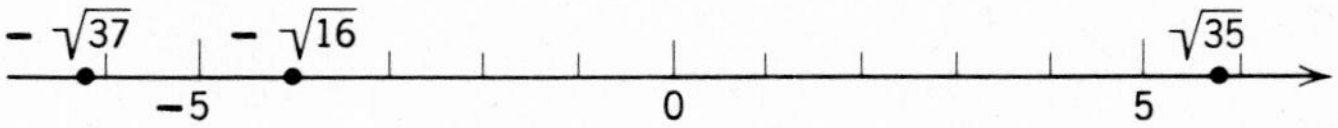

EXERCISES 8.3 (page 211)

1. $2\sqrt{2}$ **3.** $3\sqrt{2}$ **5.** $-2\sqrt{5}$ **7.** $-6\sqrt{2}$
9. 8 **11.** $12\sqrt{2}$ **13.** $5\sqrt{5}$ **15.** $-6\sqrt{30}$
17. $\pm 12\sqrt{5}$ **19.** $\pm 18\sqrt{6}$ **21.** $x\sqrt{x}$ **23.** $y^3\sqrt{y}$
25. $-x^5\sqrt{x}$ **27.** x^3 **29.** $\pm x^4$ **31.** x^6
33. $2x$ **35.** $3x\sqrt{x}$ **37.** $-2y^2\sqrt{6y}$ **39.** $8y^2$
41. $7x^3\sqrt{x}$ **43.** $\pm 4x\sqrt{2x}$ **45.** $4x\sqrt{5}$ **47.** $-8\sqrt{x}$
49. $\pm y\sqrt{5y}$ **51.** $4x\sqrt{3y}$ **53.** $5xy\sqrt{x}$ **55.** $-3a^2\sqrt{5b}$
57. $3x$ **59.** $8xy$ **61.** $-3\sqrt{cd}$ **63.** $yz\sqrt{7}$
65. $\dfrac{3}{4}xy$ **67.** $\pm\dfrac{5}{6}y$ **69.** $2x$ **71.** $6\sqrt{x}$

73. $49y\sqrt{y}$ **75.** $2x^2\sqrt{y}$ **77.** $-a\sqrt{a}$ **79.** $\pm 2x^3$
81. $\sqrt{x}$ **83.** $b^2c\sqrt{ac}$ **85.** 10.39 **87.** 16.59
89. 14.18 **91.** -12.32 **93.** -31.18 **95.** 36.59

EXERCISES 8.4 (page 214)

1. $3\sqrt{3}$ **3.** $\sqrt{5}$ **5.** 0 **7.** $5\sqrt{3}$
9. $-3\sqrt{2}$ **11.** $-4\sqrt{3}$ **13.** $4\sqrt{2} - 4\sqrt{3}$
15. $5\sqrt{3} + 3\sqrt{2}$ **17.** $8 + 6\sqrt{6}$ **19.** $5\sqrt{a}$
21. $12\sqrt{x}$ **23.** $2b\sqrt{b}$ **25.** $4\sqrt{3} + 4$
27. $-30 - 5\sqrt{7}$ **29.** $4\sqrt{2} - 4\sqrt{3}$ **31.** $3 + 9\sqrt{a}$
33. $2\sqrt{2} - 6\sqrt{3} - 10$ **35.** $-\sqrt{a} - \sqrt{b} + \sqrt{c}$
37. $x\sqrt{x} + 2x\sqrt{y}$ **39.** $xy^2\sqrt{x} + 2xy$
41. $2(1 + \sqrt{3})$ **43.** $4(\sqrt{2} - 3)$ **45.** $4(\sqrt{5} + 2)$
47. $8(1 - 4\sqrt{5})$ **49.** $6(1 + 4\sqrt{2})$ **51.** $3(1 + \sqrt{2})$
53. $4(1 - \sqrt{2})$ **55.** $3(7 + \sqrt{2})$ **57.** $4(1 + \sqrt{y})$
59. $2 + 3\sqrt{3}$ **61.** $3 - \sqrt{5}$ **63.** $-1 + \sqrt{2}$
65. $2 + 3\sqrt{3}$ **67.** $\frac{1 + \sqrt{2}}{2}$ **69.** $\frac{1 - \sqrt{3}}{2}$
71. $\frac{2 + \sqrt{2}}{3}$ **73.** $\frac{\sqrt{3} - 1}{5}$ **75.** $\frac{2\sqrt{10} - \sqrt{3}}{3}$
77. $\frac{\sqrt{11} + 1}{a}$ **79.** $\frac{5 + 6\sqrt{2}}{4}$ **81.** $\frac{3 + \sqrt{3}}{6}$
83. $\frac{6 + 5\sqrt{3}}{15}$ **85.** $\frac{3\sqrt{3} - 4}{12}$ **87.** $\frac{4\sqrt{3} - 3\sqrt{2}}{6}$
89. $\frac{8 + 3\sqrt{2}}{2}$ **91.** $\frac{\sqrt{2} + 5}{5}$ **93.** $\frac{3\sqrt{3} + 6}{2}$

EXERCISES 8.5 (page 217)

1. $\sqrt{15}$ **3.** $\sqrt{30}$ **5.** $3\sqrt{2}$ **7.** 4
9. $x\sqrt{6}$ **11.** $2xy\sqrt{3y}$ **13.** $12a$ **15.** $5x\sqrt{6y}$
17. $\sqrt{30}$ **19.** 10 **21.** $6\sqrt{6}$
23. $6x\sqrt{x}$ **25.** $abc\sqrt{abc}$ **27.** x^4
29. $3\sqrt{2} + \sqrt{6}$ **31.** $3\sqrt{2} + 2\sqrt{3}$ **33.** $4\sqrt{5} + 5\sqrt{2}$
35. $\sqrt{6} + 3\sqrt{2}$ **37.** $3 + \sqrt{6}$ **39.** $2\sqrt{5} - 2$
41. $1 - 2\sqrt{2}$ **43.** $2 + 2\sqrt{5}$ **45.** 1
47. -11 **49.** $-4 - 3\sqrt{15}$ **51.** $12 + 5\sqrt{35}$

EXERCISES 8.6 (page 220)

1. 3 **3.** 5 **5.** 3 **7.** $2\sqrt{a}$
9. $\sqrt{3b}$ **11.** $\sqrt{b}$ **13.** $2\sqrt{7b}$ **15.** 1

17. a **19.** $b\sqrt{3}$ **21.** $\frac{5\sqrt{2}}{2}$ **23.** $\frac{2\sqrt{x}}{x}$

25. $\frac{a\sqrt{b}}{b}$ **27.** $\frac{\sqrt{3}}{3}$ **29.** $\frac{\sqrt{2a}}{a}$ **31.** $\frac{\sqrt{3ab}}{b}$

33. $\frac{3\sqrt{x}}{x}$ **35.** $\frac{5\sqrt{y}}{y}$ **37.** $\sqrt{2a}$ **39.** $\sqrt{6x}$

41. $4\sqrt{a}$ **43.** $2\sqrt{3xy}$ **45.** $\frac{2\sqrt{6}}{3}$ **47.** $\frac{3\sqrt{2}}{2}$

49. $\frac{6\sqrt{10}}{5}$ **51.** $\frac{5\sqrt{x}}{x}$ **53.** $\frac{2\sqrt{2x}}{x}$ **55.** $\frac{x\sqrt{y}}{y}$

57. 0.38 **59.** 2.12 **61.** 1.34 **63.** 1.73

CHAPTER 8 REVIEW (page 223)

1. $\sqrt{6}, -\sqrt{\frac{2}{3}}$

2. a. 9.644
b. 13.712
c. 1.414

3.

8 $\sqrt{80}$ $\frac{19}{2}$

8 9 10

4. a. $6\sqrt{2}$
b. $-3\sqrt{10}$
c. $5\sqrt{7}$

5. a. y^2
b. $x^7\sqrt{x}$
c. $xy^3\sqrt{xy}$

6. a. $3x\sqrt{3}$
b. $3x\sqrt{3}$
c. $x\sqrt{3}$

7. a. $4\sqrt{7}$
b. $2\sqrt{6}$
c. $5\sqrt{2}$

8. a. $-8\sqrt{3}$
b. $\sqrt{x}$
c. $2x\sqrt{y}$

9. a. $3\sqrt{y} - 18$
b. $y\sqrt{xy} - 2y^2$
c. $\sqrt{6} - 3\sqrt{2}$

10. a. $2(1 - \sqrt{2}$
b. $3(\sqrt{3} - \sqrt{5})$
c. $x^2(3 - \sqrt{5})$

11. a. $1 - 2\sqrt{3}$
b. $2 - 2\sqrt{3}$
c. $1 - \sqrt{2}$

12. a. $\frac{3 - \sqrt{3}}{2}$
b. $\frac{3\sqrt{15}}{5}$
c. $\frac{4 - \sqrt{3}}{6}$

13. a. $\frac{2\sqrt{5} - 9}{6}$
b. $\frac{2\sqrt{3} - 5}{5}$
c. $\frac{3\sqrt{6} + 8}{4}$

14. a. $\sqrt{15}$
b. $16\sqrt{2}$
c. $3x\sqrt{2x}$

15. a. $5\sqrt{6}$
b. 12
c. $2x\sqrt{3x}$

16. a. $2\sqrt{3} - \sqrt{6}$
b. $\sqrt{14} - 7\sqrt{2}$
c. $4 - 2\sqrt{6}$

17. a. 1
b. -2
c. $12 - 7\sqrt{3}$

18. a. 3
b. $\sqrt{6a}$
c. $a\sqrt{3}$

19. a. $\dfrac{\sqrt{15}}{5}$
b. $\dfrac{2\sqrt{15}}{5}$
c. $\dfrac{\sqrt{6x}}{2x}$

20. a. $\dfrac{\sqrt{17x}}{x}$
b. $\dfrac{\sqrt{6y}}{2x}$
c. $2\sqrt{14}$

CHAPTER 8 CUMULATIVE REVIEW (page 224)

1. Two
2. -4
3. $-6ab^3c$
4. $\dfrac{5a - 9b}{6}$
5. $\dfrac{a - 1}{4x(2a + 5)}$
6. $x^2 + 3x - 1$
7. $(2x - 3)(2x - 3)$
8. -5
9. 6
10. $146 - n$
11. $25x$
12. Dependent
13. $\dfrac{9d}{7b}$
14. $-1/2, 3$
15. 54, 100
16. $3x^3\sqrt{7x}$
17. $8\sqrt{2} - 4\sqrt{5}$
18. 1
19. $\sqrt{6x}$
20. $\sqrt{6}$

EXERCISES 9.1 (page 227)

1. $2, -2$
3. $4, -4$
5. $7, -7$
7. $\sqrt{3}, -\sqrt{3}$
9. $\sqrt{10}, -\sqrt{10}$
11. $\sqrt{6}, -\sqrt{6}$
13. $2\sqrt{3}, -2\sqrt{3}$
15. $3\sqrt{2}, -3\sqrt{2}$
17. $3\sqrt{2}, -3\sqrt{2}$
19. $2, -2$
21. $\sqrt{5}, -\sqrt{5}$
23. 0, 0
25. $2\sqrt{5}, -2\sqrt{5}$
27. $3, -3$
29. $2\sqrt{2}, -2\sqrt{2}$
31. $\sqrt{a}, -\sqrt{a}$
33. $ab\sqrt{3a}, -ab\sqrt{3a}$
35. $\dfrac{\sqrt{30b}}{6}, -\dfrac{\sqrt{30b}}{6}$
37. $t = \dfrac{\sqrt{2gs}}{g}, t = -\dfrac{\sqrt{2gs}}{g}$
39. $r = \dfrac{\sqrt{\pi A}}{2\pi}, r = -\dfrac{\sqrt{\pi A}}{2\pi}$
41. $d = \dfrac{\sqrt{3Ik}}{I}, d = -\dfrac{\sqrt{3Ik}}{I}$
43. $25, -25$
45. $8, -8$
47. $5\sqrt{2}, -5\sqrt{2}$
49. $3, -1$
51. $7, -3$
53. 6, 4
55. $a + 5, a - 5$
57. $a + 3, -a + 3$
59. $a + b, a - b$

61. $-3+\sqrt{2}, -3-\sqrt{2}$
63. $-5+\sqrt{5}, -5-\sqrt{5}$
65. $-10+2\sqrt{2}, -10-2\sqrt{2}$
67. $5+\sqrt{a}, 5-\sqrt{a}$
69. $-1+\sqrt{b}, -1-\sqrt{b}$
71. $b+\sqrt{a}, b-\sqrt{a}$

EXERCISES 9.2 (page 232)

1. 2, −6
3. 1, 1
5. 4, −5
7. −1, −2
9. −1, 4
11. −2, 5
13. $1+\sqrt{2}, 1-\sqrt{2}$
15. $\frac{3+\sqrt{21}}{2}, \frac{3-\sqrt{21}}{2}$
17. $\frac{-1+\sqrt{13}}{2}, \frac{-1-\sqrt{13}}{2}$
19. $\frac{1}{2}, -\frac{3}{2}$
21. $\frac{1}{2}, -2$
23. $-\frac{5}{2}, 3$

EXERCISES 9.3 (page 235)

1. $a=1, b=-3, c=2$
3. $a=1, b=-1, c=-30$
5. $a=1, b=-2, c=0$
7. $a=4, b=0, c=-3$
9. $a=2, b=-7, c=6$
11. $a=6, b=-5, c=1$
13. $a=1, b=-8, c=4$
15. $a=2, b=-2, c=-1$
17. $a=6, b=7, c=-3$
19. $a=9, b=6, c=-8$
21. 1, 2
23. −2, 6
25. 3, −5
27. $\frac{-3+\sqrt{13}}{2}, \frac{-3-\sqrt{13}}{2}$
29. $\frac{3+\sqrt{17}}{2}, \frac{3-\sqrt{17}}{2}$
31. 0, 2
33. 0, 5
35. 0, 7
37. 0, 3
39. $\frac{\sqrt{3}}{2}, -\frac{\sqrt{3}}{2}$
41. $2, \frac{3}{2}$
43. $\frac{1}{3}, -\frac{1}{2}$
45. $\frac{5}{2}, -\frac{1}{3}$
47. $1+\sqrt{2}, 1-\sqrt{2}$
49. $2+\sqrt{6}, 2-\sqrt{6}$
51. $1, -\frac{1}{2}$
53. $\frac{3}{2}, -\frac{5}{2}$
55. $\frac{3}{2}, \frac{2}{3}$
57. $-\frac{3}{2}, 3$

EXERCISES 9.4 (page 239)

1. (0, −3)
3. (−1, 0)
5. (−2, 5)
7. (0, −2)
9. (1, 0)
11. (−2, 0)
13. (0, 12)
15. (2, 2)

17. (4, 0)

19.

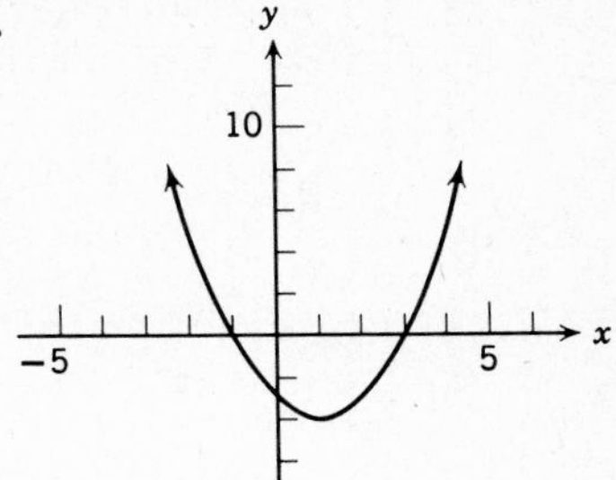

21.

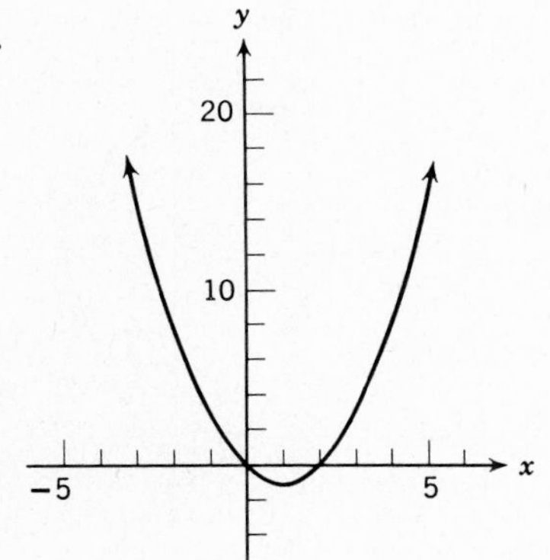

23.

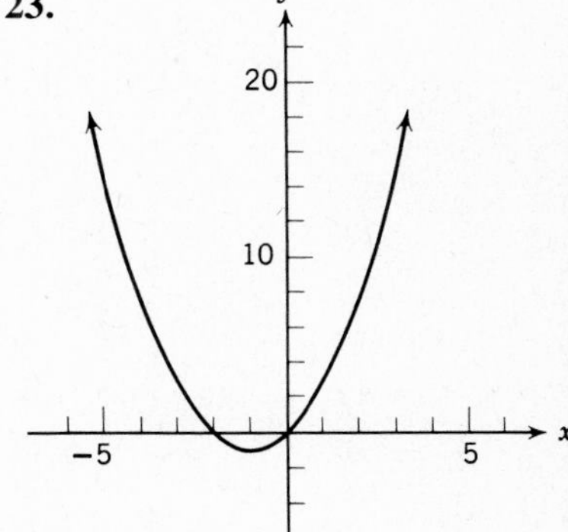

25.

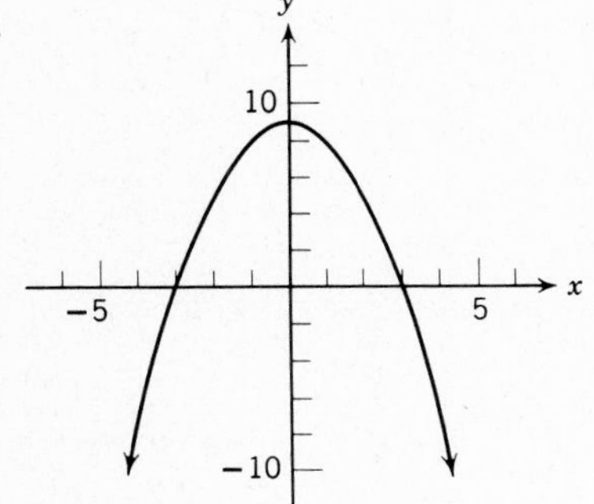

27.

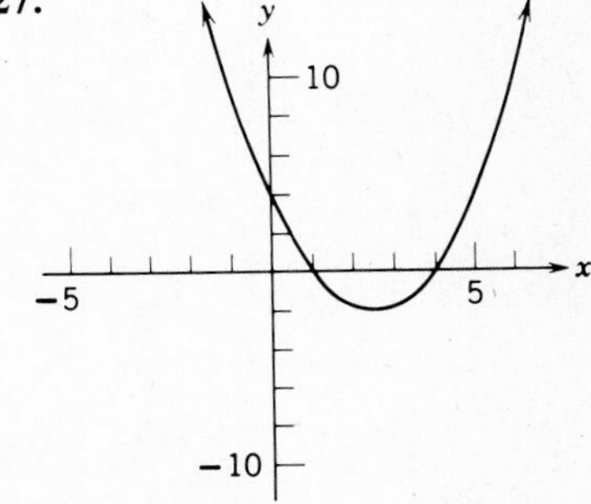

29. 0, 2; −2, 2; −2, 0; none

31.

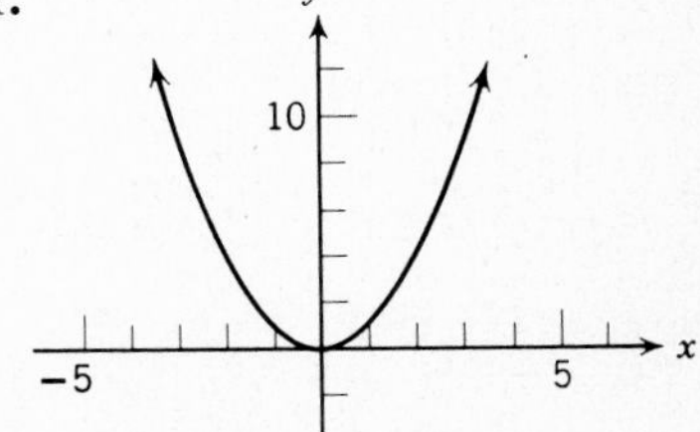

33.

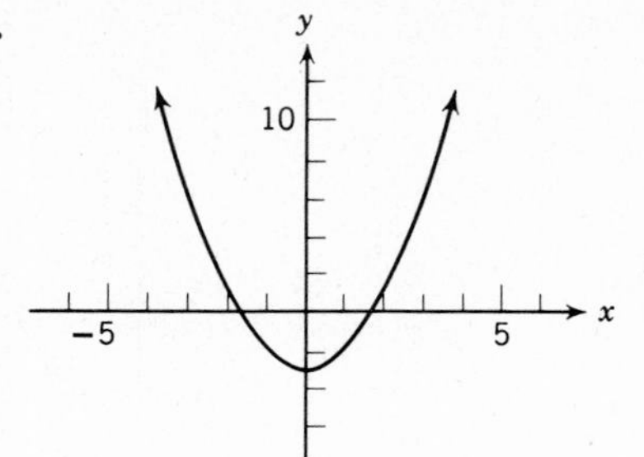

35.

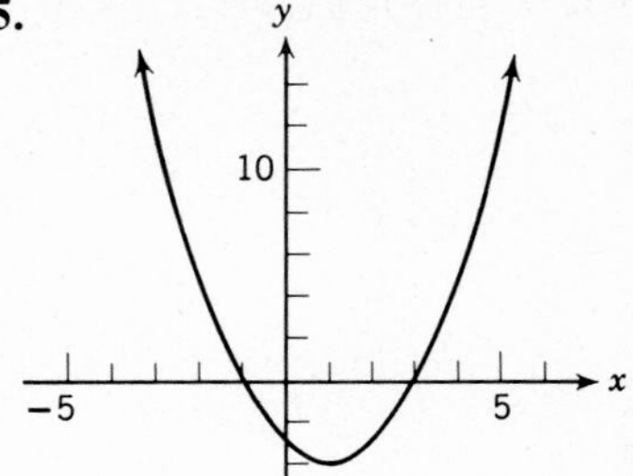

37. −2, 2

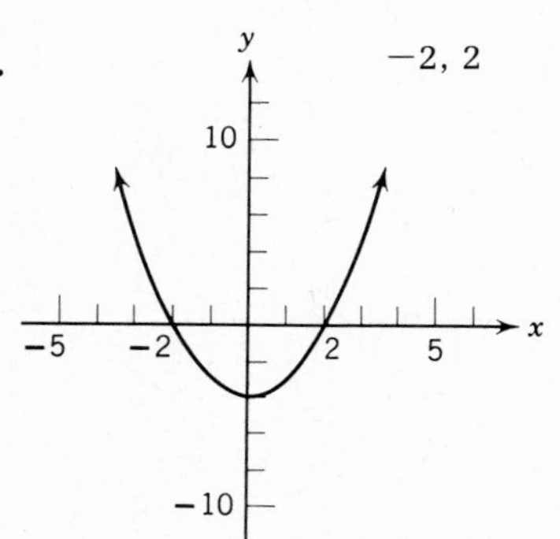

39. 3, −4

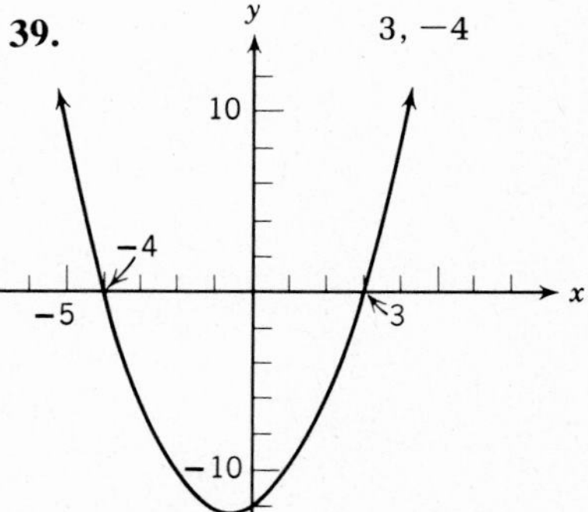

41. 1, 1

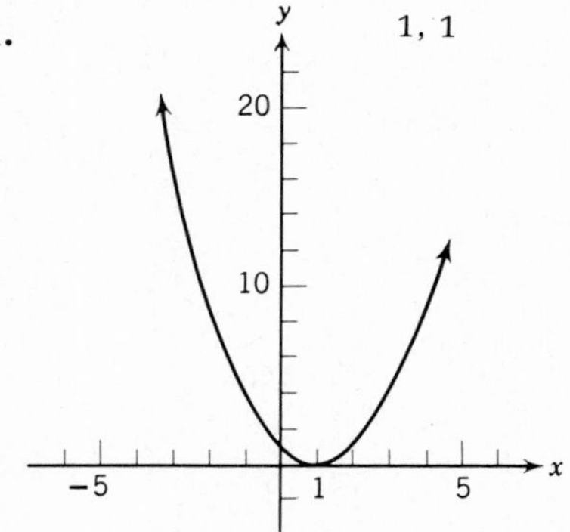

43. −2, ½

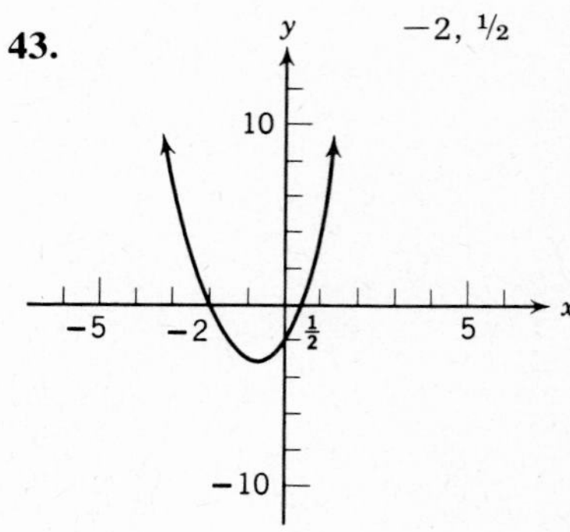

45. −3/2, 1

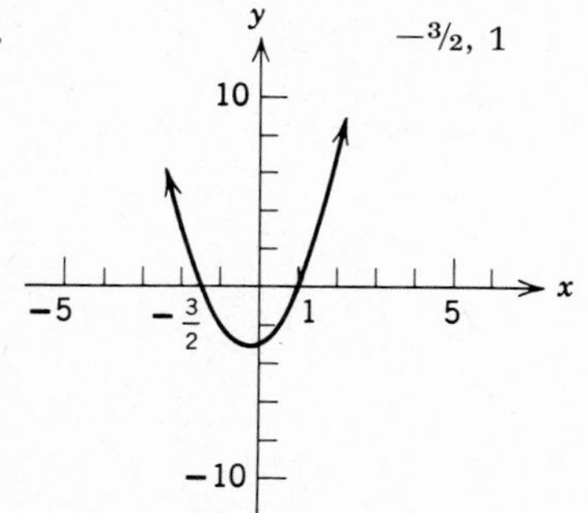

47. −1, ½

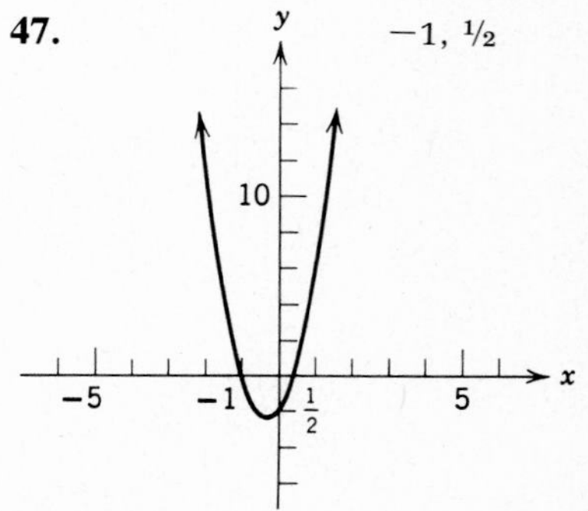

EXERCISES 9.5 (page 242)

1. 5 centimeters
3. $3\sqrt{2}$ kilometers
5. 127 feet
7. 12 inches
9. $2\sqrt{10}$ yards
11. Width: 9 meters, length: 12 meters
13. Width: 2 millimeters, length: 5 millimeters
15. Width: 3 yards, length: 6 yards
17. 50 meters
19. $30\sqrt{2}$ meters

CHAPTER 9 REVIEW (page 246)

1. a. $-5, 5$
b. $-3, 3$

2. a. $-\sqrt{7}, \sqrt{7}$
b. $-\sqrt{6}, \sqrt{6}$

3. a. $-1, 5$
b. $-4, -2$

4. a. $3, 11$
b. $a + c, a - c$

5. a. $-3 + \sqrt{a}, -3 - \sqrt{a}$
b. $2 - a, -2 - a$

6. a. $1, -4$
b. $\dfrac{3 + \sqrt{21}}{2}, \dfrac{3 - \sqrt{21}}{2}$

7. a. $-1, -1$
b. $-1, 2$

8. $-1, 5$

9. $3, -5$

10. $\dfrac{-3+\sqrt{5}}{2}, \dfrac{-3-\sqrt{5}}{2}$

11. $\dfrac{4}{3}, 3$

12.

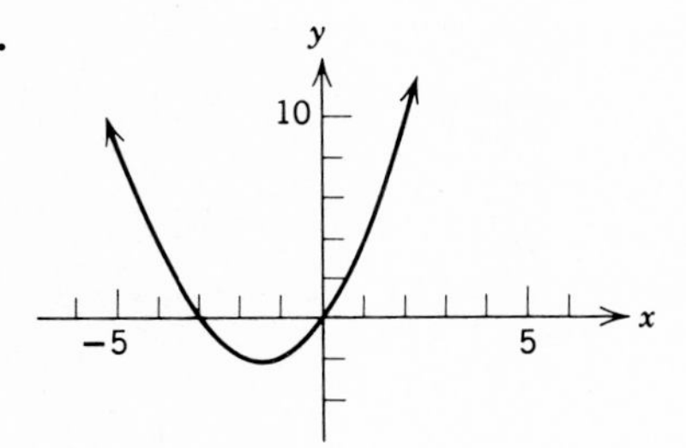

13.

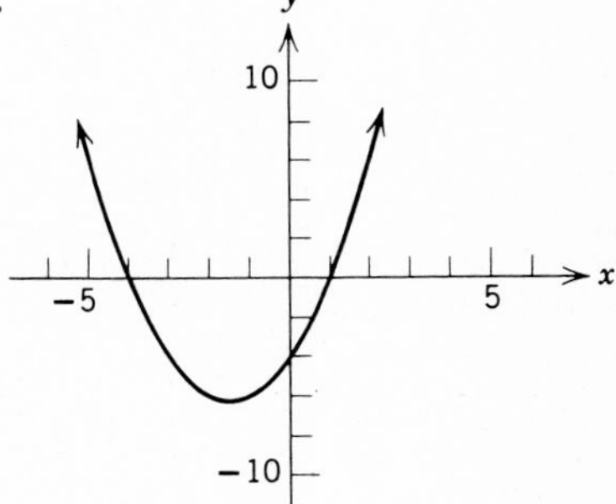

14. $-2, 3$

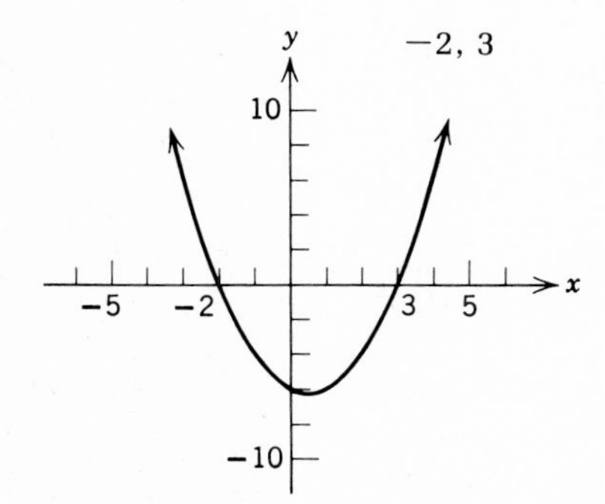

15. ± 16 **16.** $\pm 6\sqrt{2}$ **17.** $\sqrt{130}$ meters
18. $8\sqrt{2}$ centimeters **19.** 30 meters **20.** 5 meters and 15 meters

CHAPTER 9 CUMULATIVE REVIEW (page 246)

1. $2 \cdot 2 \cdot 2 \cdot 3xxxyy$ **2.** 1 **3.** 1
4. $2a^2$ **5.** 0 **6.** (1, 2)
7. $\dfrac{3x+4}{x(x+y)}$ **9.** $5x^2\sqrt{y}$ **10.** $a - x$
11. Two **12.** Either $a = 0$, $b = 0$ or both a and $b = 0$ **13.** 27
14. 8, 32 **15.** 12 inches, 8 inches, 10 inches **16.** 30°, 60°, 90°
17. Altitude: 3 centimeters base: 12 centimeters **18.** 7, 9
19. Width: 11 centimeters, length: 22 centimeters **20.** Width: 4 centimeters Length: 12 centimeters

FINAL CUMULATIVE REVIEW I (page 248)

1. 71, 73, 79, 83, 89, 97 **2.** $a + b = b + a$; $ab = ba$
3. $-4, -2, -1, 3, 5, 6$
4. $V = \pi r^2 h$ **5.** $30 - x$
6. $2a^2$ **7.** $2x + 1 + \dfrac{-1}{3x}$
8. $a = 8$ centimeters, $b = 10$ centimeters, $c = 16$ centimeters
9. $\dfrac{2b}{a+2b}$ **10.** $\dfrac{3a+3}{4a}$ **11.** $\dfrac{3ax^2y}{10}$
12. $\dfrac{4a-3b}{a}$ **13.**

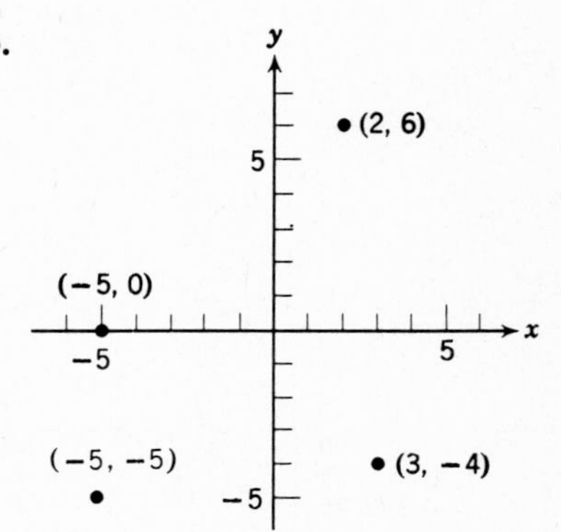

14. (1, 1) **15.** 4 **16.** $\dfrac{\sqrt{y}}{4}$ **17.** 2

18. 3.46 **19.** -3 **20.** 13 and 15

FINAL CUMULATIVE REVIEW II (page 249)

1. 2 **2.** Zero **3.** Yes **4.** $\frac{24}{b}$

5. $5b$ **6.** -5 **7.** 7 tons and 12 tons **8.** $\frac{a^2 + b^2}{ab}$

9. $\frac{3(x-3)}{x^2 - 9}$ **10.** $\frac{2}{3x}$ **11.** Proportion

12.

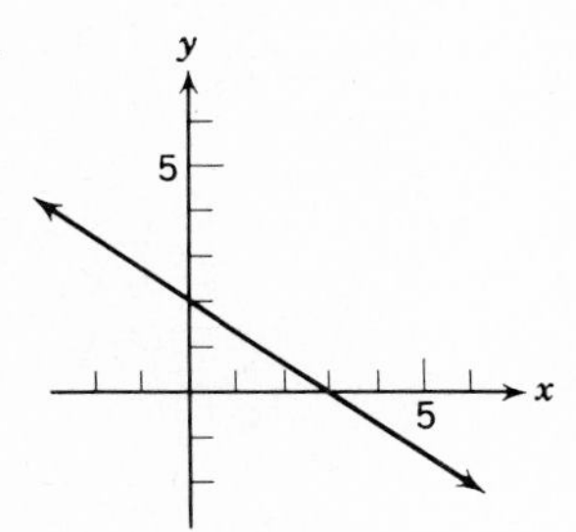

13. $x + y = 1$
14. $(-1, 3)$
15. $8, -2$
16. $\sqrt{5}, \sqrt{6}, \sqrt{7}, \sqrt{8}, \sqrt{10}$

17. $4xy\sqrt{10xyz}$ **18.** $\frac{2 - \sqrt{2}}{2}$ **19.** $\frac{-5 \pm \sqrt{17}}{2}$ **20.** $x = \frac{25}{3}$

FINAL CUMULATIVE REVIEW III (page 250)

1. $2 \cdot 2 \cdot 2 \cdot 3 \cdot 3 \cdot 5$
2. 30
3. $-a^2 - 2a$
4. No
5. 625 square millimeters
6. $ab(c - 1)$
7. $3(2x - 3)(x + 1)$
8. $5n + 10(n + 6)$ or $15n + 60$
9. $\frac{-6b}{5}$
10. $\frac{5}{6x}$
11. $x - 8$
12. $\frac{2}{a + b}$
13. Numerator: 20
14. 570 bricks
15. Linear
16. $y = \frac{4x + 2}{3}$
17. 1
18. $6a^3b^2c^2 \sqrt{c}$
19. $3\sqrt{a}$
20. $4\sqrt{10}$ meters

FINAL CUMULATIVE REVIEW IV (page 251)

1.

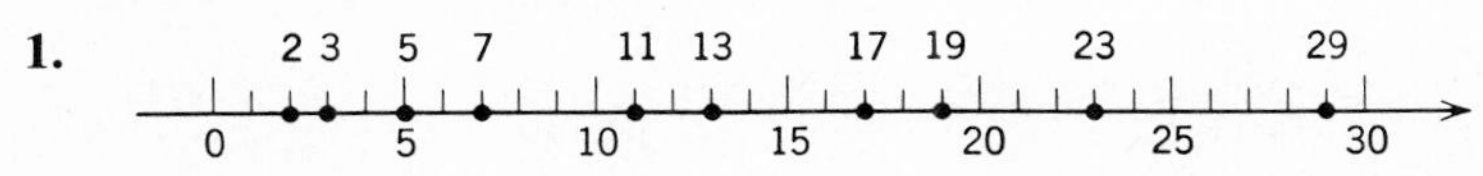

2. $r + s$; rs

3. -3

4. 5

5. $-28, -26, -24$

6. $-3bx$

7. $2(x - 2)(x + 2)$

8. 96 children's tickets, 164 adult tickets

9. $x - 1$

10. $\frac{3x + 5}{(x + 2)^2}$

11. $16x$

12. 9

13. 192 miles

14. Two

15. $\frac{4}{3}$

16. $(0, -6)$; $(2, 0)$; $(-3, -15)$

17. $\frac{a}{4}$

18. $0, \frac{1}{4}$

19. -6 and 5

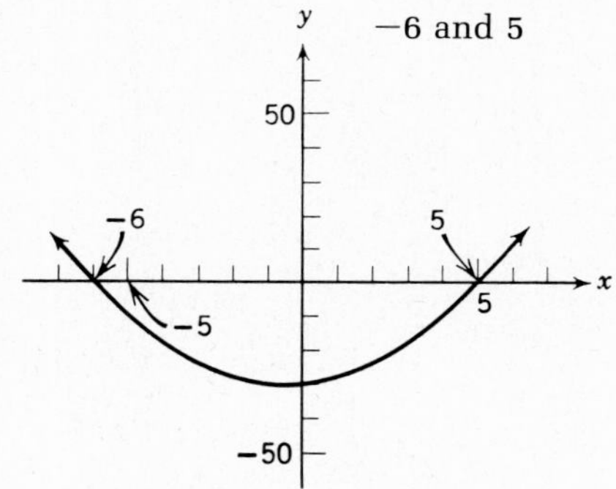

20. 12 centimeters by 16 centimeters

FINAL CUMULATIVE REVIEW V (page 252)

1. 1

2. Integers

3. 10

4. $(y - 1)(y - 11)$

5. $-8b^2$

6. $\frac{2(w + 8)}{3}$

7. $\frac{-(1 - a)}{3}$ or $\frac{-1 + a}{3}$

8. $x^2 - 2x + 1$

9. $\frac{h^2}{6 + h}$

10. $\frac{875}{t}$

11. 15

12. Means; extremes

13.

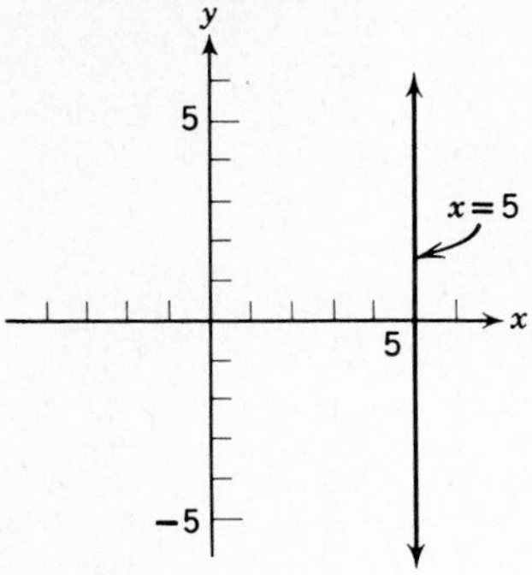

14. (1, 7)

15. −3 and 7

16.

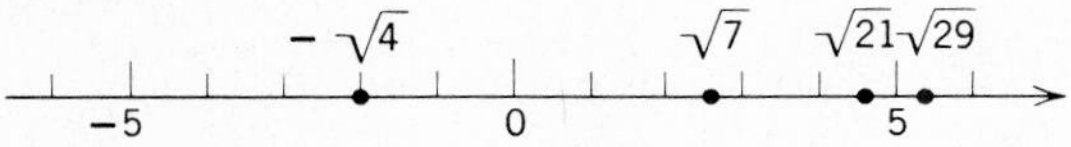

17. 9

18. $\dfrac{-5 \pm \sqrt{17}}{2}$

19. $\dfrac{2}{3}$ and −3

20. $y = \dfrac{15}{2}$

FINAL CUMULATIVE REVIEW VI (page 253)

1. −20

2. 11

3. x

4. −17°

5. $\dfrac{a + 2}{a - 1}$

6. Winner: 1844, loser: 1782

7. $3y^2$

8. $(a - 5b)(a + 5b)$

9. $5(r - 8)$ or $5r - 40$

10. 60

11. $\dfrac{11}{2}$

12. $\dfrac{-3x - 11}{6}$

13. 22

14. 2

15. $\dfrac{b(a + c)}{2}$

16. 8.46

17. (1, 2)

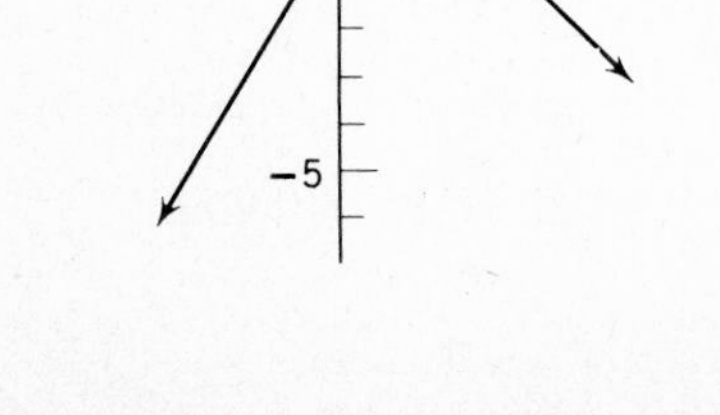

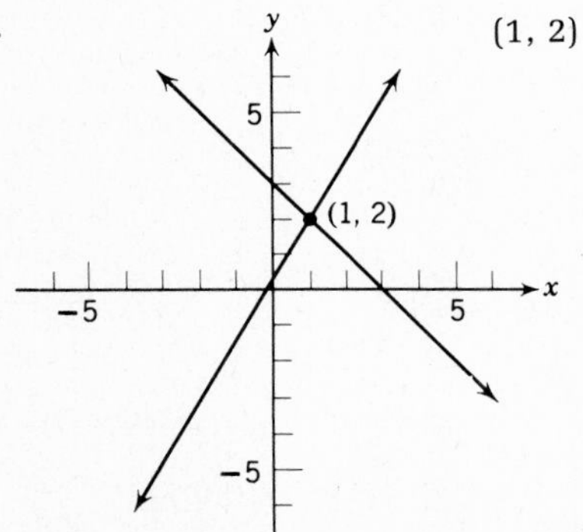

18. 0 and $\frac{-3}{2}$

19. $-1 \pm \sqrt{5}$

20. ± 12

FINAL CUMULATIVE REVIEW VII (page 254)

1. -1

2. $1 \pm \sqrt{6}$

3. $4b - 5c$

4. $x^3 \cdot x^2 = x^5$

5. $(2y + 3)(2y + 5)$

6. $6(22 - y)$

7. $\frac{3}{y}$

8. $\frac{2ac}{b}$

9. $\frac{b - a}{b + a}$

10. 20 mph, 30 mph

11. $(3, 13)$; $(-2, -2)$; $(0, 4)$; $(6, 22)$

12. 

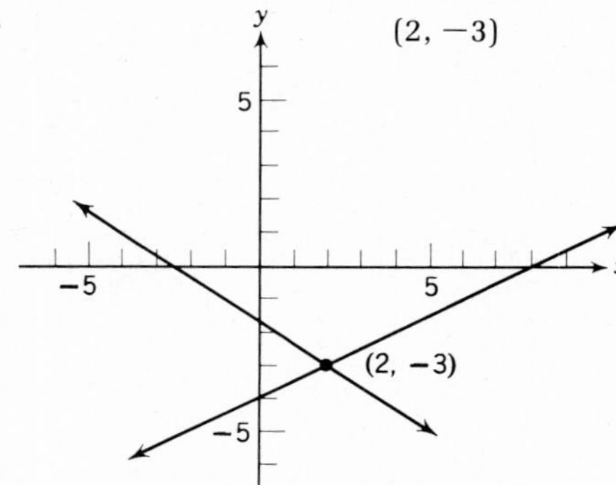

13. $(3, -2)$

14. 0, 2

15. $\frac{1}{2}, -18$

16. 16

17. $1 + \sqrt{2}$

18. $2\sqrt{3b}$

19. $\frac{\pm\sqrt{c}}{b}$

20. 4 mph

FINAL CUMULATIVE REVIEW VIII (page 255)

1. $\frac{-8}{3}$

2. 5

3. $n + 2, n + 4, n + 6$

4. $(a - 4b)(a - 2b)$

5. $-4xy$

6. $18\frac{2}{7}$ meters or $45\frac{5}{7}$ meters from one end

7. $\dfrac{x+6}{x-2}$

8. $\dfrac{2}{a+b}$

9. $b-a$

10. $x+8+\dfrac{33}{x-5}$

11. $(-1, 0)$

12. (1, 7)

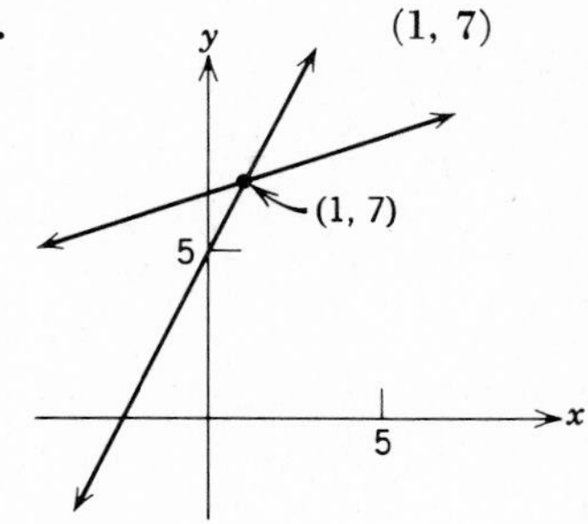

13. 67 kilograms and 79 kilograms

14. $-2, 7$

15. $10xy\sqrt{2x}$

16. $7\sqrt{5}$

17.

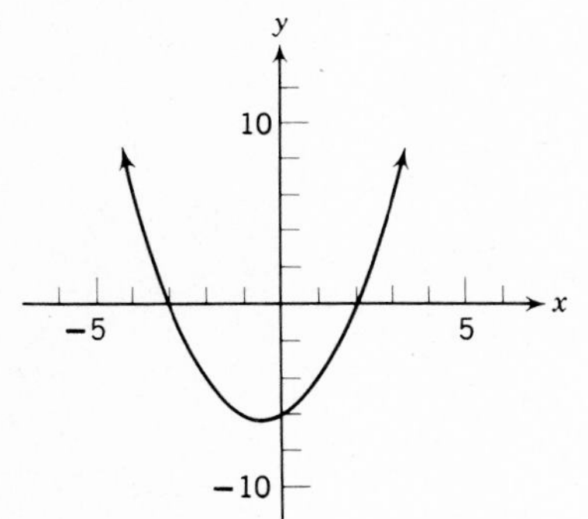

18. $\sqrt{10}$

19. $\dfrac{1 \pm \sqrt{15}}{2}$

20. Width: 12 centimeters, length: 20 centimeters

FINAL CUMULATIVE REVIEW IX (page 256)

1. 4

2. -4

3. $a^2 - 3ab$

4. 24, 26, 28

5. $2(x-11)(x-1)$

6. $n+28$

7. 5

8. $\dfrac{2y-3x}{xy}$

9. $\dfrac{1}{16b}$

10. $2x-1+\dfrac{1}{x+2}$

11. $\frac{b-a}{c}$

12.

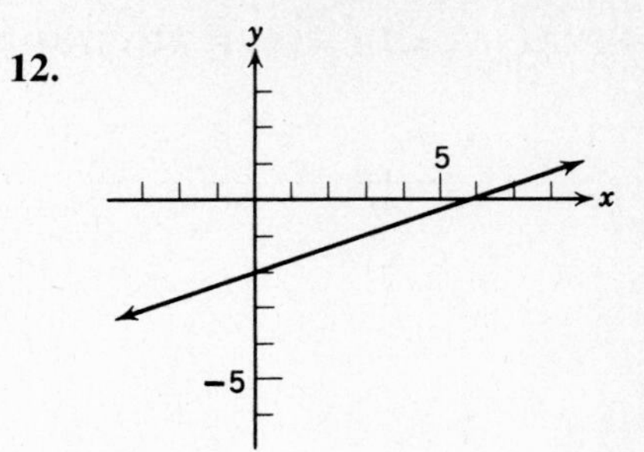

13.

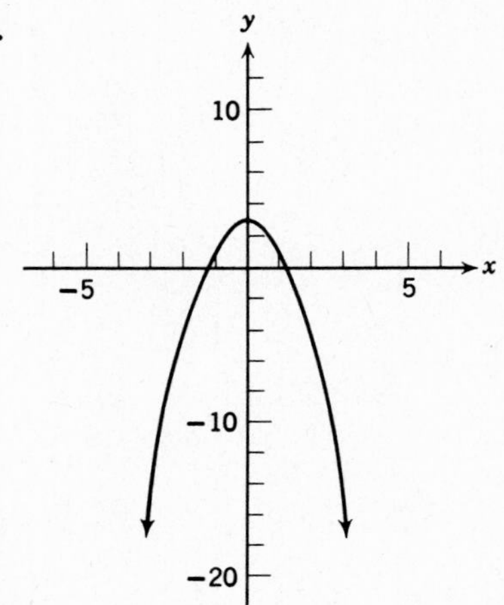

14. (1, 2)

15. 4 nickels
7 dimes
9 quarters

16. $3x\sqrt{10}$

17. $1 + 3\sqrt{3}$

18. 4.23

19. 20 kilometers

20.

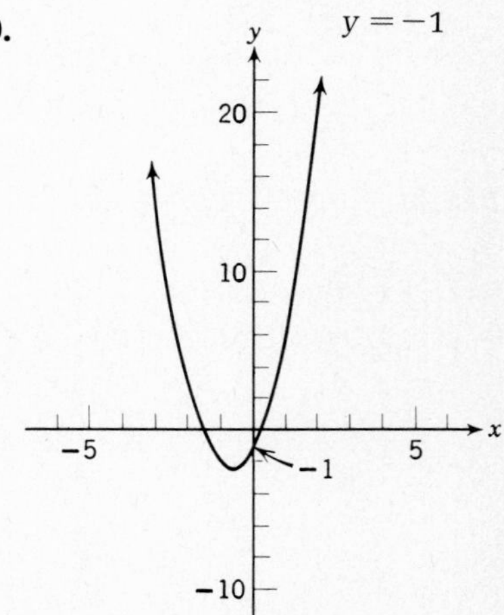

FINAL CUMULATIVE REVIEW X (page 257)

1. 23 29 31 37

20 25 30 35 40

2. $2 \cdot 2 \cdot 2 \cdot 2 \cdot 3 \cdot 5 \cdot x \cdot x \cdot y$

3. $-5, -2, -1, 1, 3, 5, 7$

4. $-2a - 5b$

5. $2ab - 2b^2$

6. $3(x - 4y)(x - 2y)$

7. $\dfrac{1}{3a + 4}$

8. $\dfrac{b}{7}$

9. $\dfrac{a - 5}{(a - 3)^2}$

10. 2

11.

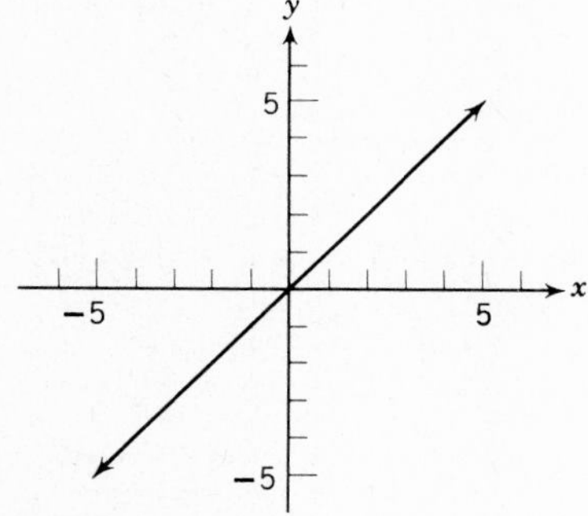

12. $(2, -1)$

13. $6\sqrt{3}$

14. $\sqrt{2} - \sqrt{3}$

15.

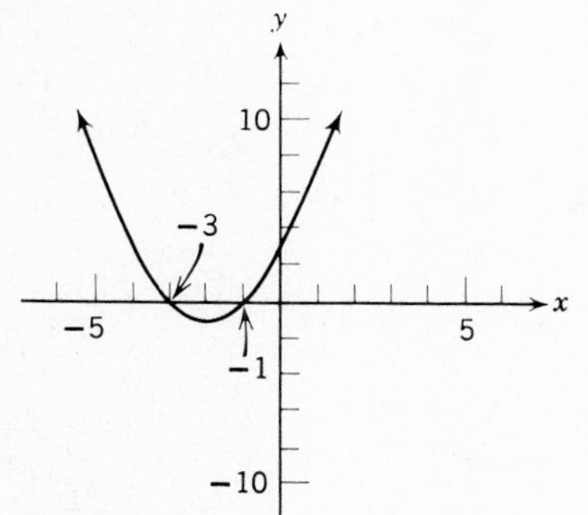

16. 8 and 9

17. 30°; 60°

18. \$600

19. Chair: \$120, desk: \$520

20. 12 meters

Metric-United States Conversion Tables

(approximate conversion factors)

Lengths

1 in. = 2.54 cm. (exact)	1 mm. = 0.04 in.
1 ft. = 0.30 m.	1 cm. = 0.39 in.
1 yd. = 0.91 m.	1 m. = 39.37 in.
1 mi. = 1.61 km.	1 km. = 0.62 mi.

Area

1 sq. in. = 6.45 sq. cm.	1 sq. mm. = 0.0016 sq. in.
1 sq. ft. = 0.09 sq. m.	1 sq. cm. = 0.16 sq. in.
1 sq. yd. = 0.84 sq. m.	1 sq. m. = 10.76 sq. ft.
1 sq. mi. = 2.59 sq. km.	1 sq. km. = 0.39 sq. mi.

Volume

1 cu in. = 16.39 cc.	1 cc. = 0.06 cu in.
1 cu ft. = 0.028 cu m.	1 cu m. = 35.31 cu ft.
1 cu yd. = 0.76 cu m.	1 cu m. = 1.31 cu yd.

Mass and Weights

1 oz. = 28.35 g.	1 g. = 0.035 oz.
1 lb. = 454 g.	1 kg. = 2.20 lb.
1 ton = 0.907 metric ton	1 metric ton = 2205 lb. = 1.10 ton

Capacity

1 pint = 0.47 liter	1 ml. = 0.002 pints
1 qt. = 0.95 liter	1 liter = 1.06 qt.
1 gal. = 3.78 liters	1 kl. = 265 gal.

Table of Squares, Square Roots, and Prime Factors

No.	Sq.	Sq. Root	Prime Factors	No.	Sq.	Sq. Root	Prime Factors
1	1	1.000		51	2,601	7.141	$3 \cdot 17$
2	4	1.414	2	52	2,704	7.211	$2^2 \cdot 13$
3	9	1.732	3	53	2,809	7.280	53
4	16	2.000	2^2	54	2,916	7.348	$2 \cdot 3^3$
5	25	2.236	5	55	3,025	7.416	$5 \cdot 11$
6	36	2.449	$2 \cdot 3$	56	3,136	7.483	$2^3 \cdot 7$
7	49	2.646	7	57	3,249	7.550	$3 \cdot 19$
8	64	2.828	2^3	58	3,364	7.616	$2 \cdot 29$
9	81	3.000	3^2	59	3,481	7.681	59
10	100	3.162	$2 \cdot 5$	60	3,600	7.746	$2^2 \cdot 3 \cdot 5$
11	121	3.317	11	61	3,721	7.810	61
12	144	3,464	$2^2 \cdot 3$	62	3,844	7.874	$2 \cdot 31$
13	169	3.606	13	63	3,969	7.937	$3^2 \cdot 7$
14	196	3.742	$2 \cdot 7$	64	4,096	8.000	2^6
15	225	3.873	$3 \cdot 5$	65	4,225	8.062	$5 \cdot 13$
16	256	4.000	2^4	66	4,356	8.124	$2 \cdot 3 \cdot 11$
17	289	4.123	17	67	4,489	8.185	67
18	324	4.243	$2 \cdot 3^2$	68	4,624	8.246	$2^2 \cdot 17$
19	361	4.359	19	69	4,761	8.307	$3 \cdot 23$
20	400	4.472	$2^2 \cdot 5$	70	4,900	8.367	$2 \cdot 5 \cdot 7$
21	441	4.583	$3 \cdot 7$	71	5,041	8.426	71
22	484	4.690	$2 \cdot 11$	72	5,184	8.485	$2^3 \cdot 3^2$
23	529	4.796	23	73	5,329	8.544	73
24	576	4.899	$2^3 \cdot 3$	74	5,476	8.602	$2 \cdot 37$
25	625	5.000	5^2	75	5,625	8.660	$3 \cdot 5^2$
26	676	5.099	$2 \cdot 13$	76	5,776	8.718	$2^2 \cdot 19$
27	729	5.196	3^3	77	5,929	8.775	$7 \cdot 11$
28	784	5.292	$2^2 \cdot 7$	78	6,084	8.832	$2 \cdot 3 \cdot 13$
29	841	5.385	29	79	6,241	8.888	79
30	900	5.477	$2 \cdot 3 \cdot 5$	80	6,400	8.944	$2^4 \cdot 5$
31	961	5.568	31	81	6,561	9.000	3^4
32	1,024	5.657	2^5	82	6,724	9.055	$2 \cdot 41$
33	1,089	5.745	$3 \cdot 11$	83	6,889	9.110	83
34	1,156	5.831	$2 \cdot 17$	84	7,056	9.165	$2^2 \cdot 3 \cdot 7$
35	1,225	5.916	$5 \cdot 7$	85	7,225	9.220	$5 \cdot 17$
36	1,296	6.000	$2^2 \cdot 3^2$	86	7,396	9.274	$2 \cdot 43$
37	1,369	6.083	37	87	7,569	9.327	$3 \cdot 29$
38	1,444	6.164	$2 \cdot 19$	88	7,744	9.381	$2^3 \cdot 11$
39	1,521	6.245	$3 \cdot 13$	89	7,921	9.434	89
40	1,600	7.325	$2^3 \cdot 5$	90	8,100	9.487	$2 \cdot 3^2 \cdot 5$
41	1,681	6.403	41	91	8,281	9.539	$7 \cdot 13$
42	1,764	6.481	$2 \cdot 3 \cdot 7$	92	8,464	9.592	$2^2 \cdot 23$
43	1,849	6.557	43	93	8,649	9.644	$3 \cdot 31$
44	1,936	6.633	$2^2 \cdot 11$	94	8,836	9.695	$2 \cdot 47$
45	2,025	6.708	$3^2 \cdot 5$	95	9,025	9.747	$5 \cdot 19$
46	2,116	6,782	$2 \cdot 23$	96	9,216	9.798	$2^5 \cdot 3$
47	2,209	6.856	47	97	9,409	9.849	97
48	2,304	6.928	$2^4 \cdot 3$	98	9,604	9.899	$2 \cdot 7^2$
49	2,401	7.000	7^2	99	9,801	9.950	$3^2 \cdot 11$
50	2,500	7.071	$2 \cdot 5^2$	100	10,000	10.000	$2^2 \cdot 5^2$

INDEX

Abscissa, 162
Absolute value, 31
Addends, 5
Addition, associative law of, 7
  commutative law of, 5
  of fractions, 125, 129
  of irrational numbers, 213–215
  of signed numbers, 33–34
Algebraic expressions, 13, 16
  evaluation of, 13–14
Algebraic solution of systems, 176, 179
Angles, acute, 73
  complementary, 73
  obtuse, 73
  right, 73
  straight, 73
  supplementary, 73
Area, of a circle, 74
  of a rectangle, 72
  of a square, 72
  of a triangle, 73
Associative law, of addition, 7
  of multiplication, 7
Axes, coordinate, 162
Axioms, 57
  addition or subtraction, 57
  division, 61
  multiplication, 63

Base of a power, 10
Binomials, 16
  products of, 84, 90
Brackets, 7
Building factor, 119
Building fractions, 119

Cartesian coordinates, 162
Checking solution of equations, 56
Circle, 74
  area of a, 74
  circumference of a, 74
  diameter of a, 74
Coefficient, 16
  numerical, 16
Combining radicals, 213–214
Combining terms, 20
Commutative law, of addition, 5
  of multiplication, 5
Completing the square, 231–232
Complex fraction, 141
Components of an ordered pair, 159
Constant, 159

of variation, 170
Coordinates, Cartesian or rectangular, 162

Degree of term, 16
Denominator, lowest common, 117
rationalization of a, 220
Dependent equations, 175
Dependent variable, 159
Descending powers, 188
Difference, of integers, 37
of variables, 21
Direct variations, 170
Distributive law, 19, 80
Dividend, 6
Division, 6
of fractions, 138
law of exponents, 25
long, 113
of monomials, 25
of polynomials, 112
of radicals, 219–220
of signed numbers, 47
by zero, 7
Divisor, 6

Equality, of numbers, 1
symmetric property of, 58
Equals, 1
Equations, 54, 158
dependent, 175
equivalent, 57
first-degree, *see* First-degree equations
graphing, 174
inconsistent, 175
involving fractions, 144
involving parentheses, 95
linear, 174
literal, 67
members of, 54
quadratic, *see* Quadratic equations
roots of, 56
solutions of, 56, 57, 60, 63, 65
systems of, 176, 179
in two variables, 158
Equivalent equations, 57
Equivalent expressions, 20
Equivalent fractions, 107, 109, 120
Equivalent systems, 176–177, 179
Evaluation, numerical, 14, 49
Even numbers, 2
Exponential notation, 9
Exponents, 10
laws of, 10, 23, 25
Expression, algebraic, 13
Expression, radical, 205
Extraction of roots, 226–227
Extremes of a proportion, 151

Factor, 5, 80
binomial, 84
monomial, 82
prime, 9
Factoring, 82
difference of two squares, 94
monomials from polynomials, 82
quadratic equations, 188, 191, 195
trinomials, 86, 91
First-degree equations, 54
graphing, 164
number of roots, 56
solution of, 56
by addition, 57
by division, 61
by multiplication, 63
Formula (s), 14
for distance problems, 14, 149, 170
from geometry, 72–74
for literal equations, 67
for Pythagorean relationships, 241–242
for solving a quadratic equation, 234
Fractions, 106
addition of, 125, 129
building, 119
building factor for, 120
changing signs of, 107
complex, 141
division of, 138
equivalent, 107, 109, 120
equations containing fractions, 144
fundamental principle for, 107, 120
graphical representation of, 106
multiplication of, 134
reducing, 107
signs of, 107
standard forms for, 107
Fundamental operations, order of, 12

Geometry, formulas from, 72–74
Graphing, intercept method, 168–169
Graphs, of first-degree equations, 164
of fractions, 106
of integers, 30
of linear equations, 174

Graphs *(Continued)*
of ordered pairs, 162
of quadratic equations, 174
of whole numbers, 2
Greater than, 2
Grouping numbers, 7

Hypotenuse of a right triangle, 241

Inconsistent equations, 175
Independent variable, 159
Integers, 30
differences of, 37
graphical representation of, 30
products of, 43
quotients of, 47
sums of, 33
Intercept method graphing, 168–169
Irrational numbers, 207

Less than, 2, 31
Like terms, 18–19
addition of, 19
subtraction of, 21
Linear equation in two variables, 165
Line graph, 2, 18, 19
Literal equation, 67
Literal numbers, addition of, 5
Lowest common denominator, 117
Lowest terms of a fraction, 109–110

Means of a proportion, 151
Metric units, 15
Monomials, 16
addition of, 18–19
division of, 24–25
multiplication of, 22–23
square root of, 210
subtraction of, 21
Multiplication, 5
associative law of, 7
of binomials, 84, 90
commutative law of, 5
of fractions, 134
of irrational numbers, 210
law of exponents for, 23
of monomials, 5
of radical numbers, 217

Natural numbers, 1
Negative numbers, 30
graphical representation of, 31

Number line, 2
Numbers, absolute value of, 31
equal, 1
irrational, 207
natural, 1
negative, 30
positive numbers, 30
prime, 1
rational, 205
real, 208
signed, 30
Numerical coefficient, 16
Numerical evaluation, 13, 14, 49

Odd numbers, 2
Ordered numbers, 2
Ordered pairs, 159
graphs of, 162
Order of operations, 12
Ordinate, 162
Origin, 2, 162

Parabola, 239
Parentheses, 7
in equations, 95
removing, 95
Perimeter, of a rectangle, 72
of a square, 72
of a triangle, 73
Polynomials, 16
addition of, 16, 18–19
division of, 112
multiplication of, 80, 84
subtraction of, 21
Positive number, 30
Powers, descending, 188
of a number, 10
products of, 23
Prime factors, 9
Prime number, 1
Principal square root, 205
Product, 5, 80
of fractions, 134
of integers, 43
of powers, 23
of radical numbers, 217
of variables, 22
Proportion, 150–151
extremes, 151
means, 151
terms of a, 151
Pythagorean theorem, 241

Quadrant, 162
Quadratic equations, 188, 226
  complete, 188, 194
  graph of, 239
  incomplete, 188, 191
  solution of, by completing the square, 231–232
    by extraction of roots, 226–227
    by factoring, 191, 194
    by formula, 234–235
    by graphing, 238–239
  standard form for, 188
Quotient, 6
  of fractions, 138
  of integers, 47
  of polynomials, 112
  of powers, 25
  of radical expressions, 219
  of variables, 24

Radicals, 205
  adding or combining, 213–215
  division of, 219
  multiplication of, 217
  simplification of, 210, 213
Radicand, 205
Ratio, 150
Rationalizing denominators, 220
Rational number, 205
Real numbers, 208
Reciprocal, 138
Rectangle, area of a, 72
  perimeter of a, 72
Rectangular coordinate system, 162
Relationship, 158
  graphical representation of a, 162
Right triangle, 241
  hypotenuse of, 241
  legs of, 241
Root, of an equation, 56
  square, 205

Sentence, symbolic, 54
Sign, of operation, 34
  of quality, 34
Signed numbers, 30
  addition of, 33–34
  division of, 47
  graphical representation of, 31
  multiplication of, 43–45
  slope, 171
  subtraction of, 38
Solution of, linear equations, 56, 57, 60, 63, 65, 174, 176, 179
  literal equations, 67
  quadratic equations, 188
  systems of linear equations, 174, 176, 179
Solving equations, 95
Square, area of a, 72
  of a number, 10
  perimeter of a, 72
Square roots, 205
  principal, 205
  product of, 210
  quotient of, 217
Standard form, for fractions, 107
  for quadratic equations, 188
  for systems of equations, 179
Substitution, in equations, 56
  in expressions, 14
  in formulas, 14
Subtraction, of fractions, 125, 129
  of like terms, 21
  of signed numbers, 38
Sums, 4
  of fractions, 125, 129
  of integers, 33
  of variables, 18
Symbolic sentence, 54
Symbols, of grouping, 7
  of operation, 7
Symmetric property of equality, 58
Systems of equations, 174, 176, 179
  equivalent systems of, 177
  solution of, 174
    by addition or subtraction, 176–177, 179
    by graphing, 174
  standard form, 179

Terms, algebraic, 16
  coefficients of, 16
  combining like, 20
  degree of, 16
  like, 18
  of a proportion, 151
  unlike, 18
Triangle, area of a, 73
  equilateral, 73
  isosceles, 73
  right, 73
Trinomial, 16
  factoring a, 86, 91

Unlike terms, 18

Variable, 5
dependent, 159
differences of, 21
independent, 159
products of, 22
quotients of, 24
sums of, 18
Vertical axis, 162

Whole numbers, 1
Word problems, 69–72, 97–103, 148–150, 154–155, 182–185, 199–203

x-intercept, 169

y-intercept, 169

Zero, division by, 7